MW01413536

ETHICAL ISSUES IN AVIATION

The aviation industry is unique in two major ways: firstly, it has a long history of government involvement dating back to the early days of aviation; and secondly, its primary concern is the safety of its passengers and crew. These features highlight the importance of ethical decision-making at all levels of the industry. However, well-publicized problems such as the disappearance of Malaysia Airlines Flight 370 highlight the need for ethics to take a more prominent role in the field.

Ethical Issues in Aviation focuses on both past and current topics in aviation, providing the reader with an overview of the major themes in aviation ethics that cover a broad range of subjects. Contributors include academics who do research in the field as well as professionals who provide first-hand accounts of the ethical situations that arise in the aviation industry. This second edition has been thoroughly revised throughout to bring it up to date, and features several new chapters that cover recent events and topics.

This book enhances student learning by providing faculty, students, and those interested in aviation with discussion of the most pressing ethical issues that continue to impact the industry.

Elizabeth A. Hoppe is Instructor of Philosophy at Loyola University Chicago. Specializing in applied ethics, she conducts research in aviation and business and is a member of the Association for Professional and Practical Ethics.

ETHICAL ISSUES IN AVIATION

SECOND EDITION

Edited by Elizabeth A. Hoppe

Routledge
Taylor & Francis Group
LONDON AND NEW YORK

Second edition published 2019
by Routledge
2 Park Square, Milton Park, Abingdon, Oxon, OX14 4RN

and by Routledge
711 Third Avenue, New York, NY 10017

Routledge is an imprint of the Taylor & Francis Group, an informa business

First edition published by Routledge 2011

British Library Cataloguing-in-Publication Data
A catalogue record for this book is available from the British Library

Library of Congress Cataloging-in-Publication Data
Names: Hoppe, Elizabeth (Elizabeth Anne) editor.
Title: Ethical issues in aviation / edited by Elizabeth Hoppe.
Description: Second Edition. | New York : Routledge, 2019. |
Revised edition of Ethical issues in aviation, c2011. |
Includes bibliographical references and index.
Identifiers: LCCN 2018025672| ISBN 9781138348080 (hardback) |
ISBN 9781472470867 (pbk.) | ISBN 9780429436789 (ebook)
Subjects: LCSH: Aeronautics, Commercial—Moral and ethical aspects. |
Aeronautics—Social aspects.
Classification: LCC TL553 .E84 2018 | DDC 174/.93877—dc23
LC record available at https://lccn.loc.gov/2018025672

ISBN: 978-1-138-34808-0 (hbk)
ISBN: 978-1-4724-7086-7 (pbk)
ISBN: 978-0-429-43678-9 (ebk)

Typeset in Bembo and Stone Sans
by Florence Production Ltd, Stoodleigh, Devon, UK

Dedicated to my mother, M. Teresa Hoppe,
and in memory of my father, Harley H. Hoppe

CONTENTS

ILLUSTRATIONS

Figures

Tables

CONTRIBUTORS

Gail L. Bigelow is a long-serving flight attendant for a major US carrier. She has been active in her union—the Association of Flight Attendants–CWA, American Federation of Labor and Congress of Industrial Organizations (AFL-CIO)—since the mid-1980s, serving on almost every committee. Ms. Bigelow was President of the Master Executive Council at her carrier for nine years, successfully leading the flight attendants through the first CHAOS(tm), campaign which resulted in an industry-leading agreement. She currently serves as Master Executive Council Grievance Chairperson.

Robert Breidenthal is Professor in the Aeronautics and Astronautics Department at the University of Washington. He received his PhD from the California Institute of Technology in 1979. A frequent consultant for industry on problems of fluid dynamics and aerodynamics (working with companies such as Boeing, Learjet, and ABB), he has also published many articles on turbulence, including "Elements of Entrainment" in *Turbulencia, Escola de Primavera em Transicao e Turbulencia.*

Paul Foster is currently the Los Angeles Chapter President of the National Black Coalition of Federal Aviation Employees (NBCFAE). He is also assigned to the Federal Aviation Administration (FAA), Western Pacific Riverside Flight Standards District Office, as an Aviation Safety Inspector (Airworthiness-Maintenance), serving as the Safety Team Program Manager. In addition, he is an adjunct professor with Embry–Riddle Aeronautical University teaching courses in aviation safety, management, and aircraft maintenance. He holds an EdD in Organizational Leadership, a Specialist in Education (EdS) degree in General Education Administration, an MA in Management, a Masters in Aeronautical Science, and a Bachelor's degree in Professional Aeronautics from Pepperdine

University, Troy State University in Montgomery, Webster University, and Embry–Riddle Aeronautical University, respectively.

Kenny Frank is a graduate of Lewis University, where he received a BS in Aviation Administration in 2006. While in college he was employed by the former airline ATA and Signature Flight Support. After graduating he worked at Chicago's O'Hare International Airport as a Ramp Supervisor for the ground handling company Servisair. He was an employee of Southwest Airlines at Chicago's Midway International Airport, where he has worked as a Ramp Supervisor and Operations Supervisor. He currently works for United Airlines as a Senior Analyst for Turn Strategy and Optimization.

Bruce Hoover received his MS degree from Oklahoma State University (OSU) and has completed hours toward a PhD. Mr. Hoover holds an Airline Transport Pilot certificate as well as a Gold Seal Flight Instructor certificate. Bruce received a Teacher of the Year award at OSU in 1986. As an adjunct faculty member in Aviation Science at Saint Louis University, he teaches undergraduate courses in applied ethics, air carrier operations, and air carrier economics. Bruce was formerly been Director of Training for a large regional air carrier and Flight Training Director for a university. He also served in the US Air Force. He has recently completed a term on the University Aviation Association Board of Trustees as an officer, and has been a consulting editor for the *International Journal of Applied Aviation Studies.*

Elizabeth A. Hoppe is Instructor of Philosophy at Loyola University Chicago. Having taught Business Ethics courses at DePaul University from 1995 to 1996, in 2003 she created an elective course in Aviation Ethics at Lewis University. She is a member of the Association for Professional and Practical Ethics and has presented papers on the ethics of airline deregulation, pilot training, and regulatory capture. She has also served as editor on several projects, and coedited an anthology on critical race theory: *Fanon and the Decolonization of Philosophy* (2010).

Warren Jensen received an MD from the University of California San Francisco and an MS in Aerospace Medicine from Wright State University. He is a professor in the Department of Aviation at the University of North Dakota and a director of aeromedical research. He teaches human factors and aerospace physiology in the undergraduate and graduate programs.

Eric B. Kennedy is Assistant Professor of Disaster and Emergency Management at York University (Toronto). He received his PhD from Arizona State University in the Consortium for Science, Policy & Outcomes. His research focuses on expertise, the management of controversial issues, and decision making under uncertainty. In addition, he studies crisis response and emergency management, with a focus on wildfires. His most recent book, *Citizen Science* (2016) is co-edited with Darlene Cavalier as part of the Rightful Place of Science series.

Steven A. Kolmes is Chair of the Environmental Science Department and occupant of the Reverend John Molter, CSC, Chair in Science at the University of Portland, Oregon. Dr. Kolmes has degrees in Zoology from Ohio University and the University of Wisconsin–Madison. His interests are in the areas of salmon recovery planning, combining ethical and scientific analyses in environmental policy discussions, water and air quality issues, and the sublethal effects of pesticides. He has also served on government scientific advisory panels.

Ted Ludwig worked in aircraft maintenance for the US Air Force (1980–1988), and as an aircraft technician at Comair (1988–1989) and Northwest Airlines (1989–2005). He was also President of Local Aircraft Mechanics Fraternal Association (AMFA) 33 in Minnesota. In 2005 he was suspended from his job at Northwest for two months for bringing to light problems the airline was having with faulty brakes from their vendor. A Department of Transportation (DOT) report released in July 2010 found that Northwest had continued to violate the same type of safety orders that Ludwig had raised concerns about and for which he had been suspended.

Gerardo Martinez, a US Navy veteran (Aviation Machinist Mate Petty Officer Third Class), has over 35 years' experience as an FAA Aviation Safety Inspector. He rose through the ranks from inspector to supervisor, and is currently Manager of the Chicago Flight Standards District Office, a post held since 2004. He is also a Vocational School Graduate of the Aviation Maintenance Technician program at the Chicago Vocational School (CVS), and holds an AAS in Aviation Maintenance Technology from Richard J. Daley College, Chicago, and a BA from Northeastern Illinois University as well as an MA in Political Science/Public Administration.

Russell W. Mills is an associate professor in the Department of Political Science at Bowling Green State University, Ohio. Prior to this appointment he worked as a policy analyst at the FAA in Washington DC, where he developed and evaluated agency-wide reauthorization proposals for political and policy implications, and liaised with internal and external stakeholders to ensure communication of proposals to the DOT, the Office of Management and Budget, and Congress. His research focuses on improving the effectiveness of aviation policy in the United States and improving governance, accountability, and performance in regulation and networks. He has received grants from the Airport Cooperative Research Program and the IBM Center for the Business of Government, and his research has appeared in publications including the *Journal of Public Policy*, *Public Administration Review*, *Regulation and Governance*, *Administration & Society*, and the *Journal of Benefit–Cost Analysis*.

Peter Nekola is Visiting Assistant Professor of Philosophy at Luther College, Iowa. He was formerly the Assistant Director, Hermon Dunlap Smith Center for the History of Cartography at the Newberry Library in Chicago. He received PhDs in both Philosophy and History from the New School for Social Research, New York. His primary area of specialization is the history of cartography.

Michael S. Nolan is a professor in the Department of Aviation Technology at Purdue University, West Lafayette, Indiana. There he leads the air traffic control program and teaches introduction to aviation, aviation infrastructure, air traffic control, and control tower operator courses. His textbook *Fundamentals of Air Traffic Control*, published in 1990, appeared in a fifth edition in 2011.

Brian W. Ohm is a professor in the Department of Urban and Regional Planning at the University of Wisconsin–Madison. He is also an affiliate professor in the Nelson Institute for Environmental Studies there, and holds a joint appointment with the University of Wisconsin Extension, where he is the state specialist in subjects such as land use law and environmental regulation. Professor Ohm previously worked as an attorney for the Minneapolis–St. Paul Metropolitan Council, and was involved in numerous planning initiatives involving land use and aviation. He holds a law degree from the University of Wisconsin Law School, an MA in History from the University of Wisconsin–Madison, and a BA in History and Political Science from St. Olaf College, Minnesota.

Michael Oksner began his career as a US Navy pilot in 1967. In 1976 he left active duty for a position with Braniff Airways as a flight engineer and first officer until the airline's bankruptcy in 1982. In 1984 Captain Oksner was hired by Southwest Airlines as a B-727 flight engineer, and in the ensuing 20 years advanced to B-737 first officer and then captain. He retired in 2004 as a consequence of the FAA's Age 60 Rule. During the last few years with Southwest, he acted as liaison for the Southwest Airlines Pilots Association with various groups and individuals opposing the Rule. Captain Oksner was a plaintiff in the *Yetman*, *Butler*, *Oksner*, and *Adams* suits seeking waivers to fly past 60 and, subsequently, retirement compensation.

Bill Parrot is Professor of Aviation at Lewis University, Illinois, where he teaches courses such as Issues and Trends and Air Traffic Control Systems in both graduate and undergraduate programs. Prior to his teaching career, Professor Parrot was a pilot for American Airlines as well as an air traffic controller.

Dawna L. Rhoades, PhD, is currently Chair of the Department of Management, Marketing, and Operations and was formerly Associate Dean for Research and Graduate Studies in the College of Business at Embry–Riddle Aeronautical University in Daytona Beach, Florida. Her research interests include airline strategy,

NextGen systems, commercial space operations, and intermodal and sustainable transportation. Her work has appeared in such journals as the *Journal of Air Transport Management*, *Journal of Managerial Issues*, *World Review of Science, Technology, and Sustainable Development*, and the *Handbook of Airline Strategy*. She is the author of *Evolution of International Aviation: Phoenix Rising* (3rd edition, 2014) and editor of the *World Review of Intermodal Transportation Research (WRITR)*.

Nathan Ross is Professor of Philosophy at Oklahoma City University. He received his PhD in Philosophy from DePaul University, Chicago, and has taught business ethics throughout his career. He has published books on the Hegel and critical theory, as well as articles in journals such as *Epoché*, *Philosophy Today*, and *Graduate Faculty Philosophy Journal*. His most recent book, *The Philosophy and Politics of Aesthetic Experience* (2017), examines the role of art in enabling critical reflections on society. He flies frequently, and makes it a habit to note areas of improvement for the American aviation industry.

Joseph P. Schwieterman, PhD, a professor in the School of Public Service at DePaul University, is a noted authority on transportation and a long-standing contributor to the Transportation Research Board (TRB), a unit of the National Academy of Sciences. He spent eight years working in yield management for a Chicago-based airline. His work appears in such scholarly journals as *Regulation*, *Transport Reviews*, and the *Journal of the Transportation Research Board*. Dr. Schwieterman has testified on transportation issues on three occasions before subcommittees of the US Congress.

James E. Sulton, III is a Supervisory Air Traffic Controller at Chicago TRACON in Elgin, Illinois. He holds a BS in Aerospace Studies and an MS in Aeronautics from Embry–Riddle Aeronautical University. Dr. Sulton received his doctorate in Educational Leadership, Administration, and Policy from Pepperdine University, California. Prior to assuming roles with the FAA, he led Oakland Aviation High School, developed the aviation curriculum for the Southern California Regional Occupational Center, and founded Aviation Ed, an organization focused on introducing at-risk youth to aviation careers. Dr. Sulton lectures frequently about the importance of education and career opportunities for inner-city youth in the field, including at the Better Boys Foundation and Upward Bound.

Mark H. Waymack is Chair and Associate Professor of Philosophy at Loyola University Chicago. His primary research interests have been in applied philosophy, including research and teaching in the areas of medical ethics, biomedical research, and organizational and business ethics. In addition to numerous articles, he has co-authored several books, including *Medical Ethics and the Elderly* (with George Taler, MD, 1988) and *Ethics, Aging, and Society: The Critical Turn* (with Jennifer Parks and Martha Holstein, 2010).

Flint Whitlock is an author and historian who has published numerous books and articles on a wide range of topics. His first book was entitled *Soldiers on Skis* (1992), and some of his more recent publications include *The Depths of Courage* (2007) and *Turbulence before Takeoff* (2009). The latter work focuses on racial discrimination against the pilot Marlon Green, as well as other aspects of racial discrimination, and was a finalist in the biography category at the 2009 Colorado Book Awards, sponsored by the Colorado Endowment for the Humanities.

Richard L. Wilson is currently an instructor at Towson University, Maryland, teaching ethics in both Philosophy and Computer Science, after having taught at the University of Maryland Baltimore County (UMBC) for many years. He has presented papers on anticipatory ethics and stakeholder theory at numerous conferences, including the Association for Practical and Professional Ethics, Interdisciplinary Studies, and the Society for Business Ethics.

PREFACE

In the academic discipline known as ethics one of the recurring issues concerns the relationship between theory and practice. While ethical theory attempts to provide frameworks and principles for the correct way for human action, problems arise when putting theory into practice. It is one thing to know what one should do in general, but how should one act in specific situations, especially when the morally correct action is unclear? In the 20th century a new area of study, known as applied ethics, arose to address questions of practice, most notably in health care, business, and the environment. Conferences and publications have addressed specific ethical questions pertinent to their field. Using the field of applied ethics as a foundation, this second edition provides unique insights into the relationship between ethical theory and the aviation industry. Because aviation is so concerned with safety, it seems natural that a text should be devoted to key ethical problems that arise.

Since the publication of the first edition in 2011, numerous ethical issues have continued to confront the aviation industry. Some prominent cases include the disappearance of Malaysia Airlines 370 and the pilot suicide of Germanwings Flight 9525, among other accidents. In addition, controversies surrounding the impact of aviation on the environment and ways to offset carbon emissions have been a focus of debate. The United Airlines route known as the Chairman's Flight led to the resignation of CEO Jeff Smisek amid a federal investigation. Regarding air traffic control in the US, debates over privatization continue today.

This new edition is updated to cover these issues, along with revising and updating chapters from the previous version of the book. One new feature includes the appearance of textboxes within certain chapters. These boxes provide specific scenarios meant to be used for further discussion of the chapter's contents. As with the first edition, this volume is not intended to provide an exhaustive account of all ethical issues that arise in aviation. It will instead provide an overview of important topics, focusing mostly on current issues and trends. However, it also addresses several past events that continue to affect the industry, such as the history of racial discrimination in the United States, the air traffic controller strike of 1981, and the ValuJet 592 accident. These historical pieces continue to provide important lessons for the ethical issues that continue to confront aviation today.

The text is divided into six parts and begins by addressing key theoretical concepts in both ethics and capitalism as they relate to the field of aviation. Part I is followed by chapters that focus on some general topics, especially concerning the business side of aviation, including the regulation of air space, outsourcing maintenance, as well as airline ticket pricing and finally the ethics of whistleblowing. Part III provides an analysis of issues in moral responsibility by examining FAA inspectors and regulatory capture, pilots, flight attendants, and airline management. In Part IV the authors examine different types of diversity issues that arise in the US, both from historical and contemporary grounds. This part includes discussions not only on race and gender but also on the former Age 60 Rule. Moving to environmental concerns, Part V addresses health, especially concerning pilots, followed by the impact of the industry on the environment. Finally, Part VI addresses airports, including land use and planning strategies, aviation mapping, and the past and future of air traffic control.

The contributors to this volume include academics doing research in the field as well as professionals who provide accounts of the ethical situations that one may encounter in the workplace. Because of the variety of viewpoints offered in this text the reader should be able to gain a broad understanding of the ethical dimensions of the aviation industry, and in turn discover ways to overcome the challenges that one may face. The book does not offer a singular moral point of view or promote a specific policy. Instead it is my hope that it will serve as a springboard for further ethical discussion on the key issues that continue to confront and challenge aviation today.

Elizabeth A. Hoppe
Loyola University Chicago
April 22, 2018

ACKNOWLEDGEMENTS

The first edition of the aviation ethics book would not have been possible had it not been for Mike Streit, former Chair of the Aviation Department at Lewis University. I first approached Mike in 2002 with the concept for a course in aviation ethics, and he immediately approved the idea. He also actively recruited students for the first course offered in the fall semester of 2003. Thanks to Mike the class was a success and led to the creation of the first edition.

For the second edition of this project, several people provided invaluable insight into the various topics. They include: Charles Barnett, III, Jalal Haidar, Ronald Kluk, Oliver Lagman, Bob Lavender, Maggie Ma, Robert Mark, David Pritchard, Joseph Ryan, and Jon P. Simon. Finally, I would like to remember Alan MacPherson, who wrote a chapter for the first edition and was published posthumously thanks to approval from his widow, Valerie Randall.

I would like to thank friends and family who not only helped me locate contacts, but also supported the overall project. In particular, I would like to acknowledge Maureen Anderson, Leslie Brissette, Stephanie Giggetts, Fran Gordon, Gary Gordon, Graham Harman, Amy Hoppe, Susan Hoppe, Teresa Hoppe, Mitchell May, Denise Taylor, Jeff Tilden, Tama Weisman, and Barbara Zorich.

Some people were instrumental in other ways—those who helped me stay alert and focused—and the acknowledgments would be incomplete without thanking the following baristas: Doug, Gary, Jacqueline, Jay, and Ricardo.

This publication would not exist without the support of Routledge Publishing, and I am especially thankful for the assistance provided by Maria Anson, Amanda Buxton (Ashgate), and Emma Redley. I would also like to especially thank Guy Loft, who worked closely with me for the first edition along with the content for the second edition. Last but not least, I would like to express my gratitude to Matthew Ranscombe for his tireless efforts in seeing this edition to fruition. Because of Matt's guidance, I was able to prepare this manuscript to the best of my abilities.

ABBREVIATIONS

AAC	Airline Attraction Committee
AAIASB	Air Accident Investigation and Aviation Safety Board
A and P	Airframe and Powerplant
ACA	Atlantic Coast Airlines
ACARS	Aircraft Communications Addressing and Reporting Systems
ACLU	American Civil Liberties Union
ACY	Atlantic City International Airport
AD	Airworthiness Directive
ADEA	Age Discrimination in Employment Act
AFA–CWA	Association of Flight Attendants–Communications Workers of America
AFGE	American Federation of Government Employees
AIM	Aviation Institute of Maintenance
AIP	Airport Improvement Program
AIT	Advance Imaging Technology
ALJ	Office of Administrative Law Judges
ALPA	Air Line Pilots Association
AOPA	Aircraft Owners and Pilots Association
ARB	Administrative Review Board
ASAP	Aviation Safety Action Program
ASD	Air Service Development
ASI	Aviation Safety Inspector
ATA	Air Transport Agreement
ATAA	Air Transport Association of America
ATC	Air Traffic Control
ATI	Anti-Trust Immunity
ATPC	Airline Tariff Publishers Company

AWAM	Association for Women in Aviation Maintenance
BEA	Bureau d'Enquêtes et d'Analyses
BFOQ	Bona Fide Occupation Qualification
CAA	Civil Aviation Authority (UK)
CAB	Civil Aeronautics Board
CAL	Continental Airlines
CAMI	Civil Aeromedical Institute
CBDRRC	Common but Differentiated Responsibilities and Respective Capacities
CFI	Certified Flight Instructor
CJEU	European Court of Justice
CLT	Charlotte Douglas International Airport
CRDA	Casino Reinvestment Development Authority (Atlantic City)
CRM	Crew Resource Management
CRS	Child Restraint System
CMO	Certificate Management Office
CSI	Customer Service Initiative
CVB	Convention and Visitor Bureau
CVR	Cockpit Voice Recorder
DHS	Department of Homeland Security
DOJ	Department of Justice
DOL	Department of Labor
DOT	Department of Transportation
ECA	European Cockpit Association
EEA	European Economic Area
EEOC	Equal Employment Opportunity Commission
EETS	European Emissions Trading Scheme
EPA	Environmental Protection Agency
ER	Emission Reduction
ERAU	Embry-Riddle Aeronautical University
ETS	Emissions Trading Scheme
EWR	Newark Liberty International Airport
FAA	Federal Aviation Administration
FAB	Functional Airspace Block
FAIR	Canadian Federal Accountability Initiative for Reform
FAR	Federal Aviation Regulation
FBO	Fixed Base Operator
FBW	Fly-by-Wire
FOQA	Flight Operations Quality Assurance
FWA	Fort Wayne International Airport
GAO	General Accountability Office
GCRI	Georgetown Clinical Research Institute
GHG	Greenhouse Gas
GPS	Global Positioning System

HFCS	Human Factors Analysis and Classification System
IAC	Internal Assistance Capability
IASA	International Air Safety Administration
IATA	International Air Transport Association
ICAN	International Commission on Air Navigation
ICAO	International Civil Aviation Organization
ICT	Information Computer Technology
ICTSD	International Centre for Trade and Sustainable Development
ICTZ	Inter-Tropical Convergence Zone
IDB	Involuntary Denied Boarding
ILEAV	Inherently Low-Emission Airport Vehicle Program
IMW	International Map of the World
IOM	Institute of Medicine
IPCC	Intergovernmental Panel on Climate Change
IRT	Independent Review Team
JAA	Joint Aviation Authorities
JAL	Japan Airlines
JFK	John F. Kennedy International Airport
JTSB	Japan Transportation Safety Board
LCA	Large Commercial Aircraft
LCC	Low Cost Carrier
LGA	LaGuardia Airport
MAS	Malaysia Airlines System
MH	Malaysia Airlines
MOU	Memorandum of Understanding
MRG	Minimum Revenue Guarantees
MSP	Minneapolis–St. Paul International Airport
MSPB	Merit Systems Protection Board
NASA	National Aeronautics and Space Administration
NATCA	National Air Traffic Controllers Association
NHTSA	National Highway Transportation Safety Administration
NIA	National Institute on Aging
NIH	National Institutes of Health
NIPP	National Infrastructure Protection Plan
NPRM	Notice of Proposed Rulemaking
NRC	National Research Council
NTSB	National Transportation Safety Board
O&D	Origin and Destination
OAA	Office of Aviation Analysis
OAK	Oakland International Airport
OBAP	Organization of Black Aerospace Professionals (formerly Organization of Black Airline Pilots)
OEM	Original Equipment Manufacturer
ORD	Chicago O'Hare International Airport

OSC	Office of Special Counsel
OSHA	Occupational Safety and Health Administration
PAFI	Piston Aviation Fuel Initiative
PAH	Polycyclic Aromatic Hydrocarbons
PATCO	Professional Air Traffic Controllers Organization
PF	Pilot Flying
PHL	Philadelphia International Airport
PMI	Principal Maintenance Inspector
PNF	Pilot Not Flying
PRO	Profile Aging Ratio
QAS	Quality Assurance Standard
RPN	Responsible Purchasing Network
RPV	Remotely Piloted Vehicle
SARS	Severe Acute Respiratory Syndrome
SCASD	Small Community Air Service Development
SES	Single European Sky
SFO	San Francisco International Airport
SMF	Sacramento International Airport
SOX	Sarbanes–Oxley Act
STS	Sonoma County Airport
SWA	Southwest Airlines
TCAS	Traffic Alert Collision Avoidance System
TSA	Transportation Security Administration
UAT ARC	Unleaded Avgas Transition Aviation Rulemaking Committee
UAV	Unmanned Aerial Vehicle
UAS	Unmanned Aircraft System
UDHR	Universal Declaration of Human Rights
UFP	Ultrafine Particulate
UN	United Nations
UNEP	United Nations Environment Programme
USCCB	United States Council of Catholic Bishops
VDRP	Voluntary Disclosure Reporting Program
VFR	Visual Flight Rules
VOC	Volatile Organic Compound
WAC	World Aeronautical Chart
WAI	Women in Aviation, International
WHO	World Health Organization
WPA	Whistleblower Protection Act
WTO	World Trade Organization
XDR TB	Extensively Drug-Resistant Tuberculosis

PART I

Theoretical frameworks

In the aviation industry a variety of ethical theories are utilized even without one necessarily being aware of them. For example, the rules and regulations that the FAA creates are based on a cost–benefit approach to decision-making. Such an approach to regulation was first introduced in Executive Order 12866 of September 1993 and reaffirmed in Executive Order 13563 of January 2011. Each agency is directed to use present and future benefits and costs as accurately as possible (White House 2011). While this type of analysis is economic rather than moral, the ethics of consequentialism parallels the same method for determining morally right and wrong actions.

Prior to examining the various ethical theories and dilemmas that arise in the aviation industry, this opening part of the text examines important theoretical frameworks, not only of ethics but also of capitalism. Some of the key concepts that arise in Part I can help provide a means for assessing the issues that arise throughout this text.

Chapter 1 by Mark H. Waymack investigates and critiques two forms of ethics that emphasize how we determine the choices we make: consequentialism and deontology, or duty-based reasoning. Both consequentialist and deontological ethics emphasize the importance of decision-making. Questions such as are consequences what matter in ethics or should morality be based on universal laws are important issues for the aviation industry. Waymack's analysis reveals the fundamental aspects of consequentialism in order to assess some of its main strengths and weaknesses. He follows this with an account of Kantian deontology, the type of ethics that is concerned with adherence to universal moral laws rather than the consequences of an action.

In focusing on how to live an ethical life, in Chapter 2 Waymack reveals two other directions that ethics may take: rights-based reasoning and virtue ethics, a form of ethics which provides ways to develop a person's character. The hope is that a virtuous person would make good decisions, and thus the focus should be on one's character rather than one's choices. Questions about rights include issues of whether or not rights are something natural or if they arise due to a contract made by people in society. Although Waymack points out that no ethical theory is perfect, he also argues against a skeptic who would assert that ethics is useless. Instead Waymack advocates for the value of studying ethics, and provides a seven-step reasoning process that utilizes the different ethical theories analyzed in Chapters 1 and 2.

The third and final chapter in this section investigates historical and contemporary assessments of capitalism in order for the reader to reflect on its strengths and limitations. As with any other industry, in their attempts to maximize profits airlines may cut corners. The point is to ask when the drive for profits crosses a line such that it becomes unethical. Nathan Ross first addresses historical proponents of capitalism—such as John Locke on property rights, David Hume's defense of luxury, and Adam Smith on the division of labor—in order to develop the merits of the capitalist system. Ross follows this analysis with an account of some of the main critics of capitalism, most notably the nineteenth-century theorist Karl Marx and his critique of surplus labor value. Marx argues that workers are exploited by the capitalist system in that management does not pay workers according to the value of what they produce. Ross then ends the chapter with a discussion of contemporary European critics, most notably Walter Benjamin, Georges Bataille, and Jean Baudrillard. Some of the more troubling aspects of contemporary capitalism include overproduction and consumerism. Both of these problems relate to issues in the aviation industry today, such as the consumer's inability to have a say in choices regarding ticketing, frequency of flights, routes, etc.

Reference

White House. 2011 (Jan 18). Executive Order 13563: improving regulation and regulatory review. Available at: www.whitehouse.gov/the-press-office/2011/01/18/executive-order-13563-improving-regulation-and-regulatory-review [accessed September 15, 2016].

1

ETHICAL THEORIES I

Dilemmas and decision-making

Mark H. Waymack

On February 12, 2009, Colgan Air (operating as Continental Connection) Flight 3407 crashed near its destination in Buffalo, NY, claiming all 50 lives on board. Since the accident occurred in wintry conditions, icing on the wings was a hypothesis that immediately came to many minds in explaining the main cause of the crash. However, as the investigation methodically proceeded, what emerged was the story of a response to an impending stall that went contrary to pilot training. According to the Executive Summary of the National Transportation Safety Board's (NTSB's) aircraft accident report:

> The probable cause of this accident was the captain's inappropriate response to the activation of the stick shaker [the stick shaker warns a pilot of an impending wing aerodynamic stall], which led to an aerodynamic stall from which the airplane did not recover. Contributing to the accident were: (1) the flight crew's failure to monitor airspeed in relation to the rising position of the lowspeed cue, (2) the flight crew's failure to adhere to sterile cockpit procedures, (3) the captain's failure to effectively manage the flight, and (4) Colgan Air's inadequate procedures for airspeed selection and management during approaches in icing conditions. (NTSB 2010: x)

Actions (or inactions) on the part of the first officer exacerbated the seriousness of the captain's incorrect actions, and succeeded only in accelerating the plane's trajectory into a doomed, fatal stall.

Behind these tragic bad decisions in the final seconds of Flight 3407, however, was a series of choices and circumstances that led to this fatal accident. The NTSB notes that the captain had several FAA certificate disapprovals, both before and after his employment at Colgan Air, two of which he failed to admit on his Colgan Air application (NTSB 2010: 9–10). The NTSB also found that the air carrier's approach-to-stall training program did not adequately prepare pilots for an unexpected stall or how to recover from it (NTSB 2010: 153). Furthermore, the low wages of the first officer made it impractical for her to live in the Newark, NJ area (her base); she in fact lived at her parents' home near Seattle, WA.[1] She had traveled from Seattle the day before, and then on the day of the accident spent part of the time sleeping on a couch in the crew room at the Newark Liberty International Airport (EWR). The captain also apparently spent some time sleeping on a couch in the crew room. Regarding the possibility of fatigue as a factor in this flight, the report states that "the pilots' performance was likely impaired because of fatigue, but the extent of their impairment and the degree to which it contributed to the performance deficiencies that occurred during the flight cannot be conclusively determined" (NTSB 2010: 153). Although we cannot say to what extent fatigue directly led to the accident, we do know that it was not icing or mechanical failure that caused the fatal crash: it was pilot error. Also contributing to the flight deck errors were inadequate training, falsification or concealment of FAA certificate disapprovals, and sleep deprivation attributed, indirectly, to the first officer's low salary.

Needless to say, the crash and its loss of life are deeply regrettable. Still, no means of transportation, ancient or modern, is entirely safe. We could just say, like an accident on the highway, bad things sometimes just happen. And that may be true in a sense. But when we step onto a commercial airplane, we, as passengers, are trusting in the good faith of the airline and in the qualifications of our flight crew. In this case it looks as though both the crew and the company failed our trust. In many ways, the things that happened "ought not" to have happened. In particular, the captain, the first officer, and Colgan Air (and its contractor Continental Airlines) all made choices and worked in ways that unnecessarily contributed to the accident. As there was a serious violation of trust, we shall argue that they all also acted unethically.

It is one thing, however, to have an intuition about ethics, or in other words a gut reaction; yet, as we all know too well, gut reactions can sometimes lead us astray. Indeed the gut reaction of the captain of Flight 3407 as the stick shaker activated was to apply a 37-pound pull force to the control column, a choice that caused the plane's wing to stall (NTSB 2010: 82). It can also be quite another thing to clearly articulate and justify our moral judgments, actions, and conclusions. Yet without express deliberation and articulation, how can we, or others, be confident in our moral choices and be assured that we are not simply working with prejudiced or misguided moral judgments?

An overview of ethical theory

Moral decision-making can be a very difficult task; yet it is not a task that we can avoid. Our society has, in its own way, made this decision-making process even harder, for it is now composed of a number of different cultures that have disparate, and often conflicting, moral traditions. For example, Roman Catholicism strongly opposes abortion rights, whereas political liberalism, which emphasizes the rights of the individual, reserves abortion as the free choice of the pregnant woman. Not everyone can be morally right when there is such fundamental disagreement. But how can we determine who is choosing wisely and who is choosing poorly? How can we know whether we are choosing correctly or if we are acting out of unjustified and narrow-minded prejudice?

Suppose there were several people all arguing about how tall a certain tree is. A sensible tactic would certainly be to measure how tall it actually is. Now this requires two important steps. First, we must agree upon what the appropriate units of measure are. That is, against what standard are we going to calculate the tree's size? And, second, we must go through some process of applying the standard to the tree; that is, doing the measuring. These two steps take the guessing out of the measurement and help ensure that everyone is most likely to agree on the correctness of the answer that we reach.

How can we translate this example to our difficulties in moral decision-making? If we can rationally agree on a standard of measurement and also agree on how to apply that standard to ethically difficult cases, then our difficulties should be largely dissolved. Unfortunately, the matter is not quite that simple. In the example of the tree, it was agreed by all that we were measuring the height of the tree. That is rather straightforward. Feet, yards, meters may be different possible units of measure, but they are all commensurable; that is, they can all be translated into the terms of each other. But with moral arguments one of the serious difficulties we face is that we may often disagree over what it is we think we ought to be measuring. It would be as though some people thought that what was important when measuring trees is their width, or the circumference of their trunks, or perhaps even the size of their leaves rather than their height.

There is, in fact, just this sort of disagreement in morality. Some people think that what ought to be measured are the consequences of an action. For example, would the action produce happiness, or would it cause pain or suffering? Other thinkers focus upon adherence to certain forms of law. Does the action conform to the law or principle that one should only do what one would want everyone else to do? Still other thinkers look to moral rights. Would a certain action promote or violate someone's moral rights? And some thinkers regard virtue or good character as the proper focus of moral evaluation.

In this chapter we will study two different theories that provide standards for moral reasoning, ones that offer relatively clear decision procedures. Such ethical theories are attempts to articulate clearly what the standard of measurement should be, why it is the appropriate standard, and how that standard or reasoning process is to be applied in our moral decision-making. The theories we will discuss each utilize quite different features as the most important aspects to be measured; but we have chosen these because we think they each capture some important and plausible insights into what moral reasoning ought to be doing and how it ought to be judged. The two theories of ethical reasoning that we will discuss in this chapter are: (1) Consequentialist, or value-maximizing; and (2) Deontological, or duty-based reasoning.

Consequentialist ethics

Consequentialist reasoning

When we contemplate a case like that of Flight 3407, some of the first things that come to mind are the consequences of our possible choices. If the captain had been honest on his job application, either he would not have been hired or he might have received additional training. Thus, the accident might have been avoided. Furthermore, if the airline had more closely enforced training requirements and also paid a more reasonable salary, the crew would have been technically better prepared as well as not sleep deprived. Instead, the captain withheld information, more than likely in order to be a better job candidate; both pilots chose to fly even when they probably did not have an adequate amount of sleep; and the airline chose to save money by requiring less training and paying wages that are hardly above the poverty line.[2]

The consequentialist begins with the assumption that humans, as agents, are goal directed. We are creatures who, in our actions, make choices that aim at certain goals in order to produce certain results. Furthermore, we judge the correctness of these actions according to how effectively they produce the intended consequences. For example, we judge investment bankers according to their success at making large profits. We evaluate an automobile manufacturer by the product of its labor—does the car run smoothly, efficiently, and safely? And we critique chefs by the level of pleasure we receive from the meals they prepare. According to the consequentialist, moral judgments are in principle made no differently. We judge human actions to be morally right or wrong, morally good or bad, according to the consequences that those actions produce.

Our account of consequential ethics has thus far been somewhat vague. To say that the consequences are what matters does not tell us much. Two questions need to be brought forward: (1) consequences may be the essence of moral reasoning, but consequences for whom?; and (2) specifically what sort of consequences should we attempt to achieve? What makes a good consequence good and a bad consequence bad?

Consequences for whom?

Any action we perform is likely to have consequences for, or upon, many different people. In cases that involve the question of whether or not to follow proper procedures in aircraft maintenance, there will be consequences to the maintenance technician, the supervisor, and even the airline. Depending on the situation, the risk of significant harm to passengers and crew may greatly increase.[3] Different moral theories, though, have given different answers to the question of who counts as morally important.

Consequences for the self: ethical egoism

One possible answer is that the consequences to me are what matter. This ethics of self-interest is also known as "ethical egoism." What is morally right is the action that has beneficial consequences for me. This is an unsatisfactory answer for a number of reasons. What are we to say of other persons? If we take this type of ethics to mean that "everyone should do those and only those actions that are in their own self-interest," then what do we say to the other person when his or her interest conflicts with our own? Am I to maintain that you are morally wrong to pursue your self-interest when what you seek to do will harm my self-interest? Does it make sense to state that the same action would be morally right and morally wrong at the same time? If we try to apply this standard of right and wrong to Flight 3407, we can easily see just how confused and useless it is. Should the pilots simply pursue their own self-interests? The self-interests of the passengers and crew, however, may well dictate a different course of action. One way out of that quagmire is to suggest that "everyone should pursue my self-interest." But it is quite difficult to see what could make any one person (me) so special that the rest of the world's population should spend its life trying to create the best consequences for that one individual (me, of course!).

Perhaps more damaging than these intellectual puzzles, however, is the question of what sort of life the ethical egoist is actually likely to live. Think of how we treat other persons whom we know are out only for their own interests. Knowing what motivates them (pure self-interest), we find we are unable to trust them, for they will be willing to lie, cheat, steal, or mislead us if doing so proves to be in their own self-interest. How many of us would be willing to be passengers onboard a commercial aircraft if we thought that airlines found it ethically permissible for pilots to falsify documents whenever they thought it might serve their financial self-interest? Would we trust our car mechanics if we thought they would be willing to fake repair work or even sabotage our automobile in order to generate more revenue?

We surely all have some experience of such egoistic individuals, and know that they do not make good friends. Indeed, they seem to lack an understanding of something that is important to what it is to be a friend. It is hard to see how the egoist could even have a real friendship. Therefore, the ethical egoist would have to live a life in which that person could trust no one and no one trusted them. The egoist could not divulge to anyone who they really are, for doing so would run counter to the egoist's self-interest. All in all, it would seem that the committed egoist would lead an isolated and impoverished life. If one truly thought that what matters in the end are consequences only to the self, then in order to live an emotionally rich and happy life the first thing one should do is stop thinking and acting like an egoist!

Consequences for others: ethical altruism

Another possible answer to the "for whom" question is that what matters are the consequences for other persons. In determining moral right and wrong, we should examine how our actions will affect other persons—morally good actions have good consequences for others and morally bad actions have bad consequences for others. One's self-interests do not have any standing in the moral decision-making.

In its own peculiar way, this sort of theory runs the risk of being self-defeating. For an individual to act altruistically there has to be another individual who accepts the benefit. But if everyone were morally good, altruistic persons, then who will be around to be the recipient of all these intended altruistic actions? Furthermore, while egoism may over-emphasize the self-interested aspect of human nature, ethical altruism seems guilty of unrealistically underestimating the strength of individual self-interest.

Consequences for everyone

Another suggestion, and one that is far more plausible than the preceding two, is that one should take into account the effects of one's actions upon everyone, one's self included. A well-known example of this way of thinking is Utilitarianism, an ethical theory that we shall discuss more fully below. The basic idea of Utilitarianism is that one ought to promote the general happiness, that is, the happiness of everyone as a whole group.

One reason for the enduring appeal of ethical theories that emphasize the consequences for everyone, and not just for particular individuals, is that such theories emphasize impartiality, something that most of us think is very central to moral claims or demands. We may have personal reasons for pursuing self-interest. We may enjoy helping our close friends. But moral reasons raise the level of discussion—of reason-giving—to a higher level, a level of impartiality. We shall, therefore, focus our discussion upon consequential ethics that emphasize consequences to everyone.

What consequences?

Another question that the consequentialist must answer is "What sorts of consequences count, either as good or as bad?" Answers to this question have varied enormously. One suggestion, though one that we should surely reject as reprehensible, is that one should seek to maximize the genetic "purity" of the population. Most consequentialists have in mind something far more sensible, such as maximizing human happiness or human fulfillment. Just what these terms mean, however, can be a matter of great dispute. For example, is the enjoyment of playing poker to be considered as ethically valuable as the enjoyment of reading an artistic and uplifting piece of literature? Are the pleasures of physical sex just as morally important as the pleasures of studying philosophy?

Utilitarianism: a well-known form of ethical consequentialism

Having outlined some of the important kinds of questions that an ethical consequentialist must answer, let us now consider one of most well-known forms of consequential ethics that have been offered over the centuries: Utilitarianism. According to this theory, the standard of moral right and wrong is the extent to which actions contribute to or detract from happiness. Utilitarian-type ideas have been around for millennia but were first presented as a clearly thought out ethical theory in the late eighteenth century by Jeremy Bentham. One of Utilitarianism's most famous exponents is John Stuart Mill (1806–1873). As Mill states in his 1863 publication *Utilitarianism*:

> The creed which accepts as the foundation of morals, Utility, or the Greatest Happiness Principle, holds that actions are right in proportion as they tend to promote happiness, wrong as they tend to produce the reverse of happiness. By happiness is intended pleasure, and the absence of pain; by unhappiness, pain, and the privation of pleasure . . . [P]leasure, and freedom from pain, are the only things desirable as ends; and that all desirable things (which are as numerous in the utilitarians as in any other scheme) are desirable either for the pleasure inherent in themselves, or as means to the promotion of pleasure and the prevention of pain . . . [T]he happiness which forms the utilitarian standard of what is right conduct, is not the agent's own happiness, but that of all concerned. (Mill 1979: 6–7, 16)

These passages make several relevant points. First, what matters for Utilitarians is happiness and suffering. Happiness (pleasure) is good, and pain and suffering are bad. Second, it is the general happiness that is important. And, third, actions are judged by their consequences, that is, by how much they contribute to, or detract from, the general happiness. The morally right action, under any given circumstance, will be the action which among all the possible choices most contributes to or least detracts from the general happiness of everyone involved.

One point that should be emphasized is that many Utilitarians, particularly Mill, are adamant that human happiness should not be thought of simply in terms of physical, mindless pleasures. Many pleasures appeal to the "higher" aspects of human nature—our intellect and noble sentiments. These higher pleasures, according to Mill, deserve more consideration than the lower pleasures. As Mill maintains:

> it is quite compatible with the principle of utility to recognize the fact that some kinds of pleasure are more desirable and more valuable than others. It would be absurd that . . . the estimation of pleasures should be supposed to depend on quantity alone. (Mill 1979: 8)

Thus, Utilitarianism should not be construed as a moral philosophy of pleasure-seeking pigs, but as the moral philosophy of an intellectually and emotionally sophisticated human society.

For our purposes it is not necessary to ascertain what a true Utilitarian would demand in the case of Flight 3407. But we can consider how Utilitarians would argue about the case. Should the captain have revealed all of his certificate disapprovals on his application? Some Utilitarians might argue that the slim chance of failure needs to be realistically weighed against the very real outcome of inconvenience, loss of income, retribution, etc. But I suspect that most Utilitarians would give great weight to the numerous deaths that might occur in the event of a tragic failure on the part of the pilot. After all, dying in a crash is a much greater loss than a single person's loss of an employment opportunity. Further, if the pilot made a tragic mistake, then the airline would surely be subject to expensive wrongful death lawsuits. Even when discounted by their low probability, these harms are so great that Utilitarianism may well recommend honesty in job applications, better training, and higher salaries. The costs involved here would be outweighed by the deaths prevented by a potential mistake.

One major difficulty Utilitarianism faces is that it does not do a very good job of explaining or supporting our moral intuitions about human rights or justice. For example, suppose that society would be made enormously happy (or it might avoid great destruction) if a particular individual, who had done no wrong, were to be imprisoned or executed. Utilitarianism would seem committed to saying that the poor fellow ought to be made a prisoner or executed. The reason of course is that the only ultimate standard in Utilitarianism is the general happiness. Since human rights and the rules of justice have no intrinsic moral value in Utilitarianism, they are valuable only insofar as they contribute to overall human happiness.

Utilitarians would reply that we should not unjustly imprison someone, or violate their rights in some other way, because doing so breeds distrust and a contemptuous or callous attitude towards rights. When society takes that particular turn, each of us becomes at risk and we become apprehensive, defensive, and unhappy. We must be aware that our actions have consequences beyond the immediate participants. What we do or sanction in a particular instance may be read by others as a precedent for their behavior. If respect for the individual and one's rights can be overlooked in one instance, it can be overlooked in other cases, including perhaps our own rights. Thus, society best promotes happiness when it chooses to respect human rights and the rules of justice as a matter of general policy. Though such a response may yield the intuitively correct moral answer when we ask about punishment of the innocent for the sake of society as a whole, it is not clear that Utilitarianism gives the right answer for the right reason. Respect for human agents still has no intrinsic moral value in the utilitarian system. At best it respects human rights as a useful means to happiness.

A second major difficulty is that Utilitarianism seems to be built upon an impoverished view of human nature and human motivations. Our experience may convince us that we at times make decisions not simply upon the basis of which actions will produce the most overall happiness. Rather, we look to other sorts of ends, goals, as well.

Strengths and weaknesses of consequential ethics

There are several strengths to consequential theories of ethics. First, they appeal to a common-sense experience—that people do things to accomplish things, to produce certain results. Moral reasoning, then, is simply one particular area of this kind of practical reasoning. A second strength is that consequential ethics appeals to a criterion of moral right and wrong that is a public criterion. Its primary focus is not on the inscrutable motives that individuals might have (or not have). It concerns the results of actions, results that we and others can observe and assess. Considering our opening case of Flight 3407, a Utilitarian would probably argue that there are several ways in which narrow-minded decisions eventually caused considerable—and avoidable—harm to many persons. The pilot, the first officer, and the company all made choices that looked to serve their own particular interests at the expense of others: instead of considering the general welfare, each of them was thinking of their perceived individual best interests. Sadly, in the end they, as well as all of their passengers and one person on the ground, were on the losing side of the equation.

Nevertheless, we might feel that consequentialist reasoning, while it captures something true about human nature and motivation, is not the whole story. Indeed, we might think that consequentialism, though an accurate description of much of human reasoning and motivation, is not an appropriate characterization of moral reasoning or motivation. Morality may have to do not with consequences, but with adhering to some moral rule, a moral law. With this suggestion in mind, we now turn our attention to deontological ethics.

Deontology defined

We can easily imagine someone arguing that moral duty is not about convenience. It is not about saving money. It is not even about making people happy. Simply stated, deontology concerns doing one's moral duty. Can a maintenance technician, in good conscience, sign off on work that they believe contains false or highly misleading statements? Can a potential employee conceal pertinent information on the job application?

Deontological reasoning, as one can see, is quite different from consequentialist reasoning. Whereas consequential reasoning assesses actions according to their end results, deontological reasoning assesses actions (or more properly, human agents) based on whether or not human actions accord with a universal rule of morality. Actions that accord with a universal rule or law of morality are those that an agent ought to do, regardless of what the actual consequences might be. Such actions or agents are morally right or good, therefore, not because of the results that they actually produced but because what they did reflects the standard of a universal moral law.

There are various deontological theories of ethics. One version, for example, is a "divine command ethic." According to this theory, what makes an act morally right or wrong is its conformity to moral laws laid down by God. Without a doubt though, the most well-known and influential of deontological moralists is the German philosopher Immanuel Kant (1724–1804).

Kantian deontology

The core of Kant's moral philosophy can be found in his short, but very dense work of 1785, *Groundwork of the Metaphysic of Morals.* The fundamental principle of morality, for Kant, may be expressed in the universal and exceptionless command: "Act only on that maxim through which you can at the same time will that it should become a universal law" (Kant 1964: 88). In order to understand more clearly what Kant means by this phrase, let us consider how he arrived at this conclusion.

Step 1: What earns us a place among the community of moral agents is our possession of a rational will. Kant observes that—contrary to what classical Utilitarianism may believe—our ordinary moral intuitions tell us that moral evaluation centers on the agent's motives. We are held morally responsible for our intentions or motives that lead us to act in a certain way. When we reflect upon this philosophically we can see that it makes sense, for responsibility requires that we should have had a choice. The world is a very complicated place, however, and much that happens is beyond our understanding or control. Hence, sometimes despite the best of intentions, what we do goes wrong—not because we meant to do wrong, but simply because things happened on which we could not have counted. This shows, as Kant believes, that we are mistaken to think that we should be held morally responsible for the actual consequences of our actions. What is within our control—and hence what we can and should be held responsible for—are our intentions, our motives. Moral mistakes therefore, reside in having immoral motives, not in producing bad outcomes. Only those who can make reasoned choices can really be said to have such intentions. Thus, motives are the center of morality and only beings that possess the ability to choose motives—that is, those who have a rational will—can be said to be morally responsible.

Step 2: The rational will is what we, as persons, really are. Who we really are, as persons, is not our physical body, for the human body may change shape or lose some of its parts, and one can still be the same person. What really makes us who we are, therefore, is our mind, in particular our rational will. Thus, when Kant says we ought to respect other persons, he means that we should respect that persons have the capacity to make reasoned choices for themselves. To use a phrase suggested by Kant, persons are autonomous—that is, they are capable of placing their own rules or laws upon themselves. To ignore, overlook, or violate the ability of another person to make choices about how he or she shall live life is to disregard one's innermost being. Kant often speaks of treating other persons not merely as means-to-an-end but rather as ends-in-themselves. When we treat something merely as a means, we treat it as a tool that has no intrinsic value, something that deserves no consideration as a thing in itself. But persons, according to Kant, have intrinsic value as beings that exist in themselves. That is, our capacity to make rational choices for ourselves—what we will do, indeed, what kind of person we will be—is a capacity that has value for its own sake. This idea has been popularized as the principle of autonomy, one that states we ought to respect the capacity of other persons (as well as ourselves) to make personal choices for ourselves.

Step 3: What is true for one is true for all. Insofar as we are rational wills, we are essentially like one another. What holds true for one of us, as a moral agent, will also hold true for all of us. For example, since we want others to respect our capacity to make choices for ourselves, we ought, by the simple law of transitivity, to respect that capacity in others. In other words, what you are willing to allow yourself to do, you must (by the simplest of logic) allow all others to do as well. This is why Kant's ethics is sometimes spoken of as an ethics of universalization.

Steps 1–3 lead Kant to his Categorical Imperative, that is, a command that applies universally and without exception: "Act only on that maxim through which you can at the same time will that it should become a universal law" (Kant 1964: 88). As we can now see, this means that we should only adopt plans of action or ways of treating other persons that we would be willing for anyone or everyone to adopt.

A classic example concerns truth-telling. Suppose the NTSB interviews you concerning an aircraft incident. You are aware that you cut a corner in a required procedure, though you are convinced that what you did had nothing to do with the actual mishap. However, if you revealed your poor choice, you could well be reprimanded or perhaps even terminated. Thus, it would seem to be in your interest to lie. According to Kant, your proposed "maxim" would be something like, "Can I tell a lie when it would be to my advantage?" But what would happen if everyone told such falsehoods when it was convenient? Well, for one thing, falsehoods would become very prevalent, especially in circumstances like this one. And since everyone would feel allowed to lie in such circumstances, the NTSB would not expect you to tell the truth. Indeed, they would not expect anyone who might have knowingly done wrong to tell the truth when questioned. So the NTSB would not believe your lie. By willing the maxim, you also will the circumstances where your action would fail. Thus, Kant argues, your maxim, when universalized, becomes self-contradictory or self-defeating. Since you cannot successfully will that everyone does this, then you should not make an exception for yourself in this situation. Lying, for Kant, is thus always morally wrong, even if it is meant as a means to relieve suffering.

Clearly, in the case of Flight 3407, the captain tells a falsehood on his job application. From a Kantian analysis, if the captain is willing to allow himself to lie, then, to be consistent, he must allow everyone else to do the same. But that means that Colgan Air would never believe what applicants state on their job applications because it would be in the interest of applicants to lie. So the captain's lie becomes self-defeating. Furthermore, how would the captain feel if he thought that the flight crew taking his family on a vacation had lied to get their jobs?

We surely, would not want our own decision-making to be misled by such false information; so neither should we be willing to participate in misleading others in this way. Further, consider the case from Colgan Air's point of view. By hiring pilots at wages that are not really "living" wages, is it treating its pilots and crew as real persons, ends-in-themselves? Or rather is it not treating them merely as tools, cheap and expendable tools, to keep the airline functioning as cost effectively as possible?

Strengths and weaknesses of Kantian deontology

Kantian moral philosophy is particularly good at emphasizing the inherent dignity and worth of the moral agent. It calls into question Utilitarianism's assumption that happiness is the only ultimate value for persons. Instead, Kant highlights the value of being true to our innermost selves, meaning our nature as rational moral agents. From such an emphasis upon the dignity and value of the individual, Kant's philosophy is able to serve as a ground for strong claims to moral rights, for example, the right to be told the truth about YY or a right to ZZ.

There are, however, difficulties with Kantian moral philosophy. One is that, as many commentators have pointed out, the Categorical Imperative allows no room for exceptions. Thus, telling a lie is always morally forbidden, even when doing so could save the life of an entirely innocent individual. There can be no denying that this is something of a problem for Kantian thought. One avenue of response is to suggest that we should incorporate relevant details of the particular circumstances into the maxim that we universalize. When enough of the details are included, then we may be willing and able to will that everyone should lie under these circumstances.

Such a suggestion has appeal and can make the results of applying the Categorical Imperative more intuitively acceptable. But the danger of such a strategy is that we (and others) may then include so many specific details in our maxims that we can universalize almost anything, knowing that these particular circumstances will never arise again. In other words, we may stick to the letter of the Kantian Categorical Imperative by universalizing our maxim; but at the same time, we may violate the spirit of the imperative by rendering the universalization process so trivial as to be meaningless.

A second difficulty of the Kantian moral philosophy is that it gives us very little guidance on how to act when two or more moral duties come into conflict. The Kantian formula seems to assume that there can never be a genuine moral dilemma. The application of the Categorical Imperative should always result in one *and only one* definitive moral answer. But can we not easily imagine cases where the moral duty to tell the truth will run counter to the moral duty to keep one's promises? The textbox provides a fictitious scenario involving a potential conflict of duty for TSA agents, and thereby indicates the difficulty in applying categorical imperatives in practice.

A TSA AGENT'S CONFLICT OF DUTY

Transportation Security Administration (TSA) screeners are obligated to provide security for passengers, aircraft, and airports. Based on the campaign "If You See Something, Say Something", an agent should report any suspicious activity. Additionally, the TSA supervisor needs to ensure that the security equipment is working properly under the Original Equipment Manufacturer (OEM) manual's guidelines. Suppose the machinery should be checked every two hours per the OEM manual. However, one particular day several hours pass before one TSA screener notices that no one has monitored the equipment. After the agent reports this to the supervisor, the supervisor tells the agent to not report the incident. The chances of an equipment malfunction would have been very low, and it would be time consuming to do the extra paperwork. As an employee, a TSA agent has a duty to respect the supervisor's instructions; but, on the other hand, one has a duty to report any potential security risks. How does one decide whether to ignore the problem or report it?

Recognizing that there can be conflicting moral duties, some philosophers, such as W.D. Ross, have argued that deontological moral reasoning is the right kind of reasoning for morality; it is just that Kant was wrong to imagine that such reasoning can be reduced to any one value or one moral law. Rather, there are many basic moral obligations and duties. Unfortunately, among such philosophers there has been little or no agreement concerning the list of duties and their order of importance. As the list grows longer and the ranking of relative importance more obscure, the less helpful such a moral theory becomes in addressing our difficult moral choices.

A third point to consider is that just as we complained that Utilitarianism had a rather impoverished view of what it means to be a human being, so too Kant's philosophy has an overly narrow view. The capacity for reasoned choice might well be an important aspect of what it means to be a moral agent; it might even be a requirement for being one. But it is not clear that such reasoned choice is all one needs to be a moral agent or that it constitutes our innermost being as persons. Might it not be the case that one's feelings could be crucial to morality in a way that Kant cannot explain? Indeed, many philosophers have argued that the foundation for morality is not cold, analytical reason, but human sentiment; and that morality is best expressed in not adhering strictly to supposedly universal moral principles, but in caring for the individual person involved. Despite its philosophical problems, Kant's moral philosophy has gained wide respect and practical application. For instance, it brings to our attention the importance of respecting other persons as persons, of not treating others merely as tools for our own convenience.

Both Utilitarianism and Kantianism offer reasonably clear decision procedures of making moral choices. In the next chapter we will consider two other moral visions, ones that emphasize not so much individual choices as ways of living.

Notes

1 According to a footnote in the NTSB report, "The CVR recorded the first officer stating, about 2030:02, that she earned a gross salary of $15,800 during the previous year (her date of hire with the company was January 16, 2008) and that 'I'm just lucky 'cause I have a husband that's working.' (The CVR recorded the captain stating that he earned a gross salary of about $60,000 during the previous year.) About 2103:03, the first officer stated that her husband had earned more in one weekend of military drill exercises than she earned in an entire pay cycle." (NTSB 2010: fn 37, 12).

2 Chapter 3 provides further discussion on pilot salaries.

3 See, for example, Chapter 11 and the analysis of the ValuJet 592 accident.

References

Kant, I. 1964. *Groundwork of the Metaphysic of Morals.* Translated by H.J. Paton. New York: Harper Torchbooks.

Mill, J.S. 1979. *Utilitarianism.* Edited by G. Sher. Indianapolis: Hackett.

National Transportation Safety Board (NTSB). 2010. *Loss of Control on Approach Colgan Air, Inc. Operating as Continental Connection Flight 3407, Bombardier DHC-8-400, N200WQ, Clarence Center, New York, February 12, 2009.* Air Craft Accident Report NTSB/AAR-10/01. PB 2010-910401. Washington DC: NTSB.

2

ETHICAL THEORIES II

Rights and character

Mark H. Waymack

In Chapter 1 we considered the Utilitarian and Kantian approaches to ethics. Each of those theories offers a relatively clear decision procedure. In Utilitarianism the decision is based on the "greatest happiness" formula; whereas in Kantianism it is the universalizability test of the Categorical Imperative. In this chapter we will consider two quite different approaches to moral reasoning: first we will examine a framework of fundamental rights; and second we will explore the idea of ethics as being about virtue—being the right kind of person—rather than it being primarily about specific decision-making.

Rights-based reasoning

We all know that there is a great deal of competition to become a commercial pilot for a major airline in the US. This has been true for many years, especially since the Airline Deregulation Act of 1978. More recently downturns in the commercial industry have only exacerbated the problem. Since there are far more qualified candidates for these few positions than there are openings, some method must be used to decide who is offered a position and who is not. As one might expect, the selection process seems to be that one applies for a position and, if qualified, is put on a list. As openings become available, the airline moves down this list one by one.

However, in at least one airline there existed a "backdoor policy" for hiring new pilots.[1] If someone was currently employed with the airline, and personally knew a potential job candidate, the employee could mention the pilot's name to management, and that pilot would be hired ahead of persons who applied through the usual "front door." The pilots with connections would be able to forego the normal testing procedures for new hires. On the surface this might seem like an innocuous method: some small favor is given to those already employed and to their personal friends. And as long as no one who is unqualified is hired, no one's safety is endangered. However, is this policy fair to those who apply through the front door? Is it morally fair that they should be disadvantaged just because they do not have an insider connection? Furthermore, because of a history of racism and sexism in the US, commercial pilots hired in the past tended to be white males; and, not surprisingly, the names being put forward in the above-mentioned example all happened to be white males. So persons of color and women would seem to be at a significant disadvantage because of this informal program. Many of us would suggest that this type of policy is fundamentally unjust. But how can a moral theory articulate such an issue?

According to a rights-based theory, the essence of morality is a set of moral rights that persons have. These moral rights belong to one as a member of the moral community, and they define what sorts of behavior toward each other are forbidden, permissible, or required. Rights, then, are the basis for claims concerning what persons may not do to one another, as well as what they must do for one another. The former, what we might call "rights of non-interference," are often termed "negative rights." The latter are rights defining what one must do for someone, often designated as "positive rights."

For the rights-based theorist moral evaluation centers on whether certain actions conform to the rules of interaction prescribed by the rights of the individuals involved. Questions of motivation or consequences are not central, though they may prove to be relevant in some instances. Key questions that any rights-based theory must answer include: (1) what moral rights do we have, and (2) what is the philosophical grounding for those rights?

Natural rights

Many moral and political philosophers have argued that persons have moral rights by nature. That is, our most fundamental moral rights are natural rights. The English philosopher John Locke (1632–1704), for example, held that we are all naturally endowed with the right to life, liberty, and property. (Thomas Jefferson would change that list to life, liberty, and the pursuit of happiness.)

According to both Locke and Jefferson, since we live in a society, rather than as isolated individuals in a "state of nature," we must find ways to mutually limit or adapt our natural rights so that we may live together peaceably. This is done by means of a social contract. In order to foster mutual cooperation, we each must be willing to restrict or give up some of our natural rights. For example, while in a state of nature we might have an unlimited right to defend ourselves, as members of a society we yield much of that responsibility to law enforcement agencies. Nevertheless, there are certain basic or fundamental natural rights that no society may legitimately ask its members to give up, such as the right to life. Other rights that we might insist upon, though by no means an exhaustive list, are the liberty to pursue our own career ambitions, liberty to pursue education, and non-discrimination in housing and employment opportunities.

One problem that faces natural rights theorists is that it is difficult, if not impossible, to agree upon what counts as "natural." Not only do different cultures have different conceptions of what is natural, but different members of the same society may also have radically different views. Are homosexual practices unnatural? Do individuals have the right to engage in such practices? What does it mean to have a right to life? Should abortion and euthanasia be permissible or forbidden?

Contractarian rights

To avoid some of the quagmire surrounding natural rights theories, some philosophers have emphasized the contractarian side of rights. A highly influential contemporary proponent of such a view is the American philosopher John Rawls, whose theory may be found in his book, *A Theory of Justice* (1971). According to Rawls, principles of justice are not discovered in nature; nor are they grounded upon the calculus of Utilitarianism. Rather, to ascertain the principles of justice we should imagine ourselves as deciding upon the fundamental ground rules for society; but we must consider and decide upon these principles in a way that prevents us from showing any partiality to ourselves or to our family or friends. To do this, Rawls would have us imagine an Original Position.

In the Original Position, we do not know who we are. We do not know what position we will have in society—whether we will be rich or poor, religious or non-religious, young or old, healthy or unhealthy. In such a position we are all equals, for persons in the Original Position have no social status or social privileges that might be used for their own benefit. Not knowing who we will be in this society, Rawls argues that we will agree upon principles of justice that will minimize the possibility of a truly bad outcome. We will, therefore, arrive at principles of justice that protect the least well off in our society, since we might be one of them. In this Original Position Rawls believes we will all agree upon two fundamental principles of moral justice:

> Each person is to have an equal right to the most extensive basic liberty compatible with a similar liberty for others.
>
> Social and economic inequalities are to be arranged so that they are both (a) reasonably expected to be to everyone's advantage, and (b) attached to positions and offices open to all. (Rawls 1971: 60)

This view differs markedly from a Utilitarian notion of moral justice. As we can recall from Chapter 1, the Utilitarian's justification for rules of justice is the promotion of the general happiness. According to Rawls, however, each of us, at least while in the Original Position, would be willing to forego the maximization of social happiness if doing so would promote basic liberties for all. Thus, the rights of the individual person, no matter what his or her station in life, cannot be violated or overridden simply to promote the happiness of others more effectively. Furthermore, the strong egalitarian nature of Rawls's theory holds that inequalities are allowed only insofar as they benefit the least well off.

Clearly from a justice point of view the described backdoor hiring policy would be considered unfair and unjust. As another example, we might also consider how scarce gates or flight slots between busy and commercially attractive airports should be allocated between competing airlines; or what to do about slots that serve small community airports.[2] Since airspace is public, might our society place a limited obligation upon airlines to serve airports that are not as commercially lucrative as the major hubs? There are also justice questions that arise in terms of how airlines charge passengers: does a ticket confer a right to free baggage? Not everyone carries the same amount of baggage; some passengers carry almost none at all. Is it fair that they all be charged the same, or should baggage be charged by the piece or the pound? Similarly, some passengers weigh substantially more than the next passenger. That translates into more jet fuel consumption (an expense), as well as taking up more seat space. Should such passengers be charged extra (see textbox for an example)? But, of course, if the industry goes down that route, would children and very small passengers get to pay less? All of these questions raise some issue of fairness.[3] Would Rawls justify charging more for overweight passengers? The way to approach that question from a Rawlsian perspective would be to ask what principle of justice would you choose if you did not know whether you are going to be fat, thin, or of average weight. My guess is that Rawls himself would probably have been against such price discrimination as it singles out a minority and treats them adversely.

In terms of weaknesses and strengths, one difficulty for Rawls is that he assumes that all persons will, if they place themselves in the Original Position, agree on the same principles of justice. But it is not clear why this must be the case. Not everyone is as risk averse as Rawls supposes. Some persons are willing to gamble, to risk losing for the sake of the possibility of a big win. This sort of individual might allow for inequalities in society that benefit the well-to-do but do not benefit the least well off. It is unclear what Rawls could say to such individuals. Perhaps the strongest feature of Rawls's theory is his account of the Original Position, for this gives us a method of reasoning whereby fair, impartial principles of justice might be settled upon.

SAMOA AIR'S PAY BY WEIGHT PROGRAM

Samoa Air began operations in 2012 and has been inactive since 2015. Its original fleet consisted of two BN2A Islanders and one Cessna 172. This small airline gained global notoriety for charging passengers by weight. The airfare was determined by a passenger's body weight plus their baggage. The CEO, Chris Langton, claimed that this system is "the fairest way of traveling." As Mr. Langton pointed out, aircraft do not run on seats, they run on weight (BBC 2013). Although this small airline appears to no longer be in operation, could Mr. Langton be correct that this is the future of air travel?

Strengths and weaknesses of rights theories

An important strength of these rights-based ethical theories is that they provide for the moral rights of the individual apart from the whims of society. Such rights do not depend upon the promotion of society's happiness since using pleasure as the main criterion could create an unstable societal foundation. Rather, moral rights are founded upon much more enduring ground, such as the inherent nature of things or a compact that all rational persons would accept.

We may feel, however, that such a view of morality renders it essentially an adversarial contest—rights versus the interests of society or rights versus rights. This may strike us as somehow lowering moral standards, reducing ethics to conventions of behavior or ground rules of conduct, rather than revealing morality as an important and enriching aspect of being a person. Morality is not just a matter of following the rules, even if they are impartial rules. It is a matter of who we are as individual and social beings, and, perhaps more importantly, our moral character. This is a key concern of virtue theorists, and it is to them that we now turn our attention.

Virtue-based moral theories

The ethical theories that we have discussed thus far each put forward some universal principle (or principles) for choosing and evaluating morally laden actions. Virtue theorists argue that this approach to moral reasoning is mistaken in two very important ways. First, universal principles are not reliable decision procedures for attaining the morally correct answer for particular cases. Often even if we wanted to work through the various possible outcomes, the moral law, or the possibilities from a justice point of view, there simply is not time. We must act promptly and deliberately. And, second, the whole perspective that these theories take toward morality directs our attention away from the most fundamental center of morality: the character of the morally virtuous agent. According to the virtue theorist, we ought to return our attention to the true substance of morality, the virtues. Let us consider these points in somewhat more detail.

Virtue theorists ask us to look at the long history of moral theories that seek to reduce morality to a set of universal principles. Despite the large number of talented thinkers who, over thousands of years, have proposed various theories of this type, there has not been any one proposal that has garnered universal acceptance, even among the thinkers themselves. One reason is that universal principles, however thoughtfully applied, sometimes fail to yield the morally correct answer in specific situations. We may recall how the Utilitarian seems to have difficulty explaining why we should always be just and respect the rights of others. Kant's theory, with its universal prohibition on lying, contrasts sharply with situations where our moral instincts tell us that lying, given our other choices, is the morally appropriate thing to do.

Given this long train of theoretical failures, virtue theorists contend that no theory based upon universal principles can yield a wholly satisfactory result.

The reason is because morality cannot be condensed so neatly to universal principles; it is far too rich and complex to be so reduced without significant loss. Imagine trying to assess what makes a painting beautiful by appealing to a set of "universal principles of beautiful painting." So it is not the case that we have not yet found the correct universal theory of morality—no such theory can ever be found.

Not only are these universalist theories prone to give the wrong answers in specific situations, they also misdirect our attention. These theories focus on abstract principles, whereas the true substance of morality is the virtue of the moral agent. To borrow once more from our analogy to painting, imagine someone trying to show why a certain landscape painting by the 19th-century French artist Claude Monet is beautiful by explaining the chemical structure of the oils or what percentage of the canvas is occupied by different colors. Something central to the appreciation of such a painting would be missing in this explanation—that special and extraordinary way in which the painting resonates and lingers in our perception. So in morality, it is not the universal rules that capture the core of morality; it is the virtuous character of the moral agent—his or her courage, justice, patience, benevolence, trustworthiness, etc. The virtue theorist would therefore have us focus our attention on developing the right kind of traits of character. Once we have developed or acquired a morally virtuous character, then we will be better able to see and do what is morally requisite in any given situation.

However, if one's character does not lend itself toward morally right actions, the result can be disastrous, especially in the aviation industry. The worst commercial air disaster occurred in 1977 when a Pan Am 747 and a KLM 747 collided in Tenerife, Canary Islands—an accident in which 583 people lost their lives. Like most tragedies, what happened must be described as the result of multiple contributing causal factors. The aircraft had been diverted to Tenerife because of a terrorist incident at the larger Las Palmas airport, and the smaller Tenerife facility was being stretched by the quantity and size of the equipment. Fog and light rain had enveloped the airport. In addition, communication problems arose, both involving language and equipment. Investigators have generally agreed that at the crucial moment, KLM Captain Jacob van Zanten's impatience, overconfidence, and unquestioned authority amongst his crew all played a significant role in the accident. In an important sense, then, it was van Zanten's character traits that precipitated the accident. If he had been more patient, if he had been more open-minded to ambiguity, if he had been willing to listen to his crew's input and questioning, the accident might not have happened.

Virtue and action

Quite obviously the center of any virtue theory of ethics must include an account of virtue and how it translates into action. Virtues are admirable traits of character that dispose one to behave in a certain manner. The courageous woman acts courageously. The benevolent man does acts of kindness and charity. The just person does not cheat, steal, murder, etc. These qualities of character are developed in the individual through education, training, and exercise. Think of how one learns to play a musical instrument well. One cannot learn to play the piano, for example, just by reading books. One must practice. At first the music one creates will be somewhat mechanical, with mistakes, and played only with focus and attention. But with practice and experience, the mechanics and finger movements become more automatic as they have become more "habitual." As the fingers have "learned" through practice what they are supposed to do, and the notes are known by heart, the pianist becomes able to escape rote, or mechanical playing, and instead creates something truly "musical." Much the same can be said of the virtues. They are acquired through teaching and practice. Once firmly established, just like the fingers that can effortlessly find their correct places in a complex composition, the person with the moral virtues ingrained in their character will simply know what one should do in a given situation and will effortlessly perform appropriately. As the virtues become a part of the individual's character, they become directing or determining forces in how the individual acts.

Like our accomplished musician, the virtuous person can know, far more accurately than any abstract rule, what ought to be done in specific cases because through one's virtuous character one is able both to perceive the morality of the situation and to act upon that perception. As the medieval philosopher, St. Thomas Aquinas, argues:

> Since discourse on moral matters even in their universal aspects is subject to uncertainty and variation, it is all the more uncertain if one wishes to descend to bringing doctrine to bear on individual cases in specific detail, for this cannot be dealt with by either art or precedent, because the factors in individual cases are indeterminately variable. Therefore judgment concerning individual cases must be left to the *prudentia* [wisdom] of each person. (Aquinas 1993: 237)

Thus, perhaps the most important virtue, since in a sense all the other virtues depend upon it, is this *prudentia*, or wisdom.

In the field of aviation we can imagine a number of virtues that would be of particular importance: benevolence, trustworthiness, compassion, courage, patience, humility, commitment, and practical wisdom. This is intended only as a partial list of virtues, for there are undoubtedly many other particular virtues that could be relevant in the aviation industry. A captain more patient than Jacob van Zanten, one with a touch more humility than arrogance, would have been more attentive to air traffic control instructions and would have fostered within the flight cabin an interpersonal dynamic where the crew would have been willing to be more assertive and more helpful. Instead van Zanten's brash arrogance appears to have been a significant contributing factor to the disaster.

Strengths and weaknesses of virtue ethics

One glaring weakness of virtue theories of ethics is how to decide what counts as a virtue. One person's virtue may well be another person's vice. When we confront such a disagreement, the only answer that virtue theory seems to give is to look to the judgment of the virtuous person. But this is not very helpful, since our disagreement is precisely concerning who counts as the virtuous person!

A second difficulty is that while the formulas of the theories based upon universal principles, such as Utilitarianism or Kantianism, might yield mistaken answers on occasions, at least such theories offer reasonably clear-cut decision procedures. But with a virtue theory of ethics, it can be difficult to know sometimes just how to translate virtuous traits of character into morally correct action in particular situations. In a more ideal world, we might simply be able to perceive clearly what morality asks of us; but in the real world, we are all too often in a moral quandary. Our problem is that we are very unsure of what we ought to do. How is virtue ethics supposed to help us here?

At the minimum virtue ethics does redirect our attention to the particularity of moral situations—particular individuals face particular choices in particular circumstances. Ethics based solely upon universal principles may sometimes run roughshod over some of the awkward details of those peculiar situations. Furthermore, virtue ethics does seem to be right in suggesting that the central importance of moral character has somehow been lost in these principle-based theories. A sense of moral integrity, of moral conscience, is really quite important to our moral experience. Thus, while we might refrain from being swept off our feet by virtue theories of ethics, they do shed light on aspects of our moral experience that tend to be overlooked by those theories that focus upon the search for universal principles of morality.

A practical decision model: the seven-step reasoning process

Unlike some areas of philosophy, ethics by its nature is supposed to be practical. To an extent far greater than in many careers, those who work in the aviation industry—whether in the pilot's seat, in maintenance, in aircraft design, air traffic control, or airport management—hold in their hands and in their behavior the safety and lives of millions of individuals every year, passengers as well as crew. Sometimes our moral duty is quite clear, and what we might need is a sort of motivational reminder. At other times there may be conflicting goods and conflicting duties, in which case we may need to do some investigation and further deliberation. To facilitate that kind of deliberation, we offer a seven-step process here. We do not pretend it is a simple, foolproof, moral calculus; but we hope that it can be a method for bringing important moral considerations out into the open for deliberation:

- How would you describe the moral problem? In other words, why does it seem to be a "moral" problem?

- What are your available choices/options?
- Who would each of those choices affect, and how would they be affected?
- Would you be willing to "universalize" your choice so that it applies to everyone?
- Are there moral rights at issue? How might those rights create an obligation for you?
- If the persons closest to you knew what you were doing, would you be proud of your choice or ashamed?
- All things considered, then, what will you do?

Once again, like learning to play a musical instrument, this seven-step method may sound overly mechanical. But working through these steps can be a very helpful exercise. And do remember that even the best of musicians still take time to deliberately practice the fundamentals. Furthermore, explicitly working through these sorts of questions can also help us articulate our moral thinking in a way that we can share with others, and thereby explain and justify what sorts of choices we will make.

Conclusion

Chapters 1 and 2 have reviewed four of the most important ways in which moral reasoning has been described by philosophers. Needless to say, there are many other possible explanations that various philosophers have offered. The four that we have considered, however, have long histories of discussion and support. In the course of our examination, we have pointed out the strong points of each theory, but we have also noted where each theory fails to be satisfactory. Indeed, in some sense each of these theories is flawed.

A skeptic might argue here that the most proper inference from all of this is that, if the best of the theories are flawed, then the most justified conclusion is that there is no real moral truth—for if there was, is it not more likely that philosophers would have discovered what it is by now? And if they had discovered it, surely most perceptive, intelligent people would have recognized it as the truth.

I believe that our response to the skeptic should be something like this: we have found something that attracts us in each of these moral theories; some insight that strikes us right should count as evidence that morality is not something that we simply make up as we please. It is not a hoax. An analogy may help here. Think of two or more different artists painting the "same" scene. It is unlikely that any of them will paint it exactly alike. Nor should the differences in the paintings be attributed to the lack of skill on the part of the artists to portray the scene "correctly." Nor should we infer from the differences in the execution of the paintings that there is no reality to the landscape that they are depicting. Rather, we should think of each different painting as a view from one perspective, highlighting some particular aspect that others might not have seen or chosen. Hence, each painting can show us something about the landscape that might have escaped the inattentive or inexperienced eye.

We can think of moral theories in the same way. Each philosopher is writing from a different perspective, and may thus choose to emphasize features in his perception that do not figure the same way in the view of someone else. But we can use these characterizations of morality to enrich our own perspective, our own understanding of what morality is and what it is to be a moral agent. That a theory is flawed, therefore, does not necessarily make it entirely false or useless. Quite the contrary: each of these theories, as we shall see, can be both insightful and useful in trying to see and come to grips with the complexity of morality in the area of aviation.

Finally, for those still not convinced of the usefulness of ethics, keep in mind the variety of ethics codes that exist within the aviation industry (see textbox for examples). Codes of ethics can never, of course, replace individual responsibility and individual judgment. But they can be of great use in articulating the moral values and responsibilities that are central to a given profession. Hence, any student and every working professional should be completely familiar with the codes of ethics that define the moral character of their chosen profession.

SAMPLE LIST OF ETHICS CODES IN AVIATION

Air Line Pilots Association, Intl. (ALPA) Code of Ethics
www.alpa.org/en/about-alpa/what-we-do/code-of-ethics

American Airlines Standards of Business Conduct
www.aa.com/content/images/corporateResponsibility/standards-business-conduct.pdf.pdf

The Boeing Company Code of Conduct
www.boeing.com/resources/boeingdotcom/principles/ethics_and_compliance/pdf/english.pdf

Flight Safety Foundation Mechanic's Creed by Jerome Lederer
www.faasafety.gov/files/gslac/library/documents/2013/feb/74231/mechanics creed.pdf

Notes

1 See *Garland v. USAir, Inc.* 1991. 767 F. Supp. 715, April 25.
2 See Chapter 23 by Russell Mills for a full discussion of this topic.
3 See Chapter 5, Joseph Schwieterman's ethical analysis of airline pricing policies.

References

Aquinas, T. 1993. *Commentary on Aristotle's Nicomachean Ethics*. Translated by C.I. Litzinger. South Bend, IN: Dumb Ox Books.

Aristotle. 1985. *Nichomachean Ethics*. Translated by T. Irwin. Indianapolis, IN: Hackett.

BBC. 2013 (Apr 2). Samoa Air boss defends charging passengers by weight. Retrieved from: www.bbc.com/news/world-22001256 [accessed February 20, 2018].

MacIntyre, A. 1981. *After Virtue*. Notre Dame, IN: University of Notre Dame Press.

Rawls, J. 1971. *A Theory of Justice*. Cambridge, MA: Harvard University Press.

References

[illegible]

3

CAPITALISM AND ITS CRITICS

Nathan Ross

The previous two chapters presented different theories regarding how we can determine moral right and wrong. However, we might ask whether these theories can be applied to aviation without some further analysis. Regardless of our best intentions, our decisions in the business place must also take account of economic realities, and often economic considerations play a crucial role in shaping how we reconcile our ethical ideals with the world around us. Thus, this chapter will investigate the question of how capitalism functions and whether it is possible to pursue ethical goals within a capitalist context.

Capitalism is a form of economic activity in which the resources needed to produce valuable goods and services are privately owned by individuals or groups whose main goal is to make a profit. In present times, capitalism often coexists with a certain degree of state oversight. For example, on the way to work you might ride a bus that is owned and operated by a state-funded agency that is not out to make a profit. Even with the Airline Deregulation Act of 1978, US aviation continues to be regulated by the government, especially to promote public safety.

However, prior to the seventeenth century, in Europe capitalism was virtually non-existent, and in many parts of the world it emerged much later than in others. Two factors stood for a long time in the way of capitalism: first, resources such as land and everything that can be gained from it were the property of the nobility or the church, or were simply considered common property, and could not be readily bought and sold by those seeking a profit. Most people worked the land and paid heavy taxes to the nobility, while merchants existed only at the margins of society. Secondly, capitalism could not develop if individuals did not have the legal right to sell their labor power on the free market. In a society that permits slavery, for example, individuals do not have the right to buy and sell labor based on free, contractual agreements, but can be forced to labor based on threats or hereditary privilege. Thus, two fundamental reforms stand at the birth of any capitalist economy: the privatization of land and resources, and the abolition of slave labor.

For as long as capitalism has existed, philosophers have debated whether it represents an ethical form of society. Does capitalism improve the lives of all people, or does it benefit only the wealthy few? Do we truly benefit from the luxuries that capitalism provides, or do these luxuries rob us of our freedom and sense of community? Can we pursue happiness and ethical goals within a capitalist society, or does the profit motive always get in the way? These questions have emerged repeatedly over the last four centuries in ever more complex forms. This chapter will examine the different stances that philosophers have taken both for and against capitalism in dealing with these questions. The goal of this chapter is to give you, the reader, some of the tools needed to think critically about the meaning of capitalism, how it works, and how we can deal with the problems that it introduces.

Historical theories of capitalism

John Locke (1632–1704)

The highly influential English philosopher John Locke paved the way for modern capitalism with his theory of private property. What gives a person the right to own something? Is it better to live in a society where resources are held in common or one where they are subject to private ownership? In solving these questions, Locke takes two premises as starting points: God gave us the earth to make the best use of it that we can; and we have an innate right of ownership over our own bodies (Locke 1963). He then demonstrates that just as we own our bodies, we "own" our activities, the things we expend our time and energy doing. Thus, he argues that we should have the right to own objects if it is the case that we have "mixed our activity with them," that is, if we have spent our time and effort to remove objects from their natural state and make them in some way more useful to human enjoyment. For example, if I take the time to climb a tree and pick an apple from it, then if someone takes the apple without my permission, this person is taking from me not just the apple, but the time that I spent picking the apple. Locke argues that just as surely as we have the innate right of ownership over our bodies, we should be considered as having the innate right to own the objects of our labor. Further, Locke contends that such ownership is better for society as a whole because people will work harder if they hope to profit from their work by owning the fruit of their labor. When resources are shared by everyone in common, they tend to be underdeveloped, and no person takes responsibility for cultivating them to their highest degree. But when they are privately owned, the owner will seek to extract as much use from these resources as possible.

For this defense of private property to stand, Locke must also argue that society as a whole, and not just a few enterprising individuals, will benefit from this kind of private ownership. The property owner will work harder to improve his or her resources, Locke claims. In doing so, the owner will cultivate more useful products than he or she can use, and will seek to trade these products on the market with those offered by other cultivators. The end result is that each party benefits from the productivity of the other. Thus, Locke claims that the productivity of all members of society, though possibly driven by self-seeking motivations, will contribute to the "common stock" of products that are available for trade and sale. As this common stock grows larger, everyone who has something to contribute will be able to acquire more. With this idea, Locke formulates for the first time one of the most incisive defenses of capitalist industry. His insight is that private ownership is essential to wealth creation, and that wealth creation can benefit everyone who participates in the market.

There are, however, a few critical questions that we can raise about Locke's theory, especially if we seek to apply it to contemporary society. This theory rests on the assumption that all things gain value by transforming them from their natural state into one that conforms to human usage. This is certainly true with many examples that Locke gives: a plowed field lends more grain than a wild patch of land. But in light of modern ecological concerns, Locke's assumption loses some of its validity. Today we recognize a distinct value in clean air and water, which in nature are available in abundance but which can be robbed of their value by the effect of human industry. The question can be raised whether value is always the result of industrial cultivation, or whether there is not also value to be found in things in their natural state. In Part V on health and the environment three of the chapters will address the problems that aircraft impose on the environment. For example, in Chapter 20 Steven Kolmes addresses the issue of persistent contrails and the environmental concerns associated with them, especially in terms of the heat-trapping effect that may contribute to climate change. Because of environmental concerns, it seems that not everything human hands transform is beneficial to society.

Secondly, we might raise a critical question as to how Locke's theory of property could apply to contemporary debates regarding intellectual property. Most of Locke's analyses involve examples of the physical cultivation of a natural resource. But many current contentious debates regarding property rights focus on less tangible kinds of products, such as the melody to a song or the configuration software. Does Locke's conception of property cover these cases, and does his theory justify these kinds of property rights? A Lockian-minded person might claim that a songwriter will work harder and write better songs if one believes that one's ideas will be protected and will stand to profit from them. But intellectual creation is different from physical labor in important ways. The artist or inventor relies heavily on the prior inventions of others, and the creative process itself seems to depend on the fluid exchange of ideas. In writing his plays, Shakespeare often made use of plots that had already been used by other playwrights. In aircraft manufacturing, intellectual property is a key issue. For example, the Boeing Company's Intellectual Property Management group licenses the use of proprietary information. The company strives to create a business advantage for Boeing and its partners, in an effort to compete globally. The technology licensing program also allows Boeing to share its technology with other companies to help accelerate innovation (Boeing 2016). The above examples show that Locke's argument on private ownership leading to greater industry and a greater "common stock" of goods cannot apply to intellectual property without further analysis of how intellectual creation is different from physical labor.

David Hume (1711–1776)

The Scottish philosopher David Hume made equally significant contributions in thinking about the issue of luxury. A capitalist society produces increasing productivity and wealth, at least in the hands of some, and so the question becomes: how does this increasing wealth truly benefit human beings? Does it merely give us momentary satisfaction, or does it really lead to a better society? Hume was confronted by a series of critics, in religion as well as philosophy, who believed that the new luxuries of capitalist society were corrupting it, making people more selfish and less focused on the common good. But Hume, a great lover of the fine things in life, believed that this view of luxury falsely blames economic progress for some occasional human failures.

He argues in his writings on luxury and commerce that luxury does not weaken, but strengthens, society (Hume 2003). This leads Hume to his famous "store-house" argument, which works as follows: Hume makes a distinction between two different kinds of labor, necessary and unnecessary labor. The former is the time that we spend to procure the things that we absolutely need to survive; the latter is the labor that we spend to procure things that we do not need but desire, which is to say, luxuries. In some less-developed, pre-capitalist societies, nearly all work is "necessary labor," and once this labor is done people turn to rest and relaxation. In a developed, industrial society, by contrast, necessary labor only makes up a small portion of the total productivity of society, and the rest of the labor is unnecessary.

It might seem as if the society with only necessary labor is better off because people have less to distract them; but history has shown that industrial societies are more powerful and more capable of sustaining themselves in the face of adversity. Hume gives the reasons for this: in the society where all labor is necessary labor, if there is a crisis, such as a war or natural disaster, people do not have any extra labor power on which to fall back. In the industrial society, by contrast, where most labor power is not necessary, one can redirect a "store-house" of labor to deal with the pressing crisis and turn it into necessary labor. Thus, Hume demonstrates that although the production of luxury might distract us from the necessities of life, it also gives us a safety net for dealing with contingencies that might arise. It seems in retrospect that Hume was struggling to find an explanation for why industrial societies, such as Great Britain, were so effective in imposing their form of society on other, non-industrial societies, such as those of the New

World; and his answer is that industrial progress, although it is driven by the desire for greater luxury, ultimately strengthens the military might of a society.

Hume argues further that luxury is not necessarily to blame when people neglect the common good and become selfish. His view is that learning to enjoy more sophisticated pleasures does not make a person more likely to engage in excessive pleasures that harm someone else. Hume does not by any means think that luxury is the only good thing in life: he places far more value on things like love for family, courage in defending one's community, and wit and intelligence in conversation. But his point is that luxury does not by itself cause one to neglect these virtues. We can enjoy luxuries when we have the wealth do so, without becoming more selfish, but luxury ceases to be positive when we pursue it at the cost of these more important things. Hume writes: "When luxury ceases to be innocent, it ceases to be beneficial" (Hume 2003: 113). For instance, having an expensive car does not make me more likely to neglect my children. But if I cannot afford the car, and spend money that would go to my child's education on the car, then the luxury represented by the car ceases to be beyond reproach.

The Concorde Supersonic Transport (SST), and other related supersonic programs, provides a case in which luxury caused more harm than good, and thereby led to the end of the program. In detailing the SST projects, R.E.G. Davies states that:

> The completely imaginary notion that the average airline passenger would pay an exorbitant price to save two or three hours across the Atlantic was a fallacy matched by the almost criminal deception in the apparent ignorance of (or a deliberate avoidance of) the magnitude of world-wide first-class air travel as a proportion of the whole. (Davies 1994: 125)

Because the luxury of flying on a Concorde was outweighed by the exorbitant costs as well as the negative impact on the environment, the program was finally terminated. This example then shows the ways in which luxury is overridden by other factors such as the high operating costs and potential harmful impact on the environment. The Concorde reinforces Hume's argument, which is inspired by the idea that we can have an affluent society without neglecting the value of people and human relationships. The "store-house" argument even supports the view that the greater affluence afforded by private industry will lead to a stronger society more capable of dealing with adversity.

But Hume's philosophy does not provide an adequate solution to one of the greatest problems faced by these more affluent societies: inequality. As societies grow more affluent through private industry, some benefit more than others, and the luxuries of which Hume speaks are not evenly distributed. But Hume's most famous student, Adam Smith, saw this problem more clearly and sought to show that even if the benefits of capitalism are disproportionate, the affluence is good for society as a whole.

Adam Smith (1723–1790)

Adam Smith inherited from his teacher, David Hume, a deep respect for the benefits of the Industrial Revolution. In his famous text, *The Wealth of Nations* (1981), Smith describes the developments that lead to greater affluence, and he defends the idea that capitalism is good for everyone, even if it leads to inequalities. For Smith, one of the central advantages of a capitalist economy is that it provides the owners of resources with the incentive to use them in the most productive manner possible. This leads to the birth of an industrial system of production in which producers constantly seek to increase the productivity and efficiency of the labor process. Smith argues that one of the central innovations of such an industrial system arises due to the increased division of labor. He cites how a handful of workers in a pin factory can produce thousands of pins if they divide up their labor process so that each person in the factory is only doing a very small part of producing a single pin. This division of labor allows the workers to focus on one minute, simple task, giving them greater dexterity and saving them the time that goes along with transitioning from one task to another. Thus, Smith sees the skilled labor of the craftsperson replaced by the unskilled labor of the assembly line worker; and he welcomes this change as it will lead to unleashing great productive forces that will vastly enrich society. While the factory owner promotes such a division of labor out of the desire to make more money, Smith argues that the capitalist is inadvertently enriching all of society by flooding the market with cheaper commodities. In an industrial, capitalist society, the poorest worker lives in greater material comfort than the king of a tribal society, as Smith contends, because goods simply become more affordable as they become easier to produce.

While Smith is a great champion of the division of labor in industrial production, he also sees ethical problems in it. When workers confine themselves to a single, unskilled task within the productive process, their work is more productive, but it also impoverishes their mental life and character. In industrial societies the workers become unhealthy, unable to solve problems and unable to think about what they are doing. The consequences of this impoverishment of the industrial laborer's activity will involve large numbers of citizens incapable of scrutinizing their government or participating in the defense of society. He worries that such a working class will provide fodder for authoritarian regimes that take advantage of their lazy habits of mind. As a solution to this problem, Smith proposes limiting child labor and introducing mandatory schooling (a solution thankfully heeded in most of the industrial nations).

Prominent critics of capitalism

A central tenet of Locke, Hume, and Smith is the idea that all of society benefits from private ownership of resources and from more advanced means of production introduced by private investors. The nineteenth-century German philosopher and economist Karl Marx rejected their views, and developed a critique of capitalism suggesting that only select individuals, not all of society, benefit from capitalism. While he did acknowledge that private investors introduce great innovations into the economy that lead to economic growth, he shows through research that this growth leads to the enrichment of the investing class at the expense of the rest of society. Marx argues that class conflict between the working class and the investing class, which is definitely not a harmony of interests that benefits everyone, is the underlying law of capitalist societies. He develops this criticism through a detailed analysis of economic realities in capitalist societies along with a careful assessment of the views of prior economists.

Karl Marx (1818–1883)

Any economist would agree that capitalism motivates people to desire to enhance capital through profits. But Marx's theory is innovative in his explanation of how capitalism is profitable, namely by means of surplus labor value. Marx (1976) analyzes the difference between two basic economic activities: selling in order to buy (C–M–C) and buying in order to sell (M–C–M). In the first instance, a person has a commodity, and sells it to get the money to buy other commodities. In the second instance a person has money, and invests it in a commodity to make more money by selling the commodity. For Marx, these are two separate activities with different motivations. C–M–C plays the valuable role of allowing people to circulate different types of commodities in society. Some people have too much of one thing and others too much of another. But those who invest in commodities to sell them do so to make money from their investment. Thus, the formula for capital is M–C–M. Marx's interest is to analyze and systematically understand this kind of investment. How is it possible for people to invest in a commodity that by its very nature generates value for the investor? In some cases, one might buy land or gold or oil. But in these instances, the profitability of the investment depends on the law of supply and demand. In order to make money from an investment in land, I must hope that due to increasing demand, the land will become more valuable. Though this kind of investment is common, it is not for Marx the overriding type of investment.

Rather, Marx argues that there is only one commodity on the market that, by its very nature, can generate a profit for the capitalist, and this commodity is human labor power. Marx argues that labor creates a certain average amount of value equal to its quantity (a pair of shoes and three shirts are equivalent in value because they both represent the same total amount of labor time). But one pays the laborer not according to the value of what is produced, but according to the cost of living in a given social context. This means for Marx that it becomes possible to systematically pay workers less than the value that their labor truly embodies. The most consistent and most viable way to invest in a commodity is to invest in the labor power that produces the commodity. Marx calls this way of making profits "surplus labor value," and he argues that this demonstrates that the profitability of capitalist enterprise is predicated upon the systematic and growing exploitation of laborers. For Marx, unlike Smith, capitalism is not a "rising tide that raises all ships," but a system of ownership that bases itself on the exploitation of workers by investors. He argues that capitalist development does not tend to improve the lives of the workers, as Smith did, but tends towards an ever-increasing gap between the wealth of capital and the poverty of workers. Marx's point about surplus labor is perhaps best illustrated in the aviation industry with airline pilot and crew member starting salaries. Ever since the airline industry was deregulated in 1978, pilot salaries and benefits have been decreasing.[1]

But Marx believed that because of this fundamental injustice in capitalism, it was preparing the way for a new, post-capitalist form of society. Capitalism was drawing an ever-increasing number of workers into dense, urban environments, where they were exposed to other workers and able to perceive their common plight. He thus envisioned the need for a worker's revolution resulting in the workers seizing ownership of the means of production and designing a new form of society in which workers must not work to generate a surplus labor value for capitalist profits. He called this post-capitalist form of society a "realm of freedom" because he believed it would combine the productivity promoted by capitalist industry with shorter working days and more time for non-labor related activities.

Twentieth-century critics

In the twentieth century capitalism continued to develop and take on forms of which its early proponents and critics could not have conceived. Theorists have referred to this later stage of capitalism as "late-capitalism," "post-industrial society," or "consumer society." One of the central issues in twentieth-century capitalism is that of overproduction. While Locke and Smith were largely concerned with understanding how resources and labor could be used in the most productive way, they did not fully envision that this process of increasing productivity would lead to the potential for overproductivity. If a business creates so large an amount of commodities that they cannot be sold, then this excess productivity will present the business with just as much wasted labor as outdated or inefficient labor. Thus, business enterprises in the modern world have the need not only to regulate the production process, but also to regulate the consumption process. They do so through advertising, which manufactures demand, through branding, which establishes the desirability of a commodity by linking it with a well-known image, and through the design of increasingly seductive commercial spaces. This process of regulating consumption is familiar to anyone who lives in a late-capitalist society. But it entails new philosophical issues that were not seized on by earlier theorists of capitalism.

The philosophers Walter Benjamin (2004) and Theodor Adorno and Max Horkheimer (2007), who worked to update Marx's ideas in a twentieth-century context, provide a compelling analysis of the way in which capitalism is complicit with many of the crises of the modern way of life. Benjamin argues that in modern life, capitalism has become a religion that grows out of, and then displaces, all other forms of religion and ethics. As he claims, one of the basic features of any religion is the ability of a social group to establish a sense of duty and obligation among its members. In most religions the creation of guilt is tied to the possibility of redemption. But Benjamin argues that capitalism is unique in that it is "probably the first example of cult that produces only guilt, not redemption" (Benjamin 2004: 288). Capitalist societies create radical guilt by making sure that every human activity only has value or meaning to the extent that it generates a profit for the owners of capital. As Marx has argued, a person will only be allowed to work to the extent that their work generates an increasing degree of surplus labor for capital. But equally, we see in modern life that people's notions of enjoyment or pleasure are connected to profit: leisure activities that do not relate to capitalist industries tend to disappear and become displaced. Even the owner of capital, the investor, will only make decisions that serve to maximize profits, or be displaced by another more ruthless investor. This point shows that even the most powerful controlling interests in the economy are not really sovereign but caught in a cycle of fear and guilt. Benjamin argues that this increasing functionalization of modern life demonstrates that capitalism is built on the radicalization of the phenomenon of guilt at the root of religions, but with the important difference that it eliminates any possibility of seeking atonement within the religion. As he writes, "in the capitalist cult, even God is drawn into the nexus of guilt" (Benjamin 2004: 299). For Benjamin, the key to breaking this predominance of guilt in modern life consists in seeing the ways in which mythology resides in many of our apparently rationalized social practices.

The critical theorists call into question the very notion of progress on which modern society is founded: they demonstrate that the very effort to free human life from myth and dependency on nature has led to a new kind of mythology. Human history, and especially capitalist industry, has progressed toward greater control over nature; but this control often means destructive violence toward nature and manipulation of people to serve the goals of industry. While they do not advocate regression to a more primitive state, they argue that we need to be constantly aware of the price that we pay for growth, prosperity, and progress in human life and freedom.

Georges Bataille (1897–1962)

One of the most significant analyses of this aspect of modern capitalism appears in the works of the French philosopher Georges Bataille (1991). He proposes a "Copernican revolution" in economic thought: instead of considering the central problem for economics to be the process of creating wealth, he suggests that we study the way that we expend wealth. His argument is that human life continually confronts an excess of wealth (beginning with biological energy) that cannot all be stored and used to increase wealth. This excess wealth is doomed to waste. But he argues that the way in which people waste wealth has a rational purpose.

Bataille (1991) studies the phenomenon of "potlatch" in Native American societies as a model for this rational expenditure of excess wealth. Potlatch rituals involve families or chiefs throwing lavish feasts in which they give away or destroy a great deal of wealth in an attempt to establish social status. Bataille studies these rituals as a model for how wasting, giving away, consuming, or sacrificing can all be done with a higher goal in mind: namely, distinguishing oneself in the eyes of others. As Bataille suggests, potlatch behavior is deeply embedded in all human economies; but capitalist society, because of its emphasis on productivity and commodification, cannot perceive the ways in which potlatch shapes our behavior. If we cannot recognize the need to waste, consume, and give in a meaningful way, he suggests that we are doomed to do so in a destructive and deeply unethical way.

He investigates the wars of the twentieth century as well as the massive arms races of the post-war era as examples of a fatal necessity in capitalism: if a society cannot use its wealth for growth and does not recognize the value in pure luxury, it will seek to waste its wealth in lavish expenditures on defense and war. War is similar to a potlatch not just in its wastefulness, but also in its fundamental motivation to elevate one group or country over another in status. Bataille thinks that recognizing the need for pure luxury and unconditional charity on an international scale will be the only way that wealthy nations can divert a scenario of all-out war and global stratification between the wealthy nations and the destitute ones.

Jean Baudrillard (1929–2007)

Following Bataille's theory of potlatch, the French theorist Jean Baudrillard developed a theory of consumer behavior in modern capitalism that points towards the economic necessity of consumption. Because of the extreme development of productivity in modern capitalism, Baudrillard (1981) argues that society must develop increasingly powerful strategies for promoting intense consumption of commodities. He argues that consumption is promoted and organized by treating commodities as "signs" that are related to each other like signs in a language. Just as there is no natural relationship between objects and the words that represent them, so too there is no intrinsic relation between what we purchase and what our needs are. We choose commodities not because of how they relate to our inherent needs or wants, but rather they are a way that we communicate with other people in society. Advertising gives us associations that allow us to connect products with the ideals that we aspire to communicate about ourselves, and products are developed which can hardly be consumed in isolation, but instead require us to buy even more products.

The result is that consumers are not sovereign in making their choices and never attain a point of satisfaction, but are engaged in a never-ending cycle of consumer behavior. Just as Marx had argued that workers are exploited for their surplus labor power, Baudrillard argues that consumers are increasingly robbed of any possibility for pleasure by the demand for consumption to keep flowing. Although commercial aviation is a service industry, a parallel exists in that consumers have almost no say in the services and costs associated with air travel. In Chapter 5 Joseph Schwieterman addresses the cost of air travel along with potential ethical lapses such as collusion between airlines. Based on his account, one can readily see that airline passengers are not sovereign. The textbox provides an example of this problem from a controversial 2017 case in which a passenger was injured while being forced off of a commuter flight.

Baudrillard argues that, unlike the workers whom Marx believes will perceive their common exploitation, consumers are isolated from each other, passive and detached in relation to the organization of consumer society (Baudrillard 1975). Thus, Baudrillard seems to question the possibility of a thoroughgoing revolution of modern consumers against the economic system that organizes their behavior. However, one might wonder whether Baudrillard might join Bataille in championing the pursuit of "pure luxury," unproductive moments of enjoyment, as an act of resistance against the denial of pleasure in modern capitalism.

FORCIBLE REMOVAL OF A PASSENGER

On April 9, 2017 a United Express (Republic Airlines) passenger, Dr. David Dao, was forcibly removed from Flight 3411. The flight, which originated at Chicago O'Hare International Airport (ORD), was sold out, and the airline needed four seats for crew members who were scheduled to fly out of Louisville (Kentucky) International Airport (SDF) the next day. When no one volunteered to give up their seat, the gate agents drew four names for removal from the aircraft, one of whom was Dr. Dao. He refused to comply and was ultimately dragged off the plane by Chicago Aviation Police, and suffered injuries as a result.

Gary Leff (2017) contends that the police overreacted, and if they had not used so much force, this incident would not have become a news story. Leff maintains that a key problem stems from heightened security measures since 9/11. In other words, in the current aviation climate the assumption is that the passenger is always wrong. This incident is another indication that the consumer is anything but sovereign, especially in the commercial aviation industry.

Note

1 The ethical implications of low pilot wages are also discussed in Chapter 1 with the case of Continental Connection Flight 3407. Chapter 1, note 1 discusses the NTSB's note that on the CVR the first officer remarked that her 2008 salary was $15,800.

References

Adorno, T.W. and Horkheimer, M. 2007. *The Dialectic of Enlightenment*. Translated by Edmund Jephcott. Stanford: Stanford University Press.

Bataille, G. 1991. *The Accursed Share: And Essay on General Economy*, Vol. I. Translated by Robert Hurley. New York: Zone Books.

Baudrillard, J. 1975. *The Mirror of Production*. Translated by Mark Poster. St. Louis: Telos Press.

Baudrillard, J. 1981. *For a Critique of the Political Economy of the Sign*. Translated by Charles Levin. St. Louis: Telos.

Benjamin, W. 2004. Capitalism as Religion. In: *Walter Benjamin Selected Writings*, Vol. 1: 1913–1926. Translated by Rodney Livingstone. Cambridge, MA: Harvard University Press, 288–291.

Boeing. 2016. Licensing. www.boeing.com/company/key-orgs/licensing/index.page. [accessed May 28, 2016].

Davies, R.E.G. 1994. *Fallacies and Fantasies of Air Transport History*. McLean, VA: Paladwr.

Hume, D. 2003. *Political Essays*. Cambridge: Cambridge University Press.

Leff, G. 2017 (Apr 11). The real reason a man was dragged off that United flight, and how to stop it from happening again. *View from the Wing*. Available at: http://viewfromthewing.boardingarea.com/2017/04/11/real-reason-man-dragged-off-united-flight-stop-happening/ [accessed June 7, 2017].

Locke, J. 1963. *The Two Treatises on Government*. Edited by Peter Laslett. Cambridge: Cambridge University Press.

Marx, K. 1976. *Capital: A Critique of Political Economy*. Vol. I. Translated by Ben Fowke. London: Penguin.

Smith, A. 1981. *An Enquiry into the Nature and Causes of the Wealth of Nations*. 2 vols. Edited by R.H. Campbell. Indianapolis: Liberty Classics.

PART II

General topics in aviation

As with other for-profit enterprises in the US, the aviation industry attempts to produce profits, and does so while decreasing costs in an effort to optimize income. But since the Airline Deregulation Act of 1978, airlines have struggled to maintain profits; most startup airlines have come and gone, and employee salaries and work conditions are ongoing topics of debate. This part addresses some of the business aspects of the industry, with Part III then focusing on the duties of the individual and the corporate components of responsibility.

Chapter 4 considers the question of freedom of the skies and national sovereignty. While aviation is a global affair, as Dawna Rhoades explains it, "aviation continues to be governed by national rules and regulations that vary widely in their quality, scope, and enforcement." In this chapter, she investigates the history of the debate over international regulations and provides the current state of this important topic. Since the early aviation conventions, such as Paris and Chicago, the question of who rules the skies has still not been settled, especially in terms of national sovereignty. A key point of debate concerns cabotage, in which non-US airlines would be allowed to fly domestic routes. In other words, one of the current dilemmas under the "freedoms of the air" regards the eighth freedom, or "the right to transport passengers, cargo and mail within the territory of a State which is not the aircraft's State of registration."

Moving to controversies surrounding airline pricing in the US, Chapter 5 by Joseph Schwieterman addresses the ethical issues that confront both airline pricing personnel and the consumers who purchase the tickets. While at first glance it may appear that the ethics of airline pricing should evaluate how airlines determine fares and fees, Schwieterman also reveals the ways in which consumers at times may behave unethically. In terms of the airlines, several potential ethical issues may arise, such as collusion, predatory pricing, retaliation, and overbooking flights. Some examples of unfair customer practices include: taking advantage of bereavement fares, cancelling flights for medical reasons when those reasons are only secondary, not completing all segments of a trip in order to save money, etc. He ends the chapter by pointing out that each side must learn from their past mistakes and respect the needs of the other party, or else both sides may continue to face a variety of ethical impediments.

Chapters 6 by Ted Ludwig provides a union perspective on the reasons why the outsourcing of maintenance to third party vendors is not conducive to a safe maintenance program. Because there are more aircraft than FAA inspectors, the FAA needs to rely on the airline to follow its own maintenance programs. But when airlines outsource to cut labor costs, safety becomes compromised because there are fewer licensed technicians working for the outside vendor. Ludwig argues that with outsourcing there is less oversight. With less oversight, the level of safety becomes questionable. This scenario creates the problem of the fox guarding the henhouse. Ludwig argues in favor of unions; one key reason being that a union protects a technician from reprisals such as for refusing to sign off repairs. In the end Ludwig admits that the airline industry is quite safe, especially given the decrease in the number of fatal aircraft accidents over recent years. However, if the push to cut labor costs leads to vendors not following the maintenance programs, then flying on commercial aircraft may not be as safe as it could be.

Chapter 7 ends this part with the important topic of whistleblowing and an analysis of both its ethical and legal dimensions. Regarding the moral side of whistleblowing, Bruce Hoover examines when it is the right thing to do along with how this action impacts the whistleblower's livelihood. Hoover then turns to the legal avenues for whistleblowers, and investigates the different ways in which federal and corporate employees are legally protected. In particular he examines two statutes that aim at protecting whistleblowers in the aviation industry: the Wendell H. Ford Aviation Investment and Reform Act for the 21st Century (AIR-21) and the Sarbanes–Oxley Act (SOX). AIR-21 protects employees who expose air carrier safety violations, while SOX addresses instances of corporate fraud or accounting abuses. In the end Hoover contends that if we want to live in a just society, we must be able to provide protections for whistleblowers who act in good faith.

4

WHO GOVERNS INTERNATIONAL AVIATION?

Dawna L. Rhoades

Three events in the second decade of the twenty-first century have raised the issue of who (and how) international aviation is governed and regulated. The first event was the June 2010 start of the new US–EU (European Union) multilateral Air Transport Agreement (ATA) extending the investment and market opportunities of the 2008 multilateral agreement which granted Open Sky access—the right of an airline established in either region to fly from any city in one region to any city in the other region with notification (European Commission 2015). The second event involved an effort by the EU to include all flights to and from the EU in their Emissions Trading Scheme (ETS), an effort to curb carbon emissions through the trading of carbon permits. Since the ETS would have included carbon emitted over the entire length of an international journey and not simply on the part where carbon would be emitted over EU territory, countries around the world complained about extraterritoriality—that is, applying the laws and regulations of one's country in the territory of another nation. Several of the non-EU nations threatened legal and political action. Subsequently the EU agreed to allow the International Civil Aviation Organization (ICAO), an intergovernmental organization associated with the United Nations (UN), to develop aviation policy on carbon emissions (Keating 2013).

The third event was the 2014 disappearance of Malaysia Airlines (MH) Flight 370. The world watched as the aviation authorities from various countries in the region argued over responsibilities, jurisdictions, rules, and ways to prevent this kind of tragedy from happening again (BBC 2014). These three events highlight the fact that aviation is a global affair. Even as it brings together people and firms from around the world, aviation continues to be governed by national rules and regulations that vary widely in their quality, scope, and enforcement. The industry that brings the world together has spent most of its existence firmly rooted in the needs, wants, and politics of home country identities. It also continues to debate the role of markets and governments in setting (and regulating) fares as well as setting standards for safety, security, and technology.

Background

The history of manned flight began on December 17, 1903 with a 12-second flight by Orville and Wilbur Wright. Five years later these aviation pioneers would be working on aircraft for the US Army. Seven years after that historic flight, amid the first whispers of what would become World War I, national governments would feel the need to attend an international conference to discuss matters relating to the regulation of flight.

Early conventions

This first conference in Paris involved 19 European nations and established the key terminology of aviation, but it would not resolve the fundamental question of who controlled (and who regulated) the skies. One group favored the concept of a free sky similar to the freedom of the seas model that had been debated for many years. The other side in the Paris debate was led by the British, who argued that a nation had the right to control the air above its landmass and regulate entry as well as all other aeronautical activities into and within that airspace. Delegates left this conference without agreement on this fundamental question.

The British Aerial Navigation Act of 1911 settled the matter in the United Kingdom by declaring that the Home Secretary had the right to regulate all aviation activities within the airspace above Britain. Other nations quickly followed suit as the continent prepared for war (Sochor 1990). Following the end of World War I, a separate convention was held in Paris concurrent with the peace conference that produced the Convention Relating to the Regulation of Aerial Navigation in 1919. This Convention affirmed the right of national governments to sovereignty over their airspace while calling for: 1) national registration of aircraft; 2) rules for airworthiness and certification; 3) regulation of pilots; and 4) restrictions on the movement of military aircraft. The International Commission on Air Navigation (ICAN) was established in Paris to continue work on international aviation legal matters. Only 26 nations would ratify the Convention, with the most notable exceptions being the United States and Russia (Kane 1998; Sochor 1990).

While the international community continued to struggle over questions about aviation, domestic development was also beginning to diverge. Although the airplane held a special fascination for many people and had gained a place in the military arsenal during World War I, it struggled to find a market among the traveling public. It was freight, particularly mail, that paid the bills for early airlines, which competed for access to government contracts to deliver this service. In the US the Postal Service would play a significant role in shaping the industry by encouraging consolidation, safety, reliability, and continental reach. This informal guidance was critical as the industry struggled with what would become one of its greatest enemies, economic crisis. The Great Depression, triggered by the stock market crash of 1929, led to financial instability in the airline industry and endangered its survival. However, in 1932 the Postmaster General was charged with violating US laws against graft and collusion in using government airmail contracts to 'shape' the industry. Beginning in 1938, postal rates would be the responsibility of the newly created Civil Aeronautics Board (CAB). In Europe governments took a more direct route to shape a stable aviation industry through government ownership. Eventually most nations developed single, large, government-owned, national flag carriers with operations centered on their capital cities.

World War II represents a watershed in aviation development. During the war, several key innovations would take place that eventually made commercial passenger aviation a reality. These include radar, turbojet engines, pressurized cabins, and newer, lighter construction materials. As the world looked to life after the war, a second aviation conference took place in Chicago in 1944. The Chicago Conference and the resulting Convention still have a major influence on matters of international aviation. The Conference created the International Civil Aviation Organization (ICAO), headquartered in Montreal, Canada, to develop the standards and practices of aviation. Representatives of all nations involved in international aviation would become members. ICAO would be governed by a council responsible for its day-to-day business and report to the General Assembly convened every three years. Eventually ICAO adopted 18 annexes to the Convention outlining the standards that govern everything from personnel licensing to aeronautical telecommunications to transport of dangerous goods. ICAO members were expected to adhere to these standards unless there was a compelling reason (usually financial) why the standards could not be implemented. They would then be expected to file an exception with ICAO.

It should be noted that nothing prevents a country from adopting standards that exceed those adopted by ICAO (see Sochor 1990 for a history of ICAO and the annexes). It should also be noted that this organization became affiliated with the UN and functions in a very similar manner. This means that issues of safety and security brought to ICAO are subject to considerable political pressure in several ways. First, the composition of committees to address aviation issues is influenced by national regulatory agencies that nominate members. Second, the standards and recommended practices that come out of these committees must be acceptable to a majority of country members before any vote is taken. If there are significant objections to the standard or recommendation, then the committee revises it (usually to lower the requirements) until a majority indicate that they can and will comply. This is the reason that many developed nations adhere to standards higher than those set by ICAO. Third, like the UN, ICAO has no enforcement powers; it cannot make a nation adopt and institute any of its standards or recommendations (Rhoades 2014; Sochor 1990).

The Chicago Conference also took up the issue of freedom of the skies and national sovereignty. This time it was the US that argued the side of "open skies," asking that the delegates "not dally with the thought of creating great blocs of closed air, thereby tracing in the sky the conditions of future war" (Sochor 1990: 8). Their chief ally in the war, Great Britain, would continue to argue for tight regulation and national control. Once again, there was no agreement on the issue of international aviation regulation. While the public debate focused on the philosophical issues, private concerns centered on economic issues. The US would emerge from World War II with its manufacturing and aviation industries intact while the airlines and aviation infrastructure in Europe were devastated by years of war. Opening their markets to the US would have almost certainly given US firms an insurmountable advantage. If there was to be no multilateral, international body regulating the terms of international aviation, then nations would have to fall back on a time-honored mechanism governing international relations between nations: the treaty.

Bilateral air service agreements

In 1946 the two key aviation nations of the Atlantic, the US and Great Britain, met in Bermuda to discuss aviation issues. The resulting bilateral treaty, the Bermuda Agreement, would become the model for future treaties. It established four key features of such agreements: 1) designated routes with possible size and frequency restrictions, 2) carrier designation, 3) reciprocity of rights, and 4) separation of passenger from cargo rights. This agreement would also give the International Air Transport Association (IATA), an airline trade organization

TABLE 4.1 The freedoms of the air

Freedom	*Description*
First	The right to fly over the territory of a contracting State without landing
Second	The right to land on the territory of a contracting State for non-commercial purposes
Third	The right to transport passengers, cargo, and mail from the State of registration of the aircraft to another State and set them down there
Fourth	The right to take on board passengers, cargo, and mail in another contracting State and to transport them to the State of registration of the aircraft
Fifth	The right to transport passengers, cargo, and mail between two other States as a continuation of, or preliminary to, the operation of the third or fourth freedoms
Sixth	The right to take on board passengers, cargo, and mail in one State and to transport them to a third State after a stopover in the aircraft's State of registration and vice versa
Seventh	The right to transport passengers, cargo, and mail between two other States on a service which does not touch the aircraft's country of registration
Eighth	The right to transport passengers, cargo, and mail within the territory of a State which is not the aircraft's State of registration (full cabotage)
Ninth	The right to interrupt a service

formed after the Chicago Convention, the right to establish airline and air cargo rates for the international system. In effect, airlines would be allowed to collude to set the fares and rates that would be charged by all member airlines. These bilateral agreements would remain in effect until the parties adopted a new agreement or renounced the existing one, an action that would halt air travel between the two countries involved.

The British government formally notified the US in 1976 that it was terminating Bermuda I (the 1946 agreement) because it felt that the original agreement gave US carriers a disproportionate share of air traffic. The British were particularly concerned with US fifth freedom rights (Table 4.1). Fifth freedom rights were viewed as direct competition for a national flag carrier as they allowed a foreign carrier to pick up and transport passengers from a non-home country to a third nation. Following the negotiations, a new agreement, Bermuda II, would govern air traffic between the US and Britain. It would be far more restrictive than the previous agreement and would provoke a major policy debate within the US.

Deregulation and liberalization

Within the US economists had been discussing the question of industry deregulation since the 1960s. The whole transport sector had received particular attention because of its role in linking suppliers and manufacturers, buyers and sellers. The airline industry became the first mode of transportation to be deregulated with the passage of the Airline Deregulation Act of 1978. This same year the US Congress passed the Policy for the Conduct of International Air Transportation, which declared the US intention to "trade competitive opportunities rather than restrictions" in aviation in an effort to expand competition and reduce fares (95th Congress 1978). CAB issued an order to IATA asking them to "show cause" why they should not be viewed as an illegal cartel under US laws and informing US airlines that they must withdraw from any rate (fare) setting activities. The International Air Transportation Competition Act of 1979 would set out three goals for US policy: 1) multiple carrier designation with no restriction on size and frequency; 2) fares set by market forces; and 3) elimination of discriminatory practices such as exclusive airport contracts and government user fees that favored domestic over foreign carriers. It should be noted that the new US-style open skies agreement would not include eighth freedom (cabotage) rights; there would be no access to US domestic markets for foreign carriers.

To gain international support for their liberalized aviation policies, the US used two levers. The first lever was to reward the airlines of nations willing to sign the so-called open skies agreement with immunity from anti-trust laws designed to prevent collusion. Anti-Trust Immunity (ATI) would allow a foreign carrier to coordinate prices, schedules, and marketing with their US alliance partners. The second lever would punish nations unwilling to enter open skies agreements by attempting to divert traffic from their country to surrounding countries. The Encirclement Strategy was based on the belief that open skies agreements would lower fares to these countries and cause consumers to bypass higher-fare, restricted markets. Two countries targeted by this strategy were the UK and Japan, the gateway destinations to Europe and Asia for US consumers. Deregulation did lower the fares of US carriers, which placed growing pressure on European carriers to violate IATA-established fares. Eventually the rate-setting system of IATA did collapse, allowing the "market" to determine fares; but neither the UK nor Japan raced to adopt open skies (Toh 1998).

Unlike the US, the nations of Europe decided to deregulate their markets in a slower, more deliberative fashion. In a series of three packages, the aviation market opened up, with the final step coming into effect on April 1, 1997. As of this date, an air carrier established in any one of the European Union countries could fly anywhere with the EU, including between cities within a member nation (cabotage). From the European perspective, this single sky now meant that US fifth freedom rights in Europe were cabotage, and a growing number of airlines and associations began calling for a change in US–EU policy that would grant similar rights in the US domestic market. After years of negotiation and setbacks, the US and the EU reached a multilateral agreement in 2008 to open up the Atlantic and are continuing to discuss a single sky concept that would extend the European concept to the US; but there is no firm timeline for the end of cabotage in the US.

A new era?

The goal of the EU in their US negotiations has been what they believe is a truly open sky that would allow EU carriers to fly between two points in the US (cabotage). While the 2008 and 2010 agreements have not yet achieved this goal, negotiations continue as European airlines push for the same rights as US carriers have in Europe. If they ever achieve this goal, then a new era of aviation will dawn for the US, which has always seen itself as the global advocate for free markets. Although ICAO is well known in aviation circles, its standards are generally only relevant in developing nations where they form the basis for national law. ICAO has not fulfilled the vision of its founders to be a global force for safety, security, and development. In the preamble to the Convention that established ICAO they cite the following goals as desirable outcomes: avoiding friction and promoting cooperation among nations and the establishment of an air transport system that is based on "equality of opportunity and operated soundly and economically" (Sochor 1990: 226). The carbon emissions issue and the disappearance of Flight MH 370 could change the perception and the role of ICAO. ICAO could become the forum for resolving the carbon issue and establishing new standards for safety, security, and global aviation cooperation and coordination. This would mark a new era for the organization that first brought the aviation world together. It would take the political and financial commitment of national governments, but the benefits could be significant for many stakeholders.

AN OPEN SKIES CONTROVERSY

One particular controversy with open skies involves three US airlines (United, Delta, and American, known as the US3) that have alleged violations by Qatar and the United Arab Emirates involving the airlines Qatar, Etihad, and Emirates (or the ME3). The US3 contend that the ME3 receive heavy government subsidies, thereby creating unfair competition. The ME3 have received direct subsidies of $52 billion since 2004. However, Neiva (2017) points out that aviation has long been prone to government support as well, and notes several examples, including US airlines receiving a $15 billion bailout after 9/11. But, as he states, two wrongs do not make a right, and the US3 might be correct in this case. One potential problem is that renegotiating the agreements may lead to negotiations of dozens of other agreements, which could then lead to rigid restrictions. Neiva instead advocates for transparency: "What can be done to strengthen these agreements is for all parties—including the United States—to demand more transparency about the aviation industry in each country. That would include a serious assessment of the amount and type of subsidies that are available to the aviation industry."

References

BBC. 2014 As it happened: MH 370 search widens. Available at http://www.bbc.com/news/world-asia-26695830 [accessed February 7, 2014].

European Commission. 2015. Air—International aviation: United States. Available at http://ec.europa.eu/transport/modes/air/international_aviation/country_index/united_states_en.htm [accessed February 7, 2014].

Kane, R.M. 1998. *Air Transportation.* 13th ed. Kendall/Hunt, Dubuque, Iowa.

Keating, D. 2013. EU offers retreat on aviation emissions. Retrieved from www.europeanvoice.com/articles/imported/eu-offers-retreat-on-aviaTION-EMISSIONS/ [accessed February 7, 2014].

Neiva, R. 2017 (Aug 9). Open skies, subsidies, and the need for transparency. *Eno Transportation Weekly*. Available at: www.enotrans.org/article/open-skies-subsidies-need-transparency/ [accessed February 22, 2018].

Rhoades, D.L. 2014. *Evolution of International Aviation: Phoenix Rising*. Ashgate, Farnham, UK.

Sochor, E. 1990. *The Politics of International Aviation*. University of Iowa Press, Iowa City.

Toh, R.S. 1998. Toward an international open skies regime: advances, impediments, and impacts. *Journal of Air Transportation World Wide*. Vol. 3: 61–70.

5

CERTAIN RESTRICTIONS APPLY

Pricing and marketing issues facing airline managers and consumers

Joseph P. Schwieterman

Many frequent flyers pride themselves on buying airline tickets with considerable sophistication and stretching their "loyalty program" points to the maximum extent possible. Less-seasoned travelers often approach ticket buying with much trepidation, while remaining mindful of the complexity and dizzying array of travel choice available to them. And the airlines—ever mindful of profit, loss, and bottom lines—constantly adjust their fares and number of seats offered. Like a game of chess, each player tries to anticipate the other's next move.

Airlines, ticket buyers, and the public agencies that manage airport systems must navigate vexing ethical issues. Travelers who want to spend as little as possible to reach their destination at a desirable time sometimes find their interests diametrically opposed to those of the airlines' revenue-management departments, which seek to *maximize* revenues.

Some of the ethical questions that emerge have clear answers, while others are more ambiguous. Should airlines be required to include taxes and security fees imposed by the government in their advertised fares? What about fees for such conveniences as paying with a credit card? Is it ethical for large airlines to undercut the prices of smaller "startup" airlines that have yet to gain a foothold in the market? In markets in which there is little competition, do airlines on occasion set prices so high that they violate the public trust?

Those purchasing air travel also confront ethical challenges. Should travelers be able to end their journey before reaching the destination shown on their ticket without notifying the airline and paying any applicable difference in fare? If the buyer of an airline ticket is an employer, should that employer be able to collect the frequent flyer points earned by employees on business trips for company use? If an employee selects a slightly more expensive flight in order to accumulate frequent flyer points on his or her preferred airline, does this constitute an ethical breach with her or her employer?

Finally, public agencies must grapple with ethical tradeoffs. At what point does "profiling" at airline security lines create an undue burden to certain passengers? Do "trusted traveler" programs allowing certain consumers to bypass security lines risk creating a society of "haves" and "have nots"? As this chapter describes, the answers are often not easy to determine. When making judgments, therefore, it is important to keep in mind critical details about the buying and selling of air travel, which are provided in the sections below.

The evolution of airline pricing

For the first 60 years of commercial flight in the US, most airlines had relatively small "tariff" departments to handle pricing issues. In general, carriers changed their fares infrequently, thereby limiting their need for staff devoted solely to pricing. To raise or lower fares, carriers needed to apply for and receive permission from the federal government (Holloway 2008; Winston 2010).

Starting in 1940, this process was handled by the Civil Aeronautics Board (CAB), an authority of appointed officials that originally operated as a unit under the Department of Commerce. Tariff departments typically worked closely with the company's regulatory affairs specialists to prepare "petitions" for authorization to change fares. The CAB often held public hearings, reviewed testimony, and conducted analysis before making a ruling—a process that could take months.

As airlines learned to "play the game" with regulators, however, it was not always clear whether federal oversight was helping or hurting consumers.[1] By the late 1960s, there were complaints that the CAB was acting more to protect carriers from competition than to shield consumers from high fares (Peltzman 1998). Some observers pointed to the existence of lower fares on *intrastate* routes, namely, routes confined within a single state, where CAB regulation did not apply (such as the intra-Texas routes famously served by Southwest Airlines) to support their contention that regulation was keeping fares artificially high.[2]

With airlines becoming larger, the demand for travel growing, and computerized reservation systems rapidly improving, airlines clamored for more flexibility in pricing their services. Carriers particularly wanted to experiment with steeply discounted fares that could attract new customers to fill seats that would otherwise go unused. This opportunity came when the CAB authorized the introduction of steeply discounted "excursion" fares in the early 1970s. "Super Savers" fares proved particularly popular, providing tremendous savings to passengers able to meet a Saturday night minimum stay and seven-day advance purchase requirement. Business travelers had great difficulty meeting the restrictions, which allowed carriers to offer such discounts without much dilution of existing revenue (Schwieterman 1995).

The success of Super Savers and other experiments cleared the way for the Airline Deregulation Act of 1978, which essentially gave the airlines freedom to price as they chose. Carriers moved quickly to exercise this freedom, developing vast hub-and-spoke systems and boosting frequency of flights on key routes.

The early years of deregulation were nonetheless turbulent. Smaller pricing actions sometimes escalated into full-scale "fare wars," which resulted in the demise of several legendary carriers, including Braniff International Airlines and Pan Am World Airways.[3]

By the early 2000s, the airlines that had weathered the storm had become much more adept at managing their prices. Gradually the sophisticated pricing strategies that we have today became standard. A competitive hodge-podge emerged, with high fares in some markets and low fares in others (Holloway 2008). Consumers had become accustomed to newly enacted fees, such as those on checked baggage and other restrictions. The complexity took its toll on the industry's image. Today, airlines are typically ranked last among major industries in customer satisfaction (Reuters 2014).

Ethical issues facing airline pricing personnel

When thinking about airfares, one should not associate them with particular *flights*, but with travel between particular *cities* (or stations) on the airline's route network. Airline prices apply to specific combinations of origins and destinations, often called *city pairs* or *markets*.

Personnel in airline yield management departments that manage prices must navigate a variety of difficult ethical issues. Where there is little competition, such as in markets too small to support more than one airline, a carrier can engage in *monopolistic pricing*. Although there is nothing illegal about charging high prices—and there is certainly no evidence to suggest that airlines as a whole have earned abnormally high profits in recent years—high fares are sensitive issues in many smaller and mid-size communities, whose economies depend on affordable air service.

Pricing concerns have escalated as a result of a series of mega-mergers that transformed the industry's competitive landscape. Delta Airlines merged with Northwest Airlines, American Airlines with US Airways, and Continental Airlines with United Airlines, leaving the country with just three large "legacy" carriers (Delta, American, and United), along with Southwest Airlines, which by some measures is just as large. Nipping at their bud are smaller but nevertheless profitable carriers such as Alaska Airlines and JetBlue (US General Accountability Office/GAO 2014b).

Consumers in cities served only by the legacy airlines often face higher fares than those in markets with more competitors. Many travelers in these locations have accepted the burden of driving to more distant airports in order to pay lower fares. As a general rule, allegations over monopolistic pricing by the legacy airlines tend to be greater for business travelers than for pleasure travelers, who tend to be more flexible and are more willing to divert to a more distant airport or take a no-frills alternative. Pleasure travelers are also more apt to shift to other modes of transportation—or not travel at all—if fares are high.

Concerns over monopolistic pricing are also common at large airports in which a local hub airline is so dominant that the carrier can command premium prices for passengers originating or terminating their trips in that city (GAO 2014b). After the latest round of mergers, all major airports in the US except Chicago's O'Hare International Airport (ORD) (at which both American and United have large operations) now have no more than one major hub operator. At Minneapolis-St. Paul International Airport (MSP), Delta Airlines now accounts for about 80 percent of all arriving and departing flights. On some occasions, its one-way fare from Minneapolis to other Midwestern hubs, such as Cincinnati or Detroit, are *higher* than for travel all the way across the country. Some critics have called such airports "fortress hubs" because they seem impenetrable to discount competition.

Related to this issue are allegations of "implicit collusion," which can occur when one airline effectively "signals" another about its desire to raise fares. There is nothing illegal about raising fares with hopes that competitors will match them; but it is illegal for one carrier to communicate these hopes directly with another, or to make statements to this effect through a third party, such as the news media.

The risk of cartel-like behavior is facilitated by the fact that most airlines process fares through a central clearinghouse, which allows them to see almost instantaneously the pricing actions of competitors. This makes it relatively easy to monitor each other's behavior. Many observers, however, fail to distinguish between pricing *coordination* of fares and collusion. There are compelling economic arguments why some degree of pricing coordination is necessary for the industry to remain viable, and why the public benefits when airlines can make investments in an environment of pricing stability (Telser 1987). Moreover, airfares have been trending downward for years when adjusted for inflation and the cost of jet fuel. As a result of these gradually declining fares, all major carriers except Southwest have entered bankruptcy at least once over the past several decades.

Even so, allegations of collusion have given rise to lawsuits, including a Department of Justice (DOJ) investigation in 2015, and embarrassing mishaps that have increased the public's cynicism about airline pricing practices. An investigation into whether airlines engaged in collusion to restrict the supply of air service, thereby driving up price, was particularly well publicized (Perez 2015). Federal officials argued that the airlines colluded by routinely discussing their plans for expansion—or to lower the number of seats offered in certain markets—in public forums. Even though this practice was common in most industries, which feel a need to provide data-hungry investors a sense of a company's plans, federal officials argued it had crossed a line with the airlines, which were doing this discretely to "signal" their competitors about the benefits of restricting capacity. Airlines had previously been accused by engaging in price fixing and "bait and switch" strategies. This investigation, however, took the antitrust process into unchartered territory.

That same year, federal watchdogs launched a second investigation several months after the deadly May 12th derailment of Amtrak Train Number 188 near Philadelphia. Airlines were alleged to have "gouged" consumers by exploiting their pricing power during the Amtrak service disruption, in some cases raising one-way prices in the Northeast Corridor to abnormally high levels exceeding $700 (USDOT 2015). The airlines vigorously denied these allegations. US Transportation Secretary Anthony Foxx drew attention to the investigation by noting: "The idea that any business would seek to take advantage of stranded rail passengers in the wake of such a tragic event is unacceptable" (USDOT 2015). Foxx added that the investigation will determine whether airline pricing during this period was "beyond the pale" (Schwieterman 2015).

The combined effect of these investigations, together with other federal actions, imposed a significant administrative burden on the airlines. Defenders of the airlines argued that federal officials were cynically playing on public wariness of air travel and were ignoring the heightening competition that airlines face from other modes of travel on short-distance routes (Schwieterman 2015). On top of the negative publicity from these investigations, United Airlines found itself embroiled in a complex "pay to play" controversy involving slots and other facilities at Newark International Airlines provided by the Port Authority of New York. As discussed more fully in Chapter 23, several senior officials at United, including the CEO, lost their jobs as a result.

Due to concerns over collusion, however, major airlines now operate under a voluntary agreement ("consent decree") with the DOJ to avoid certain practices that could facilitate cartel-like behavior (US Department of Justice 1994). It was once common for airlines to announce fare increases that had not yet taken effect, making them mere proposals. Airlines would "file" to raise fares, and if competitors did not match they could postpone the date that those fare increases took effect, hoping that the competitor would eventually reconsider. In effect, an airline could send a signal to competitors that it was ready to raise fares anytime. Critics argued that this technique provided carriers a risk-free way to solicit industry-wide fare increases and, as part of the voluntary agreement, airlines agreed to discontinue this practice.

Other types of price signaling have similarly cast the industry in a negative light. On some occasions, when a carrier opted not to match a nationwide increase announced by another airline, the latter would reportedly announce a "sale" in the competitor's hub in retaliation (Nomani 1990). The purpose of the "sale" was apparently to show a carrier that its refusal to match a price increase came at a heavy price.

Such allegations captured headlines in 1990 when an executive at Northwest Airlines (now part of Delta) testified to federal authorities about an internal memo he had written that likened pricing to an "atomic bomb" that airlines must threaten to use when a competitor did things to thwart efforts to raise fares. Apparently, the metaphor was intended to suggest that an airline must be prepared to retaliate against an unruly competitor, even if such action would mean mutually assured destruction in the form of extreme discounting. As part of the voluntary agreement, airlines adopted the practice of having all fare increases take effect instantaneously. Until other airlines match these increases, the airline that is raising fares must accept the fact that it will be noncompetitive.

Another issue is *predatory pricing*, or the practice of lowering fares with the hope of driving competitors out of a market (or out of business entirely).[4] Among the ways that airlines can—at least in theory—do this is by matching or undercutting the fares of a relatively weak startup airline and "dumping" new capacity into that market, which can render the profitable operation of the startup carrier nearly impossible (USDOT 1998). Economists are skeptical as to whether predatory pricing has ever materially changed the competitive landscape of US aviation (Viscusi et al. 2005: 321–322). Although major carriers appeared eager to force discounter Southwest Airlines out of their markets years ago, for example, they failed in every instance.

Yet there have been several high-profile instances where allegations of such behavior have cast the industry in an unfavorable light. Out of concern over pricing practices, however, the US Department of Transportation maintains guidelines for airlines to follow in order to avoid triggering a departmental antitrust investigation (USDOT 1998). These guidelines set limits on the number of seats a larger airline could make available at the price of the smaller "startup" airline as well as the number of additional seats it could add when a new competitor launches a service. The standards also stipulate that airlines refrain from offering frequent flyer bonuses as a means of implicitly undercutting the prices of new competitors.

Enter the ultra-discounters

Concerns over cartel-like behavior among airlines have been greatly lessened in recent years by the dramatic expansion of ultra-discounters, including Allegiant Air, Frontier Airlines, and Spirit Airlines. Modeled after no-frills European carriers easyJet and Ryanair, these airlines offer "unbundled" service that involves paying extra for such conveniences that passengers once took for granted, such as onboard beverages, paying with credit cards, bringing carry-on baggage, and reserving an aisle seat. The ultra-discounters generally set prices without much regard to the actions of major carriers. Spirit Airlines, for example, dramatically undercuts Delta on major routes from MSP, albeit with lower-frill and less-frequent service than many business travelers are willing to accept.

Still the expansion of these carriers poses other ethical issues. Many consumers have not come to terms with the add-on fees they impose, such as those for carry-on baggage. Yet the fierce backlash that many felt would arise never actually occurred, prompting the federal government to focus more on making sure the fees *are properly disclosed* rather than just the fees' existence. (One wonders whether things might be different had the ultra-discounters acted upon a proposal to begin charging for use of onboard restrooms.) Acceptance of these fees has encouraged airlines to consider imposing even more surcharges, such as for printing boarding passes at the airport.

The concerns that watchdog groups have over the rising number of fees are not limited to the ultra-discounters. Consider the fees imposed on passengers

TABLE 5.1 Baggage and cancellation fees on major air, bus, and rail carriers

Carrier	*Fee for Changing or Cancelling Reservation*	*Checked Baggage (First Bag)*
Delta United American Airlines	$200	$25
Frontier Airlines JetBlue Airways Spirit Airlines	$75–125	Frontier: $25 JetBlue: $20 Spirit: $20–45
Southwest	No fee	No reticketing fees No refunds for cancellations within 10 minutes of flight time
Amtrak	Depends on ticket	Free
Greyhound	$20	Free
Megabus	$2.50 Must be done online 24 hours prior to trip	Free
BoltBus	$4.50 Must be by phone up to 24 hours prior to trip	Free

changing their reservations. The three legacy airlines impose $200 penalties, plus any difference in fare, while smaller airlines (including JetBlue and Virgin America) impose fees of $75–125 (Table 5.1). Southwest Airlines is the exception, allowing all travel funds to be applied to future travel when the passenger cancels his or her reservation at least 10 minutes in advance. In general, however, the penalties imposed by airlines are steep compared to those of Amtrak and major bus lines, which maintain fees of 10 percent of the ticket price or less. Some liken the fees to a form of "gotcha" pricing that hits unsuspecting consumers who are unaware of the severity of the penalties when buying tickets. Even so, a proposed "Airline Passenger Bill of Rights" legislation, intended partially to address this issue as well as fees for checked baggage (see middle column, Table 5.1), never gained much traction. The public seems to have begrudgingly accepted the fees.

Ethical issues facing buyers of air fares

Consumers buying airfares can sometimes find themselves on the edge of an ethical precipice. Some have exploited "hidden city" discounts, that is, the ability to buy tickets to a distant city where a low fare is available without the intention of traveling to that city. For example, a consumer might buy a ticket from Cleveland to Las Vegas but leave the airport when changing planes in Denver, taking advantage of the fact that buying a ticket to Las Vegas was cheaper than buying a ticket to Denver. The website Skipplagged.com offers low fares based on such discounts (see textbox for a fuller discussion). Courts have ruled that this practice violates the contract the consumer entered with the carrier. In a few instances, airlines have successfully required passengers to pay the difference in fare.

The practice of "bracketing" is sometimes used to circumvent an airline's Saturday-night minimum stay requirement for discount fares. By purchasing several roundtrip tickets at once, with each flight staggered to meet the Saturday-stay requirements, a customer skirts the minimum-stay rule without ever staying over a weekend. The courts have ruled that this, too, violates the customer's contract with the carrier. In this case, it violates the stipulation that travelers must finish using one ticket before beginning to use another. This gives airlines the right to collect additional fees from travelers caught engaging in this practice (Schwieterman 1995).

SKIPLAGGED.COM AND ITS PROMOTION OF HIDDEN-CITY FLIGHTS

The website Skiplagged.com offers consumers lower airfares by capitalizing on loopholes the airlines create. Most notably it sells fares based on hidden-city flights in which the passenger would disembark at the layover city rather than the final destination. Under the FAQ section, this website informs its potential customers not to check bags since the baggage would end up at the final destination. It also advises passengers not to use this service too often. More specifically, it says not to fly on the same route with the same airline dozens of times in a short period. Its final FAQ says that "you might upset the airline." In addition, the website provides a link to a 2014 article by CNN Money that discusses a civil lawsuit filed by United Airlines against Skiplagged (which was dismissed in 2015). Skiplagged claims that they are trying to make a travel company that works for the consumers' benefit, and that consumers should have more power over how they spend their money. However, there can be negative consequences to the traveling public, such as flight delays in waiting for passengers and depriving other passengers of a seat they actually need. While the practice may be legal, the question remains if it is ethical.

The website is available at https://skiplagged.com/ [accessed February 22, 2018].

Consumers need to decide for themselves the extent to which they consider the above practices to be serious ethical violations. At a minimum, they should understand that travelers who use these practices are operating on the fringes of ethical behavior and that the courts have routinely sided with the airlines.

Unfortunately, some consumers engage in more flagrant forms of airfare abuse. Airline personnel for many years did battle with consumers who used deception (or tried outright intimidation) to gain access to fares for which they were not entitled, including "bereavement fares" reserved for people traveling due to the death of a family member. As a result, all airlines but one (Alaska Airlines) have dropped these special fares. Another common deception is presenting medical-waiver forms from doctors indicating that their patient should be entitled to a refund on a nonrefundable ticket when in fact the patient's health problems were either grossly overstated or had nothing to do with the patient's decision not to travel.

Ethical issues also emerge when consumers exploit obvious errors made in pricing. Stories about passengers purchasing fares filed entirely by mistake (some of which are priced at only a few dollars for an international trip) are part of cable television news coverage. In some instances, thousands of fares have been sold (in some cases offering tickets for less than $10!) before a conscientious buyer or travel agent has notified the airline of the mistake. Legally, an airline has the right to rescind a transaction when a fare was processed in error, but they often pay a heavy price in the form of negative publicity this generates.

Managing "seat inventory"

Airlines invest heavily in capacity-control systems that optimize the number of seats they can sell at various prices (Holloway 2008). Highly priced fares, such as full-fare tickets, are typically assigned to booking classes that are generously available, while discounted fares are usually limited to a smaller number of seats, which are often depleted weeks before the departure date.

When a ticket is sold, a message is transmitted to the airline's capacity-control system to decrease the number of seats available in the applicable booking class, information that is then sent to computer reservation systems throughout the world. Unlike pricing personnel, who rarely interact with independent travel agents (as published prices are not negotiable), capacity-control analysts regularly make exceptions to booking class limits when preferred clients express a need for additional seat inventory.

Ethical issues facing capacity analysts

Capacity-control analysts must navigate a variety of ethical issues. For years, airlines have been accused of "bait and switch" strategies, namely, advertising low prices but making them exceptionally hard to obtain. Although the significance of this problem should not be understated, marketing considerations tend to discourage airlines from doing this. Since low airfares tend to be matched by competitors, airlines that do not offer seats at advertised prices often face significant losses in business. "Bait and switch" was arguably a bigger problem when the primary source of information about fares was television and newspaper advertising. Before online search tools became available, passengers often had great difficulty determining when and where the lowest fares were available. Today, search engines at sites such as Expedia and Travelocity allow consumers to quickly locate the most attractive prices.

To remain compliant with federal law, airlines need to offer at least a handful of seats on every flight and at prices that are advertised (Boyd 2007). The tendency for the most highly discounted seats to be quickly depleted on flights during peak travel times, however, has generated many complaints. Similarly, many passengers voice great frustration over their difficulty in finding available seats to redeem frequent-flyer points, which are also managed through capacity-control systems. On numerous occasions, airlines have been sued for allegedly denying account holders reasonable use of their award mileage. As a compromise move, airlines now typically offer two types of award: a "premium" award that is readily available and a "saver" award that is capacity controlled. Nevertheless, the difficult of redeeming frequent-flyer credit remains a lightning-rod issue among flyers (Boyd 2007).

Overbooking flights

Space-planning personnel face a different set of problems. Their work is directed heavily at minimizing the consequences of travelers who are "no-shows" for flights. Some no-shows are attributable to intentional acts by travelers who want to hedge their bets; but many others are no-shows because of factors beyond their control, such as missed flight connections. Although the no-show (or "booking turnover") problem has dogged transportation companies for more than a century, it was not until the early 1980s that carriers added cancellation penalties to discounted tickets to reduce the severity of the problem.

For space-planning personnel, many of the most vexing ethical issues revolve around how to plan for and handle involuntary denied boardings (IDBs); namely, ticketed passengers who have not volunteered to surrender their seats but are nevertheless denied transportation due to overbooking. Until the late 1990s, consumers seemed relatively satisfied with policies regarding IDBs: typically, there were enough volunteers willing to accept travel vouchers (typically free roundtrip tickets) to keep IDBs to a minimum. In the early 2000s, however, sentiment shifted against the airlines. Airlines became less generous in compensating "bumped" passengers, and declining airfares made the free-travel vouchers seem less valuable. Moreover, as the number of empty seats diminished, it became more difficult to accommodate bumped passengers on the next flight.

After an uptick in IDBs occurred in the early 2000s, growing numbers of consumer groups alleged that the airlines had become callous and greedy. Some called for federal intervention to assure that bumped passengers were adequately compensated. Opponents of such intervention argued that airlines vigorously competed based on the experience they provided travelers, which gave them an incentive to treat IDBs fairly. Regulation, they argued, would ultimately drive up ticket prices.

The US Department of Transportation took action in 2008, mandating that airlines substantially increase their reparation (compensation) for IDBs. Passengers who are rescheduled to arrive at their destinations on domestic trips more than two hours after their originally scheduled time, for example, are eligible for 400 percent of their one-way fare (or a maximum of $1,350) in compensation (USDOT 2018). Sadly, airline employees working in space-planning departments have been known to commit ethical lapses for purely personal reasons. Some have opened up seats in sold-out booking classes to benefit friends or family; other have closed down flights for sale to abet their own "standby" travel, hoping to ensure that an empty seat would be available. Airlines now have computer programs to detect such acts.

Fees to keep passengers safe

Air travelers today pay fees that maintain the system of checkpoints and other security measures to deter terrorism. Although security checks began in the 1970s, their role greatly increased after September 11, 2001. Since then, body-scan technology, the emphasis on scanning shoes, and the introduction of rules on the transport of liquids and gels in carry-on baggage has gradually made checkpoints more intrusive (US GAO 2014a). The fees imposed by the Transportation Security Administration (TSA) are now around $6 per one-way trip.

Determining who is responsible for maintaining the cost of this system has been controversial. Airlines argue that terrorism is fundamentally an issue of national security—and that terrorists have singled out airlines due to the fact they are an easy target and have a high level of public visibility. If air service was not available, they argue, the terrorists would simply shift their emphasis elsewhere. So why should airline customers bear the full burden of TSA spending? After all, drivers are not asked to pay special fees to keep the roads safe from carjackers or drive-by shooters.

Such arguments have been largely dismissed by federal agencies, which regard ticket surcharges as simply "user fees" needed to cover the full costs of travel. The consequences to the airlines, however, have been serious. On some routes, security fees and taxes can account for as much as 25 percent of the revenue collected. These fees and taxes, together with the "hassle factor" created by airport checkpoints, have contributed to the steady decline in airline travel over routes of 500 miles or less since 2000. Routes that were once served by large planes, such as New York–Washington, Chicago–Cincinnati, and Las Vegas–Los Angeles, are now routinely operated with smaller regional jets. Even so, shifting the entire cost of airport security to taxpayers in general would not seem fair either.

Finding a perfectly fair system may be impossible, but the airlines argue that the burden created by security should, at a minimum, be unambiguously disclosed to passengers. With this goal in mind, airlines have pushed—unsuccessfully thus far—for "airfare transparency" legislation that would relieve them of a Department of Transportation regulation that requires all quoted fares to be inclusive of government-imposed fees and taxes. The airlines would like the fees kept separate so consumers take notice of the fees that their government has chosen to impose on them. The airlines believe this would create a more robust national discussion about these fees. Critics argue that airlines are merely trying to be deceptive so they can advertise fares that are less than the total cost of travel.

The TSA techniques to screen passengers also evoke emotional responses. This agency faces the delicate task of trying to stretch its budget while also avoiding hassling to an unreasonable extent. Yet there are no agreed-upon ethical principles about when to limit the "profiling" of passengers. Should the same level of effort be expended on searching a 25-year-old passenger who has (suspiciously) purchased a ticket with cash and recently traveled from a high-risk location as on an elderly man who is traveling roundtrip from Orlando, Florida, with his grandchildren? The TSA has determined that the answer is "no" as it considers profiling critical to assuring that its limited resources are spent wisely. To this end the TSA has moved forward aggressively with "trusted traveler" programs that allow people to register for the benefit of passing through security much more quickly than others. Whether the safeguard it applies, such as random checks applying to everyone, effectively maintains a sense of fairness, however, is difficult to answer.

Managing all of this has proven to be a slippery slope. Many flyers have complained of the stress and humiliation of repeatedly being searched. Heightened profiling at airport checkpoints also may have resulted in the more aggressive use of this policy in other parts of the air-travel system. Pilots and flight attendants have also used profiling on board planes to alert security to potential risks, often with backlash from those who are singled out.

Regardless of how this issue is resolved, the public has grown skeptical of the existing security practices, suggesting that change could be in the offing. Studies have shown that the present system is far less effective than one might expect considering the enormity of the funds being spent, which adds to the public's indignation (US GAO 2014a).

Reaching the destination

Consumers and airline managers—and the governments that both represent and regulate them—must ultimately decide for themselves where to draw the line on these and other ethical issues. Herb Kellerher, former CEO of Southwest Airlines, once humorously said, "If the Wright brothers were alive today, Orville would have to fire Wilbur to cut costs" (Holloway 2008: 59). For the airlines the battle for survival means not only cutting costs but also finding new revenue, which assures that new ethical questions will emerge.

Technology will push companies toward the boundaries of ethical behavior. Some firms use software that masquerades as consumers to "scrape" the computer reservation systems of rivals, collecting vast amounts of competitive intelligence about prices and flight availability. In some instances, these programs have brought rival websites to a virtual standstill. New facial recognition software will likely raise questions about invasion of privacy at the airport and the right to be anonymous.

The consumer/traveler and revenue-management personnel of airlines, together with the government agencies that oversee them, are in a chaotic dance, each trying to anticipate the other's next move. As they attempt to find their rhythm, each party must learn from their past mistakes and respect the needs of the other, while preparing for some fancy footwork, lest they stumble over ethical impediments blocking their way.

Notes

1 The most critical factor affecting CAB rulings about price changes was the cost of providing airline service, including the cost associated with labor, fuel, and aircraft equipment. Since cost escalation could be passed on to the consumers, airlines often had little incentive to control their costs. Also, since airlines could not compete based on price, carriers turned to providing amenities—such as comfortable seating configurations and food service—as well as providing more attractive schedules to attract passengers.

2 As a general rule, state governments had a greater willingness to grant airlines latitude over pricing decisions than the federal government. This allowed carriers such as Southwest Airlines and Pacific Southwest Airlines (PSG) to aggressively expand in intrastate markets prior to deregulation.

3 In one particularly notable example, Pan Am Airways introduced a $99 fare on its domestic network in 1983. Although Pan Am was oriented principally to international routes and had few domestic flights, there was "domino effect" as carriers matched and extended the discounts to new markets. Within a few days, a full-flown "$99 sale" was underway throughout the country.

4 This practice has special significance in the airline business since carriers can quickly enter and exit markets, giving incumbents an incentive to develop a reputation for vigorously defending their market share in order to deter entry.

References

Boyd, A. 2007. *The Future of Pricing: How Airline Ticket Pricing Has Inspired a Revolution.* New York: Palgrave Macmillan.

Holloway, S. 2008. *Straight and Level: Practical Airline Economics.* 3rd ed. Burlington, VT: Ashgate.

Nomani, A. 1990 (Oct 9). Fare warning: how airlines trade pricing plans. *Wall Street Journal.*

Peltzman, S. 1998. *Political Participation and Government Regulation.* Chicago: University of Chicago Press.

Reuters. 2014 (Apr 22). U.S. airlines rank lowest in satisfaction among travel sectors: poll. Retrieved from: www.reuters.com/article/2014/04/22/us-airlines-poll-idUSBREA3L06S20140422 [accessed November 25, 2015].

Schwieterman, J.P. 2015 (Aug 18). An airline investigation that misses the bus. *Wall Street Journal.*

Schwieterman, J.P. 1995. A hedonic price assessment of airline service quality in the US. *Transport Reviews*. Vol. 15, No. 3: 291–302.

Telser, L. 1987. *A Theory of Efficient Cooperation and Competition*. New York: Cambridge University Press.

US Department of Justice. 1994 (Mar 17). Justice Department settles airline price fixing suit: may save consumers hundreds of millions of dollars. Press Release.

US Department of Transportation (DOT). 2018. Fly-rights: a consumer guide to air travel. Available at www.transportation.gov/airconsumer/fly-rights [accessed March 8, 2018].

US Department of Transportation (DOT). 1998 (Mar 5). Statement of Patrick V. Murphy, Deputy Assistant Secretary for Aviation and International Affairs, Department of Transportation, before the Subcommittee of the Senate Appropriations Committee.

US Department of Transportation (DOT). 2015 (Jul). US DOT requests information on airline pricing response to Amtrak derailment. Press Release.

US General Accountability Office (GAO). 2014a. Secure flight: TSA should take additional steps to determine program effectiveness. GAO-14-531. Available at: www.gao.gov/products/GAO-14-531 [accessed July 10, 2018].

US General Accountability Office (GAO). 2014b. Airline Competition: General Accountability Office. The average number of competitors in markets serving the majority of passengers has changed little in recent years, but stakeholders voice concerns about competition. GAO-14-515. Available as a pdf at: www.gao.gov/assets/670/664060.pdf [accessed July 10, 2018].

Viscusi, K.W., Harrington, J.E. and Vernon, J.M. 2005. *Economics of Regulation and Antitrust*. Boston, MA: MIT Press.

Winston, C. 2010. *Last Exit: Privatization and Deregulation of the US Transportation System*. Washington, DC: Brookings Institution.

6

OUTSOURCING MAINTENANCE

A union perspective

Ted Ludwig

Today's global economy has created a culture of outsourcing which benefits large corporations by reducing their labor costs. Corporations contract other companies, or third party vendors, to provide labor and benefits at a lower rate, thus saving the corporation money and improving the return to investors. While this may be an economically sound business plan, I contend that it should not be utilized in certain sectors of the economy. The aviation industry is one of those sectors.

Aircraft maintenance has been one of the areas that corporations, especially airlines, have outsourced to third party vendors that are located in the US as well as overseas. Commercial aviation has contracted as much maintenance as possible to third party vendors in the hope of saving on labor costs. The following argument is based on the commercial airline industry, as it affects the greatest number of people in our society. Airlines claim that having aircraft maintenance performed by third party vendors is just as safe and reliable as maintenance performed "in-house" at the airlines' own maintenance department, using federally licensed aircraft technicians. However, aircraft maintenance that is contracted to the lowest bidder is not conducive to a *safe* maintenance program. Once an airline has its maintenance completed by third party vendors, it dismantles the infrastructure needed to perform heavy, overhaul checks, which are the most important part of aircraft maintenance. The airline is no longer capable of pulling the work back in-house if the vendor performs poorly. Once the maintenance work is finalized outside of the airline's hangars, their technicians can no longer ensure that the aircraft are being maintained to federal standards.

Outsourcing and FAA oversight

All maintenance performed on commercial aircraft is regulated by the FAA to ensure that the aircraft we fly on are maintained to safety standards. To qualify for a mechanic's certificate the worker must be at least 18 years of age and be able to read, write, and understand English. In addition the employee "must have 8 months of practical experience with either power plants or airframes, or 30 months of practical experience working on both at the same time" (FAA 2013). One must also pass three types of exam: oral, written, and practical. It is important to note that a worker is allowed to do maintenance work without a certificate. In these cases the FAA states that one "may only perform aviation related work when supervised by a person with a valid mechanic's certificate with airframe rating, power plant rating or airframe and power plant ratings (A&P)" (FAA 2013).

Because there are more aircraft being repaired than there are FAA inspectors, the agency relies on the airline to follow its own maintenance program which has been reviewed and certified by the agency. An airline lays out in detail how it will perform the maintenance on its fleet. Due to these programs, the FAA only needs to complete spot checks, or inspections, inside the carrier's maintenance facility, thus minimizing the need for inspectors.

This system has worked properly for decades. But in the era of low-cost carriers (LCCs), it can no longer work as designed. The LCCs outsource the majority of their maintenance to the lowest third party bidder. They are called "low cost" for a reason. In order to be competitive, the major airlines have followed suit with their maintenance programs in order to reduce costs. However, these programs are not designed for a third part vendor to complete the majority of their maintenance. Thus, the major airlines have revamped their programs in order to maintain compliance with the FAA.

Such revisions make sense on paper. But the truth is that the level of safety has been reduced significantly once the airlines begin outsourcing maintenance. Even today safety standards are continuing to deteriorate. There are two reasons for this change: 1) not enough FAA inspectors have been hired to spot check the thousands of third party repair facilities that have sprung up around the world; and 2) the majority of the airlines' own licensed technicians have been eliminated. The FAA has admitted it cannot spot check every vendor each year, let alone ascertain that the unlicensed technicians hired by the outside vendors are in line with federal standards.

Although the FAA relies on the individual carriers to perform their aircraft maintenance using the certified program, the problem is that with outsourcing the airline has to ensure that every one of its vendors operates properly. The airline signs a contract with the vendor, who in turn states that its work will remain compliant with the maintenance program certified by the FAA. This means that the "cost-driven" airline is like the fox watching the hen house. The vendor needs to keep its costs down, or else it risks losing the airline's contract. Thus, the vendor performs the bare minimum required to satisfy the airline's maintenance program, with lower-paid, non-union, unlicensed technicians. Who is watching to ensure that the federal standards which have kept us flying safely for decades are being met?

Specific issues with overseas outsourcing

While the FAA does not have the staff to inspect all of the domestic vendors, when aircraft repairs are completed abroad, the problem is compounded. As former FAA safety inspector Richard Wyeroski states, "for years the FAA has allowed substandard overseas 'Third World' repair stations to start up with their approval. Little or no surveillance has been conducted at these facilities and a large increase of emergency returns has occurred" (Wyeroski 2012). For example, all licensed aircraft technicians who work at airports across the US are fingerprinted, and their names are run through a ten-year background check to ensure that they are who they say they are and not someone set on doing harm to the passengers. But, when aircraft maintenance is outsourced to overseas vendors, there is no background check or fingerprinting of the technicians.

Aircraft sent to overseas vendors generally require a "heavy check," or what would be better understood as an "overhaul." This may take up to eight weeks to accomplish because of the degree to which the aircraft has to be disassembled and inspected. Every nook and cranny is exposed to potential threats of terrorism. Terrorists could plant explosives inside the walls of the planes they have worked on, all set to detonate on the same day at the same time. For instance, a reported 2009 plot by Indonesian terrorist Noordin Top targeted commercial aviation at Jakarta's main airport, "which included assistance from a former mechanic for Garuda Indonesia" (Brandt 2011).

When aircraft maintenance is kept within the air carrier, control of who has access to the aircraft, and the quality of the maintenance performed can be guaranteed. Once the maintenance is sent to an overseas vendor, the security and quality can no longer be controlled, let alone guaranteed. In countries with few unions, working conditions can be extremely hazardous compared to US standards. There may be no equivalent of the Occupational Safety and Health Administration (OSHA) or Environmental Protection Agency (EPA) to shield the workers so that they are able to perform their jobs safely. The lack of standards is in large part why companies often prefer outsourcing work to overseas vendors; it is cheap, and the workers are expendable.

Overseas vendors utilize a handful of skilled technicians to supervise the work of hundreds of laborers. The skilled technicians are supposed to guarantee that the air carrier's maintenance program is followed, at least on paper. Meanwhile the laborers do the hazardous work, such as sanding toxic paint, spraying carcinogenic cleaning solvents, and so on. Much like laborers that travel to the US from Mexico, workers from other third world countries will travel to the vendor sites. They may not be documented, or citizens of the vendor country, and may not be able to understand written English. These workers cannot voice concerns about work conditions or improper maintenance for fear of losing their jobs. In addition, if one cannot understand the manual, then one is also unable to report improper maintenance. As one can readily see, there are very few controls on overseas vendors.

The benefits of licensed, union technicians

Hopefully the reader can understand my concern for the rapid race to outsource maintenance to the lowest bidder, whether in the US or abroad. The truth is that licensed aircraft technicians are trained and entrusted to comply with all Federal Aviation Regulations (FARs) and to report any maintenance that is not performed under these regulations. This system works wonderfully when all aircraft technicians are licensed by the FAA and work under a union contract. The contract is instrumental in ensuring that US commercial aircraft are maintained properly. But to the reader, this claim may sound rather one-sided, especially coming from a union-represented, licensed aircraft technician.

However, in a real-life scenario, the reason why a licensed aircraft technician would be able to prevent an unserviceable aircraft from leaving the gate and taking off full of passengers is because the union protects the technician from reprisals. This type of scenario arose many times in my 25-year aviation career. Licensed aircraft technicians working at a non-union vendor who have refused to sign off on repairs not done according to regulations have often found themselves blacklisted. Some have even been terminated for "performance-related issues" when they, in good conscience, would not sign for an improper repair. This type of situation corrupts the standards of the technicians, who end up failing to uphold their licenses.

Only a licensed technician can sign off a repair on an aircraft. If the manager is not licensed, that person leans on a licensed technician, especially one who has an inherent fear of being terminated, to sign off, or "pencil whip," the illegal repair. Pencil whipping is a euphemism for "when workers, supervisors, and, yes, safety managers fill out observation cards, sometimes in great numbers, without actually conducting the observation (much less providing the critical feedback)" (Ludwig 2012). Naturally, the technician that puts up the least resistance ends up pencil whipping the improper maintenance. The chance that this type of scenario may arise is magnified when the airline uses an overseas vendor. If a union is present, even the most intimidated technician can refuse to sign for an illegal repair because that employee knows that union representation is only a phone call away. I witnessed many times when a union representative intervened on the tarmac to prevent the aircraft from leaving the gate in an unserviceable condition.

Union involvement in commercial aviation maintenance is more important now than ever before. The industry has been faltering economically for many years, and cuts have been made everywhere in order to restore the industry to its former profitability. These cuts have caused maintenance to be outsourced to the lowest bidder. That in itself is an unsettling fact. But add to it air carriers that have cut staff to the bare bones while at the same time being responsible for monitoring the hundreds of vendors they have contracts with, and the situation becomes frightening. With the FAA unable to keep up with outsourced aircraft repairs, and air carriers operating under extremely tight budgets in a highly competitive, cut-throat industry, unions are the only insurance that maintenance will be performed to the same standards that have kept the flying public safe for all these years.

I am not implying that management will purposely place the flying public in harm's way; but the airlines are under an incredible amount of pressure to maintain on-time performance, and this pressure may lead to the violation of the technicians' role to abide by their license. Technicians are *responsible* and *liable* for that aircraft when they sign it off. Although rare, *liability* can result in jail time for not following the FARs. Their *responsibility* is to maintain a properly functioning aircraft that in turn will ensure the passengers' safety as they travel. Flying is not natural for humans. It is a very mechanical process that allows several tons of metal to lift off the ground. If one part of the process is not functioning, the result could be catastrophic.

If the aircraft technicians only needed to abide by turn-times or vendor contract times as they performed maintenance, safety would be compromised. The licensed aircraft technician has an attachment to the aircraft, or even the part that one repairs and signs off. The technician takes ownership of that product and ensures that it is repaired or serviced properly under federal regulations. But to an unlicensed vendor technician the work performed is simply another job to do because that person does not take ownership of either the aircraft or the part. It does not matter if the repair is done per regulations, as long as it passes through one's department within the time limits. There is no liability or responsibility placed on the unlicensed technician. The onus is on the vendor company, which is allegedly working under the air carrier's maintenance program, certified by the FAA. So who is actually responsible for and taking ownership of the aircraft you are flying on? Is it the unlicensed, non-union vendor technician, or the vendor company, or the air carrier, or the FAA? No one really needs to take ownership.

The circumstances that have resulted from outsourcing aircraft maintenance are not conducive to continued safe air travel. Because of the over-engineering that goes into the design of aircraft, it will be some time before these circumstances will show that effect. We have had a very safe travel industry for many years, and that is not because of chance. Rather, licensed technicians have maintained the system by following FAA regulations. Thus, it is important to note that in the past, union workers were the ones dedicated to making sure these regulations were upheld. Now the air carrier has removed this work from union employees. The reader may want to consider why.

References

Brandt, B. 2011 (Nov 30). Terrorist threats to commercial aviation: a contemporary assessment. *Combating Terrorism Center (CTC) Sentinel*. Available at: www.ctc.usma.edu/posts/terrorist-threats-to-commercial-aviation-a-contemporary-assessment [accessed June 4, 2016].

Federal Aviation Administration (FAA). 2013. Basic requirements to become an aircraft mechanic. Available at: www.faa.gov/mechanics/become/basic/ [accessed June 5, 2016].

Ludwig, T. 2012. The anatomy of pencil whipping. *Safety-Doc.com*. Available at: http://safety-doc.com/safety-blogs/blog/the-anatomy-of-pencil-whipp.html [accessed June 5, 2016].

Wyeroski, R. 2012 (Oct 3). The Federal Aviation Administration is the problem because of "regulatory capture." *Airnation.net*. Available at: http://airnation.net/2012/10/03/faa-regulatory-capture/ [accessed June 4, 2016].

7

WHISTLEBLOWING IN AVIATION

Bruce Hoover

Whistleblowing can be defined as the voluntary and intentional release of non-public information into the public domain, as a moral protest, by a member or former member (the actor) of an organization (the target) outside the normal channels of communication to an appropriate audience (disclosure recipient) about illegal and/or immoral conduct in the organization that is opposed in some significant way to the public interest (Boatright 1993: 133). Whistleblowers act in good faith and in the public interest to raise concerns about suspected impropriety within their place of employment. Whistleblowing is commonly distinguished as internal and external (Louw 2011). The former is defined as following recognized procedures within an organization with the intention of resolving a problem internally. It is commonly required in law to exhaust the internal avenue before steps are taken externally.

Peter Jubb (1999) takes the view that whistleblowers have more in common with the dissident than with the informer. His position supports the notion that the difference is in the stand that whistleblowers take when they make the information public. Whistleblowers are individuals who expose danger, negligence, or abuse such as professional misconduct or incompetence that exists within the organization in which they are employed. The decision to "blow the whistle" on a colleague, an associate, or an employer is never an easy one; unless there is a legal obligation to report, it should be considered a step one takes when all else has failed (Ray 2006).

Similar to a siren or fire alarm, whistleblowing alerts people to pay attention to what is happening or is about to happen, and to take action immediately. Workers who blow the whistle on prohibited or unlawful practices they discover during their employment can play an important role in enforcing federal laws. However, these whistleblowers also risk reprisals from their employers, sometimes

TABLE 7.1 Well-known US cases of whistleblowing

Case	*Year*	*Industry*
B.F. Goodrich & Air Force A7-D brake problem ("Aircraft Brake Scandal")	1968	Manufacturing—wheel & brake
Pentagon Papers	1971	Military and politics
Deep Throat	1972	Politics
Silkwood and Kerr McGee	1974	Nuclear energy
Hughes Microelectronics hybrid microchips test	Mid-1980s	Manufacturing—micro electronics
Roger Boisjoly and Space Shuttle *Challenger*	1986	Aerospace
Jeffrey S. Wigand and tobacco	1996	Tobacco
Enron: Sherron, Watkins, Ceconi, and others	2001	Energy
WorldCom	2002	Telecommunications
Alderson, HCA Inc., and Medicare fraud	2003	Healthcare
San Diego Pension	2003	Local government
Rost and Pfizer	2003	Pharmaceutical
Schering-Plough Corp.	2004	Pharmaceutical
Fishbein and the NIH	2005	Federal research
FAA inspectors and Northwest Airlines: allegations of unsafe maintenance practices	2005	Airline
FAA inspectors and Southwest Airlines: oversight of airlines	2008	Airline
Edward Snowden	2013	National Security Agency (NSA)

being demoted, reassigned, or fired. Table 7.1 provides some examples of well-known whistleblowing cases across different industries in the US.

Moral dilemmas

Whistleblowing is an ethically complicated act that requires an understanding of obligation, honesty, loyalty, and duty. Jubb notes that the whistleblower's ethical dilemmas originate in role conflict. The dilemmas take on two forms, and both interact: a conflict between personal and organizational values (virtue ethics);

and a conflict between obligations owed to an organization and to parties beyond it (divided loyalty) (Jubb 1999: 81, 83). Lindblom (2007: 415) argues that the debate on morality whistleblowing centers on the conflict between the duty of loyalty to the firm or organization in which one works and the liberty to speak out against wrongdoing.

Ethical analysis of moral conflicts often finds resolution by the consequentialist or utilitarian frameworks that appeal to the greatest good, or by attempts to rank moral obligations on some moral scale. Whistleblowers face a continuous tension of responsibility: loyalty to the institutions in which they participate, or the consequences of their silence. Whistleblowers must choose an ethical path. In aviation, like many organizational settings, we are drawn deep into our responsibility for those who would be harmed by our inaction. In *Applied Professional Ethics*, the utilitarian principle states that one should choose the course of action that produces the greatest benefit for the greatest number of people. By focusing on the benefits, utilitarianism is concerned primarily with the consequences of action (Beabout and Wennemann 1994).

There is also the question of self-interest or egoism. Whistleblowers may be exposed to charges of disloyalty, disciplinary action, freezes in job status, forced relocation, and even dismissal. It is reasonable to ask ourselves the extent to which it will endanger our own well-being. To what extent will rational self-interest override moral obligations? If an employee has an excuse to refuse to be a whistleblower, then many people may get hurt when no individual is willing to take a stand. Egoistic individuals should not use self-interest to rationalize the wrongs done by employees or companies. When you are a member of the club, it becomes much more difficult to blow the whistle; you value your membership in that club. The choice is to remain silent.

Moral conflict is immediately brought to bear when an individual considers the possibility of blowing the whistle on one's own organization. An understanding of why moral conflict may arise and a perspective for thinking about how to resolve the conflict are important. On the one hand, an individual contemplating an act of whistleblowing should ask oneself what good will result and what bad will result if one carries out the act. On the other hand, air carriers and government oversight agencies are confronted with multiple regulatory and economic challenges in today's environment. They can ill afford whistleblower claims.

The nature of whistleblowing

Whistleblowing is considered a vital tool for promoting individual responsibility and organizational accountability (Louw 2011). It does not extend to the following: (1) acts motivated by bad faith (for example, a disgruntled employee attempts to damage an organization's image); (2) prematurely uncovering matters that should primarily be handled internally and are not matters of public concerns to begin with; (3) involving structures of authority that are not recognized in law; and (4) engaging in whistleblowing on mere suspicion instead of demonstrable evidence.

Whistleblowing, by its very nature, is perceived as "rocking the boat." Indeed, it is designed to wake up the silence of "affairs as usual." The elements of whistleblowing—dissent, violation of loyalty, and accusation—create a unique atmosphere of bitterness; and whistleblowers are consequently portrayed as disloyal troublemakers and stigmatized and shunned accordingly by both employers and co-workers (Johnstone 2004a; Erlen 1999; Rosen 1999). Whistleblowing is often suggested to be used as a last resort—after all other avenues have been exhausted in attempts to remedy the situation—since morally justifiable actions come with risks and do not guarantee morally desirable outcomes (Johnstone 2004b).

Perception and reception of whistleblowers

Although whistleblowers are commonly motivated by their professional conscience, the image of the whistleblower remains tainted as a "rogue" and "troublemaker." Protected in theory, the whistleblower often remains unprotected from reprisal practices (Louw 2011). Since childhood, we learned that being a tattletale is bad. The same attitude toward tattletales holds true for adults. A whistleblower, a tattletale in the professional environment, often receives a negative response from co-workers and even neighbors. In a recent famous whistleblowing case, a full criminal investigation was launched to examine the spending habits of former privacy commissioner George Radwanski and several of his subordinates. Radwanski threatened to end the career of the "rat who squealed to the MPs" if he ever found out who it was (Louw 2011).

Paradoxes of whistleblowing

The act of whistleblowing sets off alarms. A voice is raised and action follows. Dissent, breach of loyalty, and accusation are elements that may lead to bitterness and discontent.

Dissent—Differences in opinion against the majority become very public. The dissent may be focused on negligence. There is a conflict for the whistleblower: do I conform or do I stick my neck out? The latter presents great risk.

Breach of loyalty—An air carrier mechanic who blows the whistle on his/her own team may be seen to violate loyalty to the group. Conflict arises between loyalty to work colleagues and loyalty to the public interest. The whistleblower is an insider. One alternative that may provide a level of perceived safety is for the whistleblower to exit the workplace, secure a job elsewhere, and then blow the whistle. Keeping one's job is difficult during this alarming action.

Accusation—To charge others of wrongdoing arouses superiors and may lead to strong reactions. Negligence, endangerment, or abuse may be the stimulus. Accusation signals specific persons or groups of employees as responsible for threats to public safety. The alarm may address a safety defect in a specific aircraft that threatens or will shortly threaten passengers. Whistleblowers must have identified a specific, concrete risk, rather than some prediction or vague concern. Often the danger is kept secret by the organization, or a few individuals within it. The whistleblower may wish to reveal the secret for publicity, self-aggrandizement, or revenge for perceived injustices in the workplace.

Impact of whistleblowing

The stakes in whistleblowing are high: a nurse who alleges that physicians enrich themselves through unnecessary prostate surgery; the engineers who disclose safety defects in the braking systems of a fleet of new commercial passenger-carrying aircraft; the Defense Department official who alerts members of Congress to greed and overspending. All of these individuals know that they pose a threat to those whom they denounce, and that their own careers may be at risk. The Canadian Federal Accountability Initiative for Reform (FAIR 2009) describes eight phases of the typical experience of whistleblowers, how they find themselves compelled to speak out, and the consequences for them and the rest of us: (1) awareness, (2) the decision of conscience, (3) raising concerns internally, (4) facing the initial reprisals, (5) the decision to commit fully, (6) going public and the consequences, (7) the war of attrition, and (8) the endgame.

Whistleblowers are often profoundly affected by their decisions and actions. For example, in the famous mid-1980s' Hughes Microelectronics whistleblowing case, following job losses, one of the whistleblowers and her husband had to file for bankruptcy, and their marriage eventually dissolved. The other whistleblower was on welfare for a year before she could find another job. However, both of them felt they were right in taking the action. After the final settlement, one of them commented that, despite the toll it has taken, it was the right thing to do (Bowyer 2000). A study of whistleblowers in the US found that:

- 100 percent were fired and most were not able to find new jobs.
- 90 percent reported emotional stress, depression, and anxiety.
- 80 percent suffered physical deterioration.
- 54 percent were harassed by peers at work.
- 17 percent lost their homes.
- 15 percent were subsequently divorced.
- 10 percent attempted suicide (Haines 2004).

Upon exhausting all avenues available to correct a problem with no response from the employer, there is an inherent conflict between an employee's loyalty to the organization and the employee's potential legal and ethical obligation to report apparently fraudulent activities to the proper authorities (Clarke 1999). The potential significant financial gain for the whistleblower further complicates the matter. It is recommended that an employee seek the advice of competent legal counsel before acting on the whistleblowing option.

Legislation on whistleblowing

Whistleblower lawsuits (*qui tam* lawsuits) are filed under the False Claims Act and allow private citizens, whether affiliated with the government or not, to bring lawsuits on behalf of the government (thus the Latin term *qui tam*) against companies that have allegedly engaged in fraudulent activities. The government can decide to join the lawsuit or let the person who filed it see it through. In either situation the person who filed it may receive up to 25 or 30 percent of whatever is recovered. The bulk of the damages will be reimbursed to the US government. The False Claims Act upholds a citizen's right to file claims against offenders—and protects an employee from retaliation for blowing the whistle on an employer that submitted a false government claim (for example, making false or fraudulent claims about the quality of the goods or services a US government contractor has agreed to provide). Since its amendment in 1986, according to the Justice Department, by 1997 the government recovered more than $3 billion in civil fraud actions, with more than $1 billion attributed to 153 whistleblower cases (Blechman 1997).

OSHA's Whistleblower Protection Program (WBPP)

Whistleblower protection is a patchwork of laws that are often confusing to the aviation employee. The US Department of Labor (DOL) is charged with enforcing the federal laws protecting corporate whistleblowers at publicly traded companies. Since the whistleblower program began in 1970, the number of statutes the DOL's Occupational Safety and Health Administration (OSHA) is responsible for enforcing has increased—recent additions in 2008 bring the total to 17 such statutes (US GAO 2009). Table 7.2 shows the current list of statutes.

Statutes in the OSHA Whistleblower Protection Program have different focus and target user populations. Some of them address the protection of federal employees (for example, the Whistleblower Protection Act of 1989) and others address corporate employees (for instance, the Sarbanes–Oxley Act of 2002). The Whistleblower Protection Act created the Office of Special Counsel (OSC), charged with investigating complaints from persons who work for the government. Corporate whistleblower statutes and regulations aim at providing a legal prohibition against discharging, demoting, suspending, threatening, harassing, or in another manner discriminating against any employee with respect to the employee's compensation, terms, conditions, or privileges of employment because the employee engaged in certain protected conduct (Klein 2004).

Whistleblower protection for employees in the aviation industry

The FAA offers a whistleblower protection program that provides protection from discrimination for air carrier industry employees who report information related to air carrier safety (FAR Part 121 and 135 but not 91 operations). Two statutes

TABLE 7.2 Statutes in OSHA's Whistleblower Protection Program

Agency and statute	*Year of enactment of whistleblower provision*
Federal Government Employees	
Whistleblower Protection Act (WPA)	1989
Department of Energy	
Energy Reorganization Act	1978
Department of Transportation	
International Safe Container Act	1977
Federal Railroad Safety Act	1980
Surface Transportation Assistance Act	1983
Pipeline Safety Improvement Act	2002
National Transit Systems Security Act	2007
Moving Ahead for Progress in the 21st Century Act (MAP-21)	2012
Environmental Protection Agency	
Federal Water Pollution Control Act	1972
Safe Drinking Water Act	1974
Solid Waste Disposal Act	1976
Toxic Substances Control Act	1976
Clean Air Act	1977
Comprehensive Environmental Response, Compensation, and Liability Act	1980
Asbestos Hazard Emergency Response Act	1986
Federal Aviation Administration	
Wendell H. Ford Aviation Investment and Reform Act for the 21st Century (AIR-21)	2000
Department of Labor	
Occupational Safety and Health Act	1970
Securities and Exchange Commission	
Sarbanes–Oxley Act (SOX)	2002
Consumer Product Safety Commission	
Consumer Product Safety Improvement Act	2008

aim at protecting whistleblowers in the aviation industry. They are the Wendell H. Ford Aviation Investment and Reform Act for the 21st Century (AIR-21) and the Sarbanes–Oxley Act.

In the first eight years since Congress enacted significant new protections for whistleblowers in the airline industry, more than 200 AIR-21 cases have gone before the Office of Administrative Law Judges (ALJ), some of which have resulted in significant recoveries in favor of employees (OAL 2009). In 2000 Congress passed AIR-21 to protect employees who expose air carrier safety violations. Congress began seriously contemplating a whistleblower protection program in the late 1990s after a string of commercial passenger plane crashes pushed safety concerns to the forefront. Under AIR-21, Section 519, employees who believe they have suffered adverse action for reporting air safety violations can file a complaint with OSHA within 90 days of the date on which the discriminatory decision was made and communicated to the employee. An employee may still file the complaint with an OSHA office after 90 days, but the OSHA Area Director may dismiss the complaint as untimely. The FAA may still investigate the safety information provided by the employee. More recently, air carriers have become subject to the whistleblower protection provisions contained in SOX, which imposes criminal penalties for retaliation against whistleblowers. In the FAA Modernization and Reform Act of 2012, Section 341, Congress established an Aviation Safety Whistleblower Investigation Office and a Director's position (US Congress 2012).

Wendell H. Ford Aviation Investment and Reform Act for the 21st Century

There are joint responsibilities for the FAA and OSHA under the AIR-21 program. A prevailing plaintiff is entitled to reinstatement, back pay, compensatory damages, and attorney's fees and costs.

Covered employers

Employees of air carriers, their contractors, and their subcontractors are protected from retaliation, discharge, or otherwise being discriminated against for providing information relating to air carrier safety violations to their employer or to the federal government. This includes information filed, testified, or assisted in a proceeding against the employer relating to any violation or alleged violation of any order, regulation, or standard of the FAA or any other federal law relating to air carrier safety. AIR-21's prohibition against retaliation applies broadly. It applies to air carriers—which includes any "citizen of the United States undertaking by any means, directly or indirectly, to provide air transportation" (49 U.S.C. §40102(a)(2))—and to contractors and subcontractors of air carriers. A "contractor" is defined as any "company that performs safety-sensitive functions by contract for an air carrier" (49 U.S.C. §42121(e)). It is important to note that if a person is used by an air carrier certificate holder or an air carrier that does not hold a certificate, that person may still be considered an "air carrier employee" for the purposes of whistleblower protection. Given that Congress specifically defines that an air carrier must be a citizen of the United States, §42121 does not encompass foreign air carriers, except to the extent that they may act as a contractor or subcontractor to a US one.

Protected conduct

Employees who report air safety violations in the following manners are engaging in conduct which is protected by AIR-21: (1) providing information to the employer or the federal government relating to any violation or alleged violation of any federal air safety statute or regulation; (2) filing a proceeding relating to a violation or alleged violation of air safety rules; (3) testifying in such a proceeding; or (4) assisting or participating in such a proceeding. Generally it is not difficult for a complainant to establish that he or she engaged in protected conduct. For example, the DOL decisions have held that the following activities are protected: (1) alerting the FAA that an aircraft was being flown past its maintenance threshold; (2) reporting to a supervisor that some aircraft parts in warehouse bins did not contain the FAA-required serviceable tag; and (3) alleging to management that maintenance records were falsified. The FAA's Advisory Circular on the subject notes that:

> air carrier safety is any safety concern that you believe, in good faith, is a violation of an FAA regulation, order, or standard or any other Federal law that implicates the safety and security of air carriers. The safety information you report must be related to air carrier safety (not personal safety). (USDOT 2004)

The FAA publication provides examples of information related to air carrier safety (USDOT 2004: 2). These may include:

- Falsification of records
- Noncompliance with flight and rest requirements
- Improper maintenance practices
- Security breaches
- Inadequate compliance with training requirements
- Use of suspected unapproved aircraft parts
- Improper manufacturing procedures
- Crewmember medical qualifications
- Improper production of aircraft parts
- Instruction not to document aircraft maintenance discrepancies.

If you are an employee of a US air carrier, its contractor, or its subcontractor this law protects you against discrimination by your employer for reporting air safety information. An employer cannot discriminate against you because you provided, caused to be provided, or are about to provide to the employer or federal government information that relates to any violation or alleged violation of any order, regulation, or standard of the FAA or any other provision of federal law that relates to air carrier safety. Also, an employer cannot discriminate against you because you filed, caused to be filed, or are about to file a proceeding that relates to any violation or alleged violation of any order, regulation, or standard of the FAA or any other provision of federal law that relates to air carrier safety.

Adverse action

Almost any action taken by an employer which has a negative effect on the employee's terms, conditions, or privileges of employment amounts to adverse action. This includes intimidating, threatening, retraining, coercing, blacklisting, or discharging an employee. DOL authority construing similar whistleblower protection statutes indicates that adverse action also includes demotion, reduction in salary, failure to hire, harassment, transfer to a less desirable position, and even a change of office location.

OSHA investigation

Within 60 days of the filing of a complaint, OSHA must investigate the complaint and determine whether the employer violated §519 of AIR-21. The employer has 20 days to submit to OSHA a response to the complaint. If OSHA finds that the complaint has merit, the secretary will issue a preliminary order requiring the employer to: (1) take affirmative action to abate the violation; (2) reinstate the plaintiff to his or her former position; (3) provide the plaintiff with back pay; and (4) provide compensatory damages to the plaintiff.

The employer can appeal the preliminary order by requesting a hearing before an administrative law judge (ALJ) within 30 days of the issuance of the preliminary order. Requesting a hearing will stay enforcement of the preliminary order, except for reinstatement of the plaintiff. Therefore, where OSHA finds that an air carrier violated §519, the air carrier will be placed in the difficult position of rehiring the plaintiff while he or she pursues a lawsuit against the carrier.

Hearing before an ALJ

Hearings are held before DOL ALJs and are conducted *de novo*, that is, the ALJ disregards OSHA's findings. The rules of evidence applied in hearings before ALJs are somewhat more liberal than are the federal rules of evidence. ALJs apply a broad scope of relevance, and hearsay evidence is not automatically excluded. In addition to reinstatement, back pay, and compensatory damages, a prevailing party is entitled to attorney's fees and costs.

Within 60 days of the issuance of the ALJ's decision, either party can file a petition for review before the DOL's Administrative Review Board (ARB). The ARB is authorized to award the same type of relief that an ALJ may award. Either party may appeal the ARB's decision by filing a petition for review with the United States Court of Appeals in which the alleged violation occurred or in which the plaintiff resided on the date of the alleged violation.

Settlements

The DOL will not automatically dismiss a claim that has been settled by the parties. Instead, the settlement will be reviewed to ensure that it is fair, adequate, and reasonable. The DOL is concerned primarily with "gag provisions," that is, provisions that might hinder a plaintiff from raising concerns.

FAA procedures and enforcement

An employee of an air carrier, contractor, or subcontractor may contact an FAA aviation safety inspector (ASI) with information regarding a violation or an alleged violation of an FAA order, regulation, or standard, or any other provision of federal law relating to air carrier safety. The employee may also request whistleblower protection by the DOL. In order for ASIs to adequately advise employees of their rights, inspectors should immediately advise them of the WBPP and that for a personal remedy they should contact the DOL/OSHA as soon as possible (not later than 90 days after the discrimination event) (US DOL 2008).

ASIs may conduct an investigation

The ASI will contact the complainant, ask if they have any additional supporting evidence, or can tell the ASI where to look. The ASI may inquire if the complainant has any witnesses or other employees that can corroborate the alleged violation(s), or if any other employees are having the same or similar problems. The ASI will then interview company personnel involved and/or other persons that may have knowledge. The ASI should only investigate the safety issues, and not reveal that they are investigating a whistleblower complaint. The FAA inspector will collect evidence and make copies of all relevant information, and may review company manuals and/or records for alleged violations. The inspector will then take appropriate enforcement and/or corrective action, or the inspector will close out with no action.[1]

In addition to the remedies that the DOL is authorized to provide to a prevailing plaintiff, a carrier who violates Section 519 of AIR-21, which is administered by both the DOL and the FAA, may also be subject to an FAA civil penalty. When a complaint is filed under AIR-21, OSHA will provide the FAA with a copy of the complaint and the FAA will investigate safety issues related to the complaint. A memorandum of understanding (MOU) between the FAA and the DOL provides that the two agencies will share all information they obtain relating to complaints of discrimination and will keep each other informed of the status of any administrative or judicial proceeding associated with the complaint.

The FAA has the responsibility to investigate complaints related to air carrier safety, and has statutory authority to enforce air safety regulations and issue sanctions for noncompliance with these regulations. FAA enforcement action may include certificate suspension or revocation and the imposition of civil penalties. Additionally, the FAA may issue civil penalties to US air carriers, contractors, or subcontractors for discriminating in violation of the Whistleblower Protection Program.[2]

It is very important that each complaint be thoroughly investigated. The FAA needs to investigate the safety issues regardless of whether the complaint is filed in a timely manner or not (that is, over 90 days). OSHA may dismiss the complaint of discrimination as untimely, but under certain circumstances (namely, the complainant has a valid reason for not submitting the complaint within 90 days) may accept the complaint. Figure 7.1 illustrates the whistleblowing process in aviation.[3]

The number of whistleblower claims under the Wendell H. Ford Aviation Investment and Reform Act (AIR-21) between fiscal years 2008 and 2013 totaled 462, or fewer than 80 per fiscal year. In fiscal year 2013, 60 percent of the OSHA-adjudicated claims under AIR-21 were dismissed; only 2 of the 88 claims had merit, and 20 percent were settled. Such numbers may discourage employees, who may believe they have been retaliated against, from filing a claim (GAO 2014).

Whistleblower protection provisions of SOX

In addition to the whistleblower protection provisions of AIR-21, many carriers are subject to the protection provisions in SOX: Section 806 creates a federal civil right of action on behalf of any employee of a publicly traded company who is subject to discrimination in retaliation for reporting corporate fraud or accounting abuses.[4]

Criminal provisions

In contrast to the civil remedies created by §806, the criminal provisions of §1107 of SOX are not limited to the actions of publicly traded companies; nor are they restricted in scope to matters involving corporate fraud or accounting abuses. Section 1107 imposes criminal penalties on any individual who:

> knowingly, with the intent to retaliate, takes any action harmful to any person, including interference with the lawful employment or livelihood of any person, for providing to a law enforcement officer any truthful information relating to the commission or possible commission of any federal offense. (US Congress 2002)

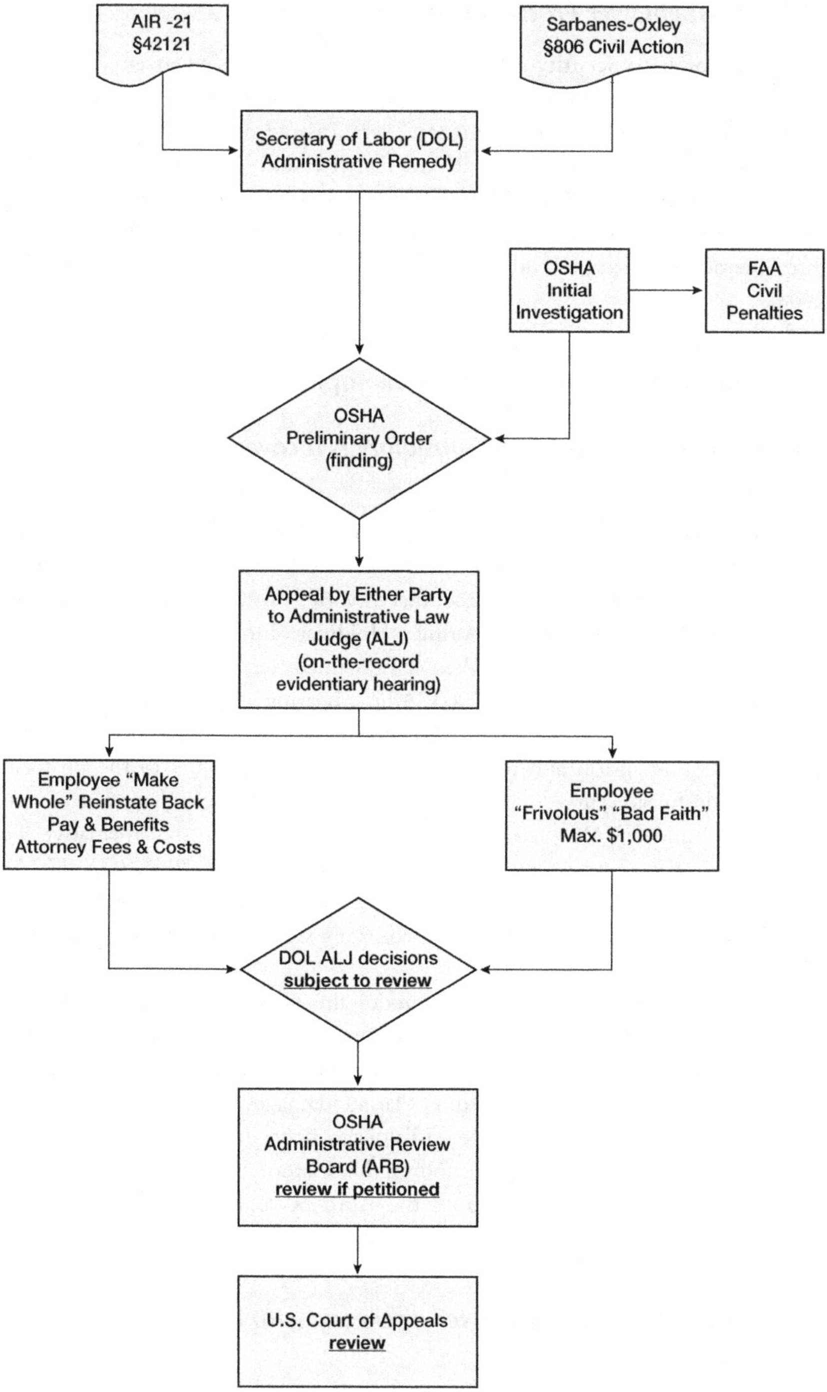

FIGURE 7.1 Aviation whistleblowing process flow chart

TSA's Whistleblower Protection Program

The Transportation Security Administration (TSA) announced an agreement with the American Federation of Government Employees (AFGE) to enhance whistleblower protection for airport screeners. Instead of being sent to the OSC for investigation, a whistleblowing TSA screener can file an appeal with the Merit Systems Protection Board (MSPB), offering an additional layer of protection. The MSPB is an independent, quasi-judicial agency in the executive branch, whose three members review and adjudicate individual employee appeal cases (Wilson 2008).

US aviation whistleblowing case examples

Whether the employee of the subsidiary is a covered "employee"

Stacy M. Platone worked as an airline pilot union communications specialist. On April 2003, she filed a complaint with the DOL under Section 806 of the Corporate and Criminal Fraud Accountability Act of 2002, Title VIII of SOX. She alleged that Atlantic Coast Airlines Holdings, Inc. violated the employee protection provision of the Act when it suspended her and later terminated her employment effective March 19, 2003. After a hearing an ALJ issued an April 30, 2004 recommended decision finding that Platone had engaged in protected activity and that her suspension and termination were causally related to the protected activity. The first inquiry—whether the employee of the subsidiary is a covered "employee" under SOX—has been consistently answered in the affirmative. For example, in *Platone v. Atlantic Coast Airlines Holdings, Inc. 2003-SOX-27* (ALJ April 30, 2004), an ALJ held that an employee of a non-publicly traded subsidiary was a covered "employee" where the company's parent/holding company was publicly traded.

The ALJ in *Platone* reasoned that, under the facts of the case, the holding company was the alter ego of the subsidiary and that it certainly had the ability to affect the complainant's employment. In a supplemental decision dated July 13, 2004, the ALJ recommended awarding Platone damages, costs, and attorney fees. ACA Holdings timely appealed the ALJ's decisions to the DOL's ARB. Because the ALJ erred as a matter of law in finding that Platone had engaged in protected activity, the ARB declined to adopt the April 30 and July 13 recommended decisions and denied Platone's complaint (*Platone v. US Dept of Labor, No. 07-1635*).

FAA manager files whistleblower disclosure with the OSC, June 2002

Is there a need to reform and amend the Whistleblower Protection Act of 1989? Rejecting repeated, explicit congressional intent, the Federal Circuit Court of

Appeals ruled in *Willis v. Department of Agriculture* (141 F.3rd 1139 (Fed. Cir. 1998)) that a disclosure made as part of an employee's normal job duties is not protected.

Gabe Bruno, an FAA employee for over 20 years, was assigned responsibility for oversight of the ValuJet–AirTran merger following the tragic 1996 ValuJet Flight 592 accident that killed all 110 on board. Bruno was to implement necessary oversight to prevent a recurrence of the tragedy. He filed an official whistleblower disclosure with the OSC in June 2002, citing the oversight problems with AirTran and the cancellation of a mechanics reexamination program. In May 2001 the FAA reassigned Bruno from his management position as a result of his expressed concerns to supervisors.

According to existing case law, Bruno's whistleblowing does not qualify as protected speech because his disclosures were made "during the course of his job duties." The resulting FAA "security investigation" could therefore not be successfully challenged as retaliatory. The case was settled in the summer of 2005 with Bruno losing his job at the FAA in a "nuisance settlement."

DOL SLAPS AIRLINE IN WHISTLEBLOWER CASE

An Administrative Law Judge, Anne Beytin Torkington, found there was "ample proof" that Alaska Airlines managers pressured Carroll Sievers and others at the airline's Portland facility to get planes back into service after they raised legitimate concerns (*Sievers v. Alaska Airlines, Inc., 2004-AIR-00028*). The ALJ found Sievers's work to be "competent and aggressive" as well as "reasonable, genuinely motivated by concern for safety, and an activity protected" under the federal whistleblower law. The ALJ ordered Alaska to pay Sievers $534,000 for past and future lost wages and $50,000 for emotional distress. Alaska appealed the decision. The ARB (Case No. 05-109) reversed the ALJ decision on January 30, 2008. The ARB found that "Seviers did not demonstrate that protected activity contributed to the termination. Rather, the weight of the evidence demonstrates that Alaska terminated Sievers because of timecard fraud." The ARB denied Seviers's complaint.

Conclusions

Anyone may find themselves in a position of making a difficult decision between their conscience and what they are being told to do by their boss. Fortunately, most of us do not have to agonize over critical issues. Whistleblowing occurs when an individual within an organization openly accuses others, usually colleagues, of professional wrongdoing through a mechanism with the federal government or through a public medium. Such action requires ethical justification, which may be found in preventing harm to the public, issues of safety, or opening people's eyes to similar wrongdoing. If the good outweighs the bad, it may be permissible; whether it is obligatory depends on federal law and the degree of negative repercussions the whistleblower is most likely going to undergo. Organizations such as air carriers must provide employees with internal mechanisms for listening and responding to their concerns. If we are to have a just society, it must protect, to the degree feasible, individuals who blow the whistle in good faith.

Notes

1 It should be noted that FAA Order 2150.3, Compliance and Enforcement Program, Chapter 13, contains FAA policy and procedures for providing immunity from enforcement action, in some cases, to persons who provide information about violations. Information regarding regulatory violations occasionally is offered to an FAA ASI or attorney along with a request that, in exchange for the information, the person making the offer be granted "immunity from prosecution" for his or her participation in the violations. The phrase "immunity from prosecution" ordinarily refers only to criminal matters (DOT FAA Order 8900.1 2007).

2 The FAA may impose a civil penalty after the Secretary of Labor's Order of a finding of a violation of 49 U.S.C. 42121 becomes final. This civil penalty is in addition to any enforcement action the FAA may impose for safety violations as well as any abatement action OSHA may impose concerning a finding of discrimination. The FAA will receive a copy of the Secretary of Labor's Order when it becomes final (i.e., all appeals by either party are exhausted). The FAA may then issue the civil penalty for a violation based upon the finding by the DOL, in addition to any FAA safety investigation enforcement action the FAA may have previously taken (see 49 U.S.C. §46301).

3 Complainants must realize records are retained. All records associated with whistleblower investigations must be kept for three years after the investigation is closed.

4 Procedures for the Handling of Discrimination Complaints under Section 806 of the Corporate and Criminal Fraud Accountability Act of 2002 are noted within Title 29 CFR Part 1980.

References

Beabout, G.R. and Wennemann, D.J. 1994. *Applied Professional Ethics: A Developmental Approach for Use with Case Studies*. Lanham, MD: University Press of America.

Blechman, W.J. 1997. Blowing the whistle on health care fraud. *Business & Health*. 15(11): 51.

Boatright, J. 1993. *Ethics and the Conduct of Business*. Englewood Cliffs, NJ: Prentice Hall.

Bowyer, K.W. 2000. Goodearl and Aldred versus Hughes Aircraft: a whistle-blowing case study. 30th Annual Frontiers in Education Conference (FIE 2000), Vol. 2, S2F/2–S2F/7.

Clarke, R.L. 1999. The ethics of whistle-blowing. *Healthcare Financial Management*. 53: 16.

Erlen, J. 1999. What does it mean to blow the whistle? *Orthopedic Nursing*. 18: 67–70.

Federal Accountability Initiative for Reform (FAIR) 2009. The Whistleblower's Ordeal. Available at: http://fairwhistleblower.ca/wbers/wb_ordeal.html [accessed April 23, 2009].

Haines, P. 2004. Famous cases of whistle-blowing. *Brock Press*, March 30. Available at: http://media.www.brockpress.com/media/storage/paper384/news/2004/03/30/Business/Famous.Cases.Of.WhistleBlowing-645453.shtml [accessed January 23, 2009].

Johnstone, M. 2004a. *Bioethics: A Nurse Perspective*. Sidney: Elsevier.

Johnstone, M. 2004b. Patient safety, ethics and whistleblowing: a Nursing response to the events at the Campbelltown and Camden Hospitals. *Australian Health Review*. 28(1): 13–19.

Jubb, Peter B. 1999. Whistleblowing: a restrictive definition and interpretation. *Journal of Business Ethics*. 21(1): 77–94.

Klein, C.A. 2004. New OSHA rules protect corporate whistleblowers. *Tribology & Lubrication Technology*. November

Lindblom, L. 2007. Dissolving the moral dilemma of whistleblowing. *Journal of Business Ethics*. 76(4): 413–426.

Louw, T.J.G. 2011. The ethics of whistleblowing: an evaluation of a South African case. In *Whistleblowing: In Defense of Proper Action*, edited by M. Arszułowicz and W. Gasparski. New Brunswick, NJ: Transaction, 2011, 61–68.

Ray, Susan L. 2006. Whistleblowing and organizational ethics. *Nursing Ethics*. 13(4): 438–445.

Rosen, L.F. 1999 (Jan). Whistle blowing. *Today's Surgical Nurse*. 21(1): 41–2.

US Congress 2002. Sarbanes–Oxley Act of 2002. P.L. 107-204, Title XI-Corporate Fraud Accountability, Sec.1107. Retaliation against Informants.

US Congress. 2012. FAA Modernization and Reform Act of 2012. P.L. 112-95, Section 341. Aviation Safety Whistleblower Investigation Office.

US Department of Labor (DOL). 2008. Information for Whistleblowers. Washington, DC: Available at: http://www.dol.gov/ejudication/whistleblowers.htm [accessed November 13, 2008].

US Department of Transportation (DOT). 2004. Federal Aviation Administration. Whistleblower Protection Program (Air Carrier), Advisory Circular AC No. 120-81.

US Department of Transportation 2007. Federal Aviation Administration. Order 8900. 1, Volume 11, Chapter 3, Section 2.

US Government Accountability Office (GAO). 2009. Report to Congressional Requesters, Whistleblower Protection Program: Better Data and Improved Oversight Would Help Ensure Program Quality and Consistency, GAO-09-106, Washington, DC.

US Government Accountability Office (GAO). 2014. Report to Congressional Committees, Whistleblower Protection Program: Opportunities Exist for OSHA and DOT to Strengthen Collaborative Mechanisms, GAO-14-286, Washington, DC.

Wilson, B. 2008 (Feb 28). New deal offers TSA screeners more whistle blower protection. *Aviation Daily*.

PART III

Issues in responsibility

While the previous part focused on the business side of aviation, this one investigates the individual players in order to examine some of the key issues in responsibility. Some of the key players include governments, flight attendants, pilots, and corporations. The question of moral responsibility is of major importance in ethical thought, especially regarding what moral duties entail, when they should be enacted, and toward whom. This part begins with the responsibilities of government employees, followed by the duties of flight attendants, pilots, and questions regarding corporate responsibility.

Chapter 8 by Gerardo Martinez provides an account of the ethical responsibilities of FAA inspectors, followed by a brief examination of possible ethical lapses that may arise. He reveals the lengths the FAA goes to in trying to eliminate any possible conflicts of interest between the inspectors and the airlines. Focusing on the duties of the Aviation Safety Inspectors (ASIs), Martinez discusses the disclosures an ASI must make, such as re-employment rights with an airline, airline pension funds, etc. He ends by stating that ethical responsibility is of utmost importance since an ASI's primary duty is to promote safety.

In Chapter 9 Elizabeth A. Hoppe takes a closer look at one of those lapses, namely, regulatory capture. Capture occurs when a regulating body, such as the FAA, works so closely with a regulated entity that the agency ends up elevating the interests of the airline over what should be the FAA's main concern, namely safety. Hoppe examines cases of capture that arose in the FAA's oversight of Southwest Airlines and Northwest Airlines (now Delta Air Lines, Inc.). The problem of capture in the two cases also entails a question of the validity of consequentialist approaches to decision-making. Investigations of Southwest showed that the Certificate Management Office (CMO) was more worried about the impact of its decisions on the airline it was overseeing than on public welfare. Hoppe ends the chapter by showing the importance of incorporating alternative ethical approaches to decision-making, such as Kantian deontology, which concerns moral obligations that cannot be overridden no matter the consequences.

In Chapter 10, Gail L. Bigelow provides her perspectives on the responsibilities of flight attendants, especially in the post-9/11 era. In tracing the basic duties that a flight attendant should follow, she also notes the ways in which ethical dilemmas may arise, such as whether or not to report possible safety issues. Bigelow also discusses passenger responsibilities and some of the more trying situations, such as parents flying with young children. In light of 9/11 she reveals the way in which the profession has changed in that it is more focused on questions of terrorism and whether or not certain passengers should be considered more suspect than others. By providing us with an overview of the complexity of the role of the flight attendant, her chapter also indirectly shows us the ways in which they can be taken for granted and thereby undervalued.

The final three chapters in this section revolve around issues concerning major aviation accidents. Chapter 11 concerns the 1996 crash of ValuJet 592 and the question of corporate responsibility. Kenny Frank first analyses some of the key aspects of the NTSB report, which states that the accident was probably caused by a fire started by oxygen canisters being transported as cargo. But Frank shows us that the NTSB report may be incomplete, and he reviews anonymous witness accounts that claim that the wiring was faulty and thereby could have been the main source of the onboard fire. In the end Frank shows us that whatever the actual cause of the crash, the real issue is corporate responsibility, or lack thereof, of ValuJet to ensure the safety of its passengers and crew.

In Chapter 12 Eric B. Kennedy investigates recent aviation disasters such as Air France 447 and Germanwings 9525 by examining the ethical questions that arise from the intersection between technology and air safety, or what is known as the sociotechnical space. The Air France accident resulted from a mechanical malfunction that rapidly incorporated human factors like disorientation, confusion, and differing actions among the crew members. While this accident led to the query if pilots still know how to fly, Kennedy demonstrates why this question does not do justice to the complexity of the issue. He also examines three issues in today's cockpit: security systems based on the Germanwings accident, inclusivity of race and gender, and reliability, such as on the pressure for on-time performance. Finally, he concludes by showing the ways in which aviation ethics is useful for addressing failures and improving outcomes, especially in terms of sociotechnical considerations.

This part ends with Chapter 13 and the ethical dimensions of the loss of Malaysia Airlines Flight 370. Richard L. Wilson first provides the reader with an overview of the Boeing 777 design, the timeline of the flight, and several possibilities of what may have occurred—ranging from intentional interference to technical failure. Although providing an ethical analysis of MH370 based on incomplete information is difficult, Wilson uses stakeholder analysis as an approach that can help to highlight some of the ethical issues that arose in the aftermath of the flight. By examining the primary and secondary stakeholders, Wilson shows that one key ethical concern is the inadequate communication reports from both the airline and the Malaysian government to the relatives and the public. In the end, whether or not a probable cause is ever determined, at a minimum airline management can be more mindful of all of its stakeholders by providing improved safety features and better communication in the event an emergency arises.

8

ETHICS AND FAA INSPECTORS

Gerardo Martinez

> I, (name), do solemnly swear (or affirm) that I will support and defend the Constitution of the United States against all enemies, foreign and domestic; that I will bear true faith and allegiance to the same; that I take this obligation freely, without any mental reservation or purpose of evasion; and that I will well and faithfully discharge the duties of the office on which I am about to enter.
>
> So help me God

Above is the oath of office that all FAA Aviation Safety Inspectors (ASIs) take when they are first hired. It means that we accept full responsibility for doing the best job possible, to the best of our abilities, all of the time. Who we work for, ultimately, is you, the public. We ensure the public's safety, and we are charged to regulate the industry in a fair and ethical manner. We represent the Federal Executive Branch of the government. We work closely with industry, which includes manufacturers, operators, repair stations and small corporate aviation departments, schools, and various other entities such as museums, aviation associations, clubs, and individuals.

The FAA assigns ASIs from three specialties (Operations, Maintenance, and Avionics) to each operator. Based on the size and complexity of the airline, we can assign anywhere from 3 to more than 100 ASIs to an operator. Each inspector has a responsibility to ensure safety and adherence to Federal Aviation Regulations (FARs):

- We have the authority granted by Congress to stop an operation or punish an operator or airman who violates one or more of these regulations.
- We have the authority to revoke or suspend certificates, including the certificates of pilots, mechanics, and operators.

Inspectors oversee the carrier's entire system, visit locations where they operate, and review and approve their policies and procedures. These reviews include, but are not limited to, training, maintenance programs, content and presentation of instructions, check lists, routes, dispatch, scheduling, maintenance control, weight and balance programs, and anything else that an operator would need to safely operate.

The FAA inspector: responsibilities and authority

To accomplish our goals we need an experienced and educated team of employees. Our ASIs often come from the industry, where they already have been exposed to the aviation skill sets that the FAA requires. Each of our employees must undergo a very thorough background investigation, which covers moral character, financial responsibility, criminal records, educational credentials, work history, military records, and reputation. We also have financial disclosure and conflict of interest requirements. Because we hire from the industry, our employees have, for the most part, vast amounts of experience and ties to former employers. We require those new hires to divest themselves of any financial ties to a carrier or employer regulated by the FAA.

Our ethics rules require that we annually sign and submit for review a financial disclosure report. We must list investments we hold, real estate, stocks, bonds, and any travel reimbursements we have received (other than government travel) for each year. The rules basically prohibit ASIs from using the public trust for personal financial gain. The FAA has significant public safety responsibilities, and we must discharge our responsibilities effectively, vigorously, and evenhandedly. We cannot do so if we have interests in the companies we oversee.

If inspectors or their spouses have employment or previous employment with an airline, or an operator then special rules apply to them:

- Previously earned pensions and are understood to be something one is entitled to, but one still has to report them.
- Re-employment rights must be disclosed.
- Flight privileges must also be disclosed, and in some instances pre-approved.
- FAA employees are prohibited from owning stocks or other interests in airlines or aircraft manufacturing companies, or even in companies predominantly in the business of supplying aircraft components, goods, services, or parts to the airlines or the aviation industry.

For example, we cannot own stock in Boeing, General Dynamics, General Electric, Goodrich, Honeywell, Lockheed Martin, Northrop Grumman, Raytheon, Rockwell Collins, Textron Inc, United Technologies, or any of the airlines. Even owning one of these stocks will require either divestiture or electing not to work for the FAA.

Our close oversight of operators means that we often end up being privy to what can be considered insider information. We find out what the operator is planning to do in the near term or long term, where it expects to extend operations, where managers plan to shrink operations, what markets the carrier plans to expand to, what fleets the carrier anticipates will be used in the future, what fleet types and sizes they hope to obtain, what fleet types and sizes they intend to eliminate, and much more. It is vital that we guard this proprietary information closely.

Ethical lapses

Have there been ethical lapses among inspectors? The answer is yes. Fortunately I am aware of only a small handful. Most are quickly and severely punished. I know of one instance where an ASI created a consulting firm, then suggested that an operator hire this consulting firm to help write the carrier's compliance manual. The final product would be submitted to the FAA for review and final approval. The only problem was that the ASI was the FAA official who would review the final product. He was terminated.

Another case involved an airframe and powerplant (A&P) school. The ASI who had oversight responsibility for the school was enticed to work for it. The school was investigated after the military noticed that a large amount of military mechanics were returning as civilian-certificated A&P mechanics, all in record time. The FAA investigated and discovered that there was no humanly possible way to test the amount and quantities of mechanics that the school was processing. The operators of this school eventually faced criminal charges, and all of its graduates had their certificates revoked but were offered the chance to retest for them. There were many innocent victims in that episode.

Capture

In *The Regulatory Craft* (2000), Malcolm K. Sparrow describes the problem of capture that can happen to regulatory agencies. Capture as described by Sparrow means that the relationship between a regulator and the regulated entity becomes so close that the regulator ends up identifying with the party he or she is supposed to be overseeing. Capture happens when one loses sight of who one works for, and the fear of damaging one's relationship with the party one is supposed to be regulating causes one to lose sight of one's responsibilities. We are not robots; we end up working with people whom we regulate, sometimes for years. An ASI can experience an interaction with a floor mechanic, and then later interact with her or him when the ASI becomes a supervisor, followed by manager, and later a director.

But it is possible that capture could be neutralized by proper interaction and oversight. An ASI who has a good professional relationship with the above-mentioned director can be effective, and the parties can agree to disagree if they have to. The danger occurs when the relationship becomes so close that the regulator ends up not wanting to fulfill his or her duties and responsibilities for fear of somehow damaging the relationship with the regulated entity.

Conclusion

The bottom line is that ethical behavior has to be at the forefront of our interactions with all of our audiences: the public, the industry, each other, and, above all, the parties we regulate. We cannot forget our primary task, which is to ensure the public's safety and regulate the industry in a fair and ethical manner. What does "ethical manner" mean? It means maintaining a professional distance that allows interaction *and* objective oversight. It means remembering our duty to serve the public and maintain the public's trust. And, above all, it means avoiding such a close relationship that "capture" becomes preferable to oversight.

Reference

Sparrow, M.K. 2000. *The Regulatory Craft: Controlling Risks, Solving Problems, and Managing Compliance.* Washington, DC: Brookings Institution Press.

9

THE FAA AND THE ETHICAL DIMENSIONS OF REGULATORY CAPTURE

Elizabeth A. Hoppe

In March of 2008 the public first learned that Southwest Airlines (SWA) had grounded 46 of its Boeing 737 aircraft due to potential structural cracks, and also that it was subject to a heavy fine by the Federal Aviation Administration (FAA). What may at first sound like a corporation engaged in unethical business practices turned out to be a much more complex, and more troublesome, matter. A year prior to this incident Southwest had reported the problem to the FAA through the voluntary disclosure reporting program (VDRP). However, the FAA's Principal Maintenance Inspector (PMI) did not require that the affected B737s be withdrawn from service. Thus, with FAA approval, Southwest continued flying aircraft that should have been grounded. An FAA Independent Review Team (IRT) found that one of the central issues involved regulatory capture. In other words, the FAA had placed its continued relationship with the airline ahead of public safety. Both the FAA and Southwest Airlines were fortunate that a tragedy did not result from their questionable decision-making. Because of dangerous situations such as this one, it is imperative to find ways to overcome capture so that public welfare and safety remain the primary focus of the FAA.

While the cases of capture occurred during the FAA's oversight of the airlines, I find that the ethical problems initially stem from the federal government's emphasis on consequentialist approaches to rule-making decisions. The FAA is required to determine its rules using the methodology of cost–benefit analysis. As mentioned in the Introduction to Part I, this type of approach parallels the ethics of consequentialism. However, one of the limitations is that this type of economic-based methodology fails to consider any non-monetary values, such as human well-being. The proneness to capture arises in a type of consequentialist thinking since the FAA's Certificate Management Offices (CMOs), which specialize in the certification, surveillance, and inspection of major air carriers, placed priority on the potential consequences of their relationship with the airline, rather than public welfare.

The two cases of capture that concern this chapter highlight the need for finding alternative ethical frameworks for both the rule-making process and the oversight of the aviation industry. One such alternative can be found in Kantian deontology, a concept that Mark Waymack addresses more fully in Chapter 1. Two aspects of Kantian ethics are important to consider: 1) the concept of Persons, in which people are treated as being ends in themselves; and 2) the idea that universal laws contain no exceptions. Regulatory capture coincides with the question of ethical responsibility, and to solve the problem, the US needs a regulatory system that includes a stronger sense of obligation toward the public.

Regulatory capture and its ethical dimensions

While regulatory capture is a complex issue, in general it arises when a regulator who works closely with the regulated entity ends up identifying with the interests of the regulated party, rather than the rules it should be enforcing. Much research has been conducted regarding regulatory capture from economic, legal, and political viewpoints. The concept is associated with George Stigler and his theory of economic regulation in which he argues that "as a rule, regulation is acquired by the industry and is designed and operated primarily for its benefit" (Stigler 1971: 3). However, Stigler also mentions that one of the alternative views is that regulation is supposed to exist for the protection and benefit of the public at large, a view that is seen today in the regulatory practices of the FAA.

Of course, regulatory capture is an immense topic, and there are variations on its meaning and impact. Ernesto Dal Bó reviews literature by academic economists and provides evidence on the causes and consequences of capture. According to Dal Bó, broadly speaking it "is the process through which special interests affect state intervention in any of its forms," while narrowly speaking it "is specifically the process through which regulated monopolies end up manipulating the state agencies that are supposed to control them" (Dal Bó 2006: 203). The main idea in both definitions concerns the ability of the industry to influence government for its own advantages. Interestingly, neither of Dal Bó's definitions encompass the sense in which the FAA's CMOs are prone to capture. It does not appear that through voluntary disclosure any of the airlines have exercised undue influence on the FAA. Rather, the FAA appears to have been a willing participant in allowing certain airlines to continue operating aircraft that violated Airworthiness Directives (ADs). In other words, rules to protect the public were already in place; the FAA simply did not follow them.

Regarding its ethical dimensions, one may argue that capture, although problematic at times, is not really an ethical issue, especially since the research tends to be addressed through the fields of political science, economics, law, and sociology. For instance, one could claim that it is a simply a phenomenon that arises in the relationship between the regulator and the regulated. However, I contend that it is one thing to be "complimentary" or "cooperative"; it is another to be captured. Cooperation could involve a positive reinforcement of regulations insofar as the business entity agrees that problems need to be addressed and resolved. In working with the regulator to find solutions, a positive outcome could be achieved by all involved. Capture differs from cooperation especially since capture can lead to harmful consequences due to the lack of concern over public safety. Of course, no system is perfect and an unforeseen negative effect could result from cooperation; but that would be not an ethical problem as much as an unfortunate incident. Additionally, cases of regulatory capture often involve conflicts of interest that in turn can be considered unethical in themselves. Conflicts of interest arise in many government agencies, especially ones where there is a "revolving door" policy in which government officials may have formerly worked for the industry and vice versa.

The potentially harmful consequences of regulatory capture further suggest the need to study the problem from an ethical framework. In examining two instances of capture at FAA CMOs, I will show that it is not only an ethical matter, but also that ethical theories, especially deontology, can provide an alternative way of creating possible alternatives to decision-making.

The role of consequentialism in federal regulations

The move toward a cost–benefit approach to rule-making arose during the 1970s and 1980s because many government officials became concerned about the cost of regulation to the industry, and thereby proposed that regulations provide net social benefits (Mills 2011). One finds the impact of this concern in President Bill Clinton's Executive Order 12866 of June 1994, which states: "in deciding whether and how to regulate, agencies should assess all costs and benefits of available regulatory alternatives, including the alternative of not regulating" (White House 1994). This order also states that:

> The American people deserve a regulatory system that works for them, not against them: a regulatory system that protects and improves their health, safety, environment, and well-being and improves the performance of the economy without imposing unacceptable or unreasonable costs on society; regulatory policies that recognize that the private sector and private markets are the best engine for economic growth.

One can clearly see that the government is concerned not only with public safety, but also with the economic impact of regulations on industry. Thus, federal regulations attempt to achieve positive results for both the public and industry by utilizing a consequentialist approach to rule-making, a methodology that was later reaffirmed by President Barack Obama's 2011 Executive Order 13563 as well.

Consequentialism in FAA rule-making

In providing rule-making decisions that are based on cost–benefit analysis, the FAA is in line with the federal government's general approach to regulations. The agency justifies this type of methodology by saying:

> For many Governmental investments and regulations, the recipients of the benefits are not those who bear the costs. From an overall perspective, society's welfare is improved as long as all accepted projects and regulations have benefits in excess of costs. (FAA 1998: 2-6)

The agency's statement typifies the view that the government should place the priority of social welfare at the forefront of its rules and decision-making process. Although the FAA does not explicitly address the meaning of "welfare," we can infer that the overall benefit of society is factored in its rationale for creating new regulations.

Although this approach to rule-making sounds plausible, it can lead to controversial decisions that in turn help reveal the limitations of consequential-based decision-making. For example, the FAA does not currently require child safety seats on commercial flights, citing the fact that the costs greatly outweigh the benefits. Since the chances of an aircraft accident are very low, by requiring child restraint systems (CRSs), the increased airfare would cause some families to drive to their destination instead of fly. The FAA (2006) claims that "the risk to families is significantly greater on the roads than in airplanes, according to FAA and National Highway Traffic Safety Administration (NHTSA) statistics." Since the chances of an automobile accident are higher than an aircraft accident, families would be in more danger if they drove to their destinations. Although an infant would fly out of a person's arms in an accident, the FAA does not find it beneficial to regulate safety seats.

However, not all government entities agree with the FAA, and this point helps show the limitations of consequentialist decision-making. One only needs to examine the 2004 analysis by the National Transportation Safety Board (NTSB) to find a counterargument to the FAA's position. Perhaps one of their most important points is that, "basing the evaluation of relative worth of a requirement for appropriate child restraints solely on the number of historic injuries and deaths overlooks the fact that every lap-held child traveler lacks adequate protection" (NTSB 2004: 2). The NTSB also points out that items such as laptops and large objects need to be secured prior to take-off and landing, especially since they could become projectiles that could in turn harm other passengers. So why are children under two not required to be secured as well? At least indirectly, the NTSB is questioning the benefit of consequentialist reasoning when they point out that rules and regulations based solely on statistics overlook the fact that a child is unprotected during a flight. This point highlights what is missing in a purely cost–benefit type methodology. Some problems, such as a child's well-being, cannot be adequately calculated in monetary terms. However, despite the controversy surrounding safety seats, the FAA does appear dedicated to its mandate of air safety.

Airworthiness Directives

Prior to examining the two cases of capture, it is important to keep in mind that the regulations that the FAA overlooked were perhaps the ones most concerned with safety: Airworthiness Directives (ADs). Under FAR 39.3 the FAA issues an AD whenever an unsafe condition exists in a product and the condition is likely to exist or develop in other products of the same design. In terms of its connection with deontology, FAR 39.11 is of particular importance. This states that "airworthiness directives specify inspections *you must* carry out, conditions and limitations *you must* comply with, and any actions *you must* take to resolve an unsafe condition" (emphasis added).

Similar to Kantian categorical imperatives discussed in Chapter 1, these rules should not be overridden since they are meant to correct unsafe conditions. Ostas likewise claims that "laws directly affecting human health and safety have a high degree of moral saliency and call upon a spirit of cooperation, not mere compliance or evasion" (Ostas 2004: 576). We can apply his point about cooperation to demonstrate that in the cases of regulatory capture, the FAA should have been enforcing the ADs rather than allowing the airlines to continue to operate potentially unsafe aircraft. If the FAA did not base its decisions solely on consequences, but also took into consideration the necessity of adhering to ADs, then this could be one of the solutions to the problem of capture. Later I will discuss deontology in more detail in order to demonstrate that the federal government's approach to regulations needs to include other factors besides the consequences.

If the ADs are so crucial for air safety, the question becomes why the FAA would ever overlook them. One place to find an answer stems from the history of the FAA. This agency arose in 1958 largely as a reenactment of the Civil Aeronautics Act of 1938. The 1938 Act helped create a firm regulatory system for the airline industry, and the 1958 Federal Aviation Act contained one key difference: the government's expanding power in air safety (Kane 2007: 116). The FAA continued the dual mandate of safety and economic growth; and it was not until October of 1996, in light of the 1996 ValuJet accident in Miami, that its mandate changed in order to center its efforts on air safety (FAA 1996: 301–302).

Despite this change, elements of economic promotion have continued to arise. In 2003 the FAA initiated a Customer Service Initiative (CSI) in order for its customers to ask for reviews of certification decisions without fear of retribution. The CSI was also meant to provide more consistency and fairness in FAA certification decisions (Saenger 2003: 1). However, one of the key problems was that the FAA "defined its customers as the people and companies requesting FAA certification" (USDOT 2008: 6). In 2009 FAA Administrator Randy Babbitt ordered the FAA to stop calling the airlines its clients, and instead changed the terminology to "stakeholders" (Lowy 2009).[1] By itself this change does not address the main issue of regulatory capture. Perhaps more importantly, because of its history in promoting the industry, FAA oversight can sometimes narrow its focus on the potential consequences for the airlines rather than the public.

Voluntary reporting programs and the problem of trust

Another important aspect of capture in the FAA concerns the agency's voluntary disclosure programs. In general these types of program are meant to ensure compliance and foster collaboration between government and industry, and thereby allow for a free exchange of ideas without fear of retribution (Mills 2011: 6). Rather than see the government as an enforcer of penalties, a business may be more willing to work with the regulator if the business is able to avoid severe penalties by self-reporting infractions. For example, the Environmental Protection Agency (EPA) has over 40 voluntary programs, some of which are meant for businesses to self-police their environmental performances (see Potoski and Prakash 2004).

In addition the relationship between the regulator and regulated entity could develop into one of trust, thereby benefitting both sides. According to the US GAO report on aviation safety, the "combination of promise of immunity for self-reporting and threat of enforcement and disciplinary action for remaining silent creates an incentive for industry personnel to participate in the voluntary reporting programs" (US GAO 2010: 17). Perhaps more importantly, as the FAA moves toward a risk management approach to safety oversight, data from the voluntary reporting programs will become an invaluable means of enhancing safety.[2]

The FAA is responsible for several voluntary reporting programs, and for the purposes of this chapter I will focus on its Voluntary Disclosure Reporting Program (VDRP), which was established in 1990 and includes entities such as air carriers and repair stations.[3] In its 2009 Advisory Circular on the VDRP, the FAA notes that civil penalties are one of the means to promote compliance with FAA regulations. However, "the public interest is also served by positive incentives to promote and achieve compliance." According to the FAA, aviation safety is enhanced by incentives for air carriers "to correct their own instances of non-compliance and to invest more resources in efforts to preclude their recurrence." Thus, the FAA will forego any civil penalties when the air carrier detects violations, reports them to the FAA, "and takes prompt corrective action."[4] Another condition is that "immediate action, satisfactory to the FAA, was taken upon discovery to terminate the conduct that resulted in the apparent violation" (FAA 2009: 3, 4). This condition in particular does not seem to have been followed in the Southwest and Northwest Airlines cases, a point that will be addressed more fully later.

In terms of the benefits of the voluntary reporting programs, the 2008 IRT's report on the FAA's safety culture reaffirms the value of these voluntary programs. The team maintains that "such programs are *more* vital to FAA, in our view, than to other regulatory agencies, given the essentially preventive nature of the residual risk-control task, and the resulting importance of learning about and learning from precursor events" (Stimpson et al. 2008: 5, emphasis in original).[5] Voluntary disclosure can be a reliable and important aid in ensuring safety and limiting the possibility of an aircraft accident. If the airlines believe that they will not be penalized for reporting problems, they can enhance safety by reducing the need for inspectors to oversee aircraft repairs. Thus, if everyone follows the rules of the maintenance programs, then safety can be optimized.

However, a major problem arises when the voluntary reporting system does not lead to beneficial results. The Inspector General's report to the DOT regarding Northwest Airlines states: "while voluntary programs can help to identify and correct safety issues that might otherwise not be known, a partnership program that does not ensure air carriers correct underlying problems is less likely to achieve benefits" (USDOT 2009: 2). Due to regulatory capture arising in some of the FAA certificate management offices, the VDRP was not enhancing public safety, but instead detracting from it. This problem indicates that trust cannot be the solution to the problems that arise in capture. Instead what is needed is a strong sense of moral obligation, one of the key ethical concepts missing in the regulatory system.

Two instances of regulatory capture in FAA oversight

On March 5, 2008 the FAA proposed a $10.2 million civil penalty against Southwest Airlines:

> for operating 46 airplanes without performing mandatory inspections for fuselage fatigue cracking. The FAA alleged that the airline operated 46 Boeing 737 airplanes on almost 60,000 flights from June 2006 to March 2007 while failing to comply with an existing FAA Airworthiness Directive that required repetitive inspections of certain fuselage areas to detect fatigue cracking. (FAA 2011c: 52231)

Due to this incident in June of 2008 the Office of Inspector General (OIG) issued a report on the FAA's oversight of airlines, and it determined that the FAA had developed an overly collaborative relationship with Southwest. The Inspector General states that: "although FAA requires air carriers to ground non-compliant aircraft and its inspectors to ensure that carriers comply, the inspector did not direct SWA to ground the 46 affected aircraft" (USDOT 2008: 2). The report also states that:

> The events at SWA demonstrated serious lapses in FAA's air carrier oversight. We found that FAA's inspection office overseeing SWA (the Certificate Management Office, or CMO) developed an overly collaborative relationship with the air carrier, which allowed repeated self-disclosures of AD violations through its partnership program. (USDOT 2008: 3)[6]

A key problem is that the "FAA's oversight in this case appears to allow, rather than mitigate, recurring safety violations" (USDOT 2008: 4). This of course is a major issue since the FAA's mandate is to promote safety.

In addition to the IG's report, a separate investigation by the IRT agrees with their findings. The then Secretary of Transportation, Mary Peters, charged the IRT with making recommendations concerning the FAA's safety culture as well as its implementation of safety management (Stimpson et al. 2008: 4). As part of its findings, the IRT concluded that regulatory capture played a role in the CMO office overseeing Southwest Airlines. The CMO seemed to be more concerned about preserving a good relationship with the airline than it was about enforcing regulations.

Northwest Airlines

However, it is important to note that the Southwest case is not an isolated incident, but rather has arisen in other FAA CMOs, such as the one for Northwest Airlines. The Inspector General's report of 2008 notes that control deficiencies exist across the voluntary disclosure programs, and that this problem is not unique to Southwest Airlines (USDOT 2008: 4). In March of 2008, responding to lapses in FAA oversight of AD compliance, the FAA initiated a Special Emphasis Validation of Airworthiness Directives Oversight in which inspectors audited the execution of ten ADs applicable to each fleet type at each carrier (USDOT 2009: 3). Based on its 4-month special emphasis review, 14 AD deficiencies were found at Northwest Airlines which varied from the 8 the CMO had identified over the previous 4 years, and the number of AD non-compliances was "one of the highest of all airlines reviewed" (USDOT 2009: 6). To indicate the seriousness of the problem, in November of 2008 Northwest Airlines had to ground 27 aircraft because the airline had not performed required inspections of landing gear parts. As the report states, "these inspections were intended to prevent the separation of the main landing gear from the wing and possible rupture of the wing fuel tank" (USDOT 2009: 6). This example reveals the seriousness of the problem of non-compliance. Just as in the case of Southwest Airlines, Northwest was fortunate that no tragic accident occurred due to their failure to resolve maintenance issues.

Perhaps more importantly, the IG indicates that regulatory capture played a role in the Northwest Airlines case. According to their report, the CMO inspectors "continued to primarily work collaboratively with the carrier to resolve AD deficiencies in 2008 and 2009" and the inspectors issued "letters of correction rather than seeking civil penalties" (USDOT 2009: 6).[7] Due to the collaboration between Northwest and FAA inspectors, it is apparent that regulatory capture also played a role in the problems with the ADs. The voluntary reporting program was once again unhelpful insofar as the inspectors allowed Northwest Airlines to use the VDRP when it was in direct conflict with the FAA's own guidelines. I do not mean to imply that there is a cause and effect relationship between the VDRP and capture, but the two seem to go hand in hand in the above-mentioned cases.

Why capture occurred

One of the key questions concerns why agencies such as the FAA continue to fall prey to capture. According to the IRT report, the relationship between the CMO and the airline allows for capture to arise since one office works with one airline over a long period of time. The IRT notes that:

> One feature of the FAA's current structure has the potential to increase this risk: the inspection teams are mostly organized around *airlines*, rather than cutting across multiple airlines and organizing around some other dimension, like geography, or type of plane. (Stimpson et al. 2008: 36, emphasis in original)

In other words, the majority of FAA airline inspectors are assigned to a specific CMO and "deal with exactly one airline, full time, and for many years at a stretch (e.g. the 'Southwest CMO' deals only with Southwest, and is responsible for Southwest Airlines' operations everywhere)" (Stimpson et al. 2008: 36). Such a relationship creates a potential for capture to arise since the CMO employees develop a long-term relationship with one airline, with both sides working closely together.

FAA responses

The government reports on regulatory capture in the FAA have not gone unheeded, and due to the 2008 reports the FAA has made changes to some of its rules and procedures. Above I mentioned that in the aftermath of the Southwest case, the 2003 Customer Service Initiative was amended in 2009 so that airlines are no longer known as "customers" but instead as "stakeholders."

But perhaps the most important change involves the revolving door issue. In response to the reports on the Southwest Airlines case, first by the Office of Inspector General (USDOT 2008) and then by the IRT (Stimpson et al. 2008), in August of 2011 the FAA issued a final rule regarding "Restrictions on Operators Employing Former Flight Standards Service Aviation Safety Inspectors." This rule "will prohibit any person holding a certificate from knowingly employing, or making a contractual arrangement with, certain individuals to act as an agent or a representative of the certificate holder in any matter before the FAA under certain conditions" (FAA 2011c: 52231). In other words any FAA inspector who had a direct responsibility to oversee the operations of an airline cannot be employed with that airline in matters pertaining to the FAA for at least two years after their FAA employment. According to the rule, its purpose is to "prevent potential organizational conflicts of interest which could adversely affect aviation safety" (ibid.). A former FAA employee may have had close relationships with some of the current inspectors, and they may have even shared oversight responsibilities with the airline. Thus, if the former inspector were to work for the airline, then the possibility of capture could be greatly increased. Regarding the Southwest Airlines case, the Southwest Airlines Regulatory Compliance Manager was a former inspector assigned to the airline who reported directly to the PMI when he worked at the FAA. He became manager just two weeks after leaving the FAA (USDOT 2008: 5).

However, even prior to the 2011 final rule, the FAA had policies in place to mitigate potential conflicts of interest. As Gerardo Martinez discussed in Chapter 8, because the FAA hires from the industry, employees are required to divest themselves of any carrier regulated by the FAA. For example, if an inspector owns stock in an airline or other related corporations, he or she must divest from that stock or risk losing employment with the FAA. If one is hired from the industry, one must disclose whether or not there is an agreement that the inspector can later return to that business.

Although this rule shows that the FAA is attempting to overcome the possibility of capture, the question remains whether or not it will adequately address the problem. The FAA received a total of five comments on the proposed rule, and one commentator "noted the challenge of trying to regulate integrity" and took a stronger stance by contending that no former FAA employees should ever be allowed to become FAA designees for an airline (FAA 2011c: 52232). The concept of integrity is one that can be addressed by ethical theories, and in the following sections this chapter shows the ways in which the application of ethics can help address the problem of capture more fully. Before moving to the next section, the textbox reveals another case of capture that indicates the fact that regulatory capture is not limited to the FAA.

REGULATORY CAPTURE AND THE MINERAL MANAGEMENT SERVICE

One may wonder whether regulatory capture is a major problem in other government agencies. While the revolving door issue has arisen in the FAA, the situation may have been much worse in the Mineral Management Service (MMS). The MMS is responsible for managing the nation's natural gas, oil, and other mineral resources. It was responsible for issuing permits to the Deepwater Horizon drilling rig which exploded on April 20, 2010. Prior to the tragic accident, the Department of Interior Office of Inspector General had investigated MMS employee relationships with the Island Operating Company in which some government employees had allegedly accepted gifts from the company. According to a memorandum from the Acting Inspector General, Mary Kendall, of great concern was the environment in which the inspectors easily moved between the industry and the government (DOI OIG 2010: 1). The individuals involved had not only worked in both places, but also often knew each other from childhood. In the Lake Charles, Louisiana district office, some employees had been doing illicit drugs, and many inspectors had inappropriate humor and pornography on their office computers. The report notes that once the MMS supervisor was terminated, the behavior of accepting gifts drastically declined. This case shows that the problem of regulatory capture is not limited to the FAA, and it indicates that much work needs to be done to eliminate the problems associated with revolving door policies.

Why the FAA needs the industry

Because capture arises when the FAA works too closely with the airlines, one may be tempted to think the solution is to eliminate the revolving door altogether. However, one needs to keep in mind that capture does not automatically occur when an agency turns to the regulated entity regarding certain matters, such as technological expertise. Regarding the FAA's type certification process, in which it evaluates new aircraft designs, the technological knowledge of the regulated entity is often far superior to the regulators who are unfamiliar with new and innovative designs. According to Downer (2010: 86):

> The complexity of modern aircraft has long passed a level where regulating it is within the FAA's budget and manpower, and yet the FAA would be ill-placed to make informed judgements, even with infinite resources; they simply lack the "tacit knowledge" to make the requisite judgements about the technologies they certify. (Downer 2010: 86)

For example, the Boeing 787 contains new technologies that the engineers would be able to assess better than the FAA would. The 787 uses composite materials in over 50 percent of its airframe, and is thought to be superior to the traditional aluminum structure (Norris et al. 2005: 50).[8] Due to innovations that arise from industry experts, it would be reasonable to assume that the FAA lacks the expertise to assess new designs and technologies. In order to address its lack of knowledge, the FAA has formalized the relationship between itself and those with expertise through Designated Engineering Representatives (DERs) (FAA 2011a). DERs are employees of the manufacturers or industry consultants, so would their work in the industry imply that capture necessarily arises in the type certification process?

According to Downer, many see this relationship as a conflict of interest that in turn becomes a problem of capture. However, as he claims, "rather than there being a conflictual relationship between regulator and regulated, [the] FAA and the manufacturer share the same interests" (Downer 2010: 91, 92). In other words, neither side would want to certify an aircraft that is unsafe and unreliable since it would not only potentially result in a tragedy, but also it would lead to revenue losses for the manufacturer. However, the famous example of the Ford Pinto case in the 1970s would indicate that this claim is problematic given the fact that Ford management apparently knew of the design flaw (concerning the fuel tank position) and elected not to fix the problem. Thus, Downer is also right to point out that the aligned-interest approach is not beyond reproach. According to Downer (2010: 93), "manufacturers certainly see value in building reliable aircraft, but they juggle other pressures as they compete in a highly demanding marketplace."[9]

Even though the system is not perfect, there does not seem to be a better alternative to certifying aircraft. According to a GAO report on the designee program, "allowing technically qualified individuals and organizations to perform 90 percent of certification activities enables FAA to better concentrate its limited staff resources on the most safety-critical functions, such as certifying new and complex aircraft designs" (US GAO 2004: 1). One of the great benefits of this program is that certification can take place in a timely manner, and this in turn reduces costs to the industry.[10] Given that the FAA often needs to employ industry workers, the solution to capture should entail other methods.

Consequentialism vs. deontology

I have indicated that the consequential approach to rule-making can lead to potentially harmful results. But one may try to defend consequentialism in that this type of ethics is a necessary tool for helping public safety. I do not mean to imply that consequentialism should be abandoned, but rather by itself it is inadequate for addressing the problem of capture. I agree that consequentialism is an important element of any agency that is concerned with public well-being and safety since one needs to anticipate the potential risks for unsafe conditions. I also do not mean to suggest that the cases of capture at Southwest and Northwest Airlines can be simply seen as "unethical" practices. After all, the 2008 IRT report finds that "the FAA's aviation safety staff to be unambiguously committed to its core mission of safety" (Stimpson et al. 2008: 5). In other words, the problem is not a clear-cut case of a few employees behaving in a blatantly immoral manner. Rather, the dilemma centers on consequentialism itself.

One of its key drawbacks is determining which end results are the most important. A consequentialist considers whether one is concerned with the more immediate and local consequences or calculating the more far-reaching and general consequences to the public. Concerning the immediate impact of decision-making, the FAA CMO would evaluate the office's relationship with the airline it oversees and the short-term impact a decision might have on the two parties' continued relationship. For example, the IRT interviewed the PMI at the Southwest office, who said: "I permitted unairworthy SWA aircraft to operate in revenue service, and I was wrong to do so. However, politically, I felt that grounding the SWA aircraft would have negative consequences for the FAA" (Stimpson et al. 2008: 25). Thus the PMI himself had been considering the consequences all along, but the problem was that he prioritized the wrong set of consequences. Regarding the more far-reaching consequences, they are more difficult to discover, and their effects may not be felt for quite some time. Thus, it is often easier to focus on the short term rather than the long term.

In the end a major drawback to consequentialist ethics is that it remains numbers based rather than human based. A consequentialist approach, such as utilitarianism, can try to counter the problems mentioned above by claiming that one should consider the greatest amount of happiness altogether, but nevertheless its overall focus concerns the net results rather than human beings themselves. If we are to eliminate regulatory capture, more diverse ethical considerations are required, especially those that focus on character and responsibility.

Deontology as an alternative

Because of its opposition to consequentialist ethics, one obvious place to find an alternative is through Immanuel Kant's concept of moral duty and respect for persons. In line with Kantian thinking, the IRT report mentions that "a regulator's job is primarily to deliver *obligations*, not *services*" (Stimpson et al. 2008: 37, emphasis in original), and in terms of the FAA one's obligations should be public safety.

In his concept of the Kingdom of Ends, in which rational beings are in union with each other through common moral laws, Kant claims that "man" and in general every rational being "exists as an end in himself and not merely as a means to be arbitrarily used by this or that will" (Kant 1993: 35). In this ideal kingdom, as Kant calls it, people are treated with dignity and as having absolute worth rather than as a means to be used for some purpose. As an example, one could draw a parallel between FAA ADs and Kantian moral laws, for the ADs would count as laws that should not be overridden. Unlike consequentialism, for Kant there should be no exceptions to following these laws. Furthermore people (Persons in Kantian terminology) would be treated as having absolute worth, and thus the airlines would need to comply with the ADs because doing so would indicate respect for human beings.

Of course, as with any ethical theory deontology has its flaws, and one problem with the Kingdom of Ends is that Kant himself makes it clear that "it is certainly only an ideal" (Kant, 1993). Thus, for some it may be unrealistic to apply Kant to government regulations. It is simply impractical to treat all human beings as ends in themselves since we often need to use people for some other purpose. However, some variation of his ideal could be quite useful in assessing regulatory oversight and enforcement. Since consequentialism leaves out the concept of treating humans as ends in themselves, a Kantian approach could supply a viable alternative by having governmental agencies focus on the public as human beings first rather than as numbers to be used in managing risks or weighing costs and benefits.

Final suggestions

A lesson we can learn from the application of ethical theories to capture is that in the end the main ethical problem may not be with capture itself as much as it is the federal government's methodology for rule-making, one that privileges the needs of business over the needs of the public. Thus, the regulatory system needs to be revised so that it truly focuses on humans as having absolute worth. In addition, to prevent future lapses in oversight, perhaps government regulators need to be held accountable for their decisions. Note that although the FAA fined Southwest Airlines, Southwest could not in turn penalize the FAA for failing to provide adequate oversight. As Shapiro and Steinzor claim, agency accountability is one way to enhance the benefits of the regulatory system, especially through using the internet to give public notice when agencies are successful and when they fail (Shapiro and Steinzor 2007–08: 1771). If there were concrete ways to hold the FAA accountable for lapses in oversight, perhaps that would make the CMOs take into consideration all of the stakeholders, not just their "customers."

Perhaps one main drawback to my position is the question of feasibility, especially when it comes to creating a system that is truly humanizing. And yet, we have already seen aspects of Kantian ethics in one area of government: namely, the NTSB. In the previously mentioned debate regarding the FAA's decision not to mandate CRSs, the NTSB criticized the FAA on the fact that any unrestrained child is unprotected. Thus, the NTSB was not concerned with the probability of an accident, but rather its focus was on the safety of the child. This point indicates that alternative views already exist within the federal government. While the NTSB's role is to make recommendations, not enforce them, the FAA could perhaps better heed the necessity for prioritizing human factors. While in the end one may argue that utilizing various aspects of ethics sounds too piecemeal, or perhaps impractical, at a minimum Kantian ethics helps reveal the limitations of decision-making based solely on the possible consequences, especially when human life is at stake.

Notes

1 The 2008 report of the IRT also mentions the CSI, and that it was poorly named. However, they find that "regulating the FAA's use of language should be unnecessary. It would be awkward in any case to try to control the use of language or attitudes through statute" (Stimpson et al. 2008: 37).

2 The report also notes several negatives, such as the problem that the accuracy of voluntarily reported data cannot always be verified, and also can be subjective based on the reporter's experiences (see US GAO 2010: 19–20).

3 Two other programs include the Flight Operations Quality Assurance (FOQA), established in 1995, and the Aviation Safety Action Program (ASAP) of 1997.Concerning the FOQA program, devices on specially equipped aircraft collect data from the aircraft's flight recorders; and regarding ASAP, members of participating aviation industry employee groups report a variety of safety events without fear of reprisals (see US GAO 2010: 5, Table 1).

4 As for evaluating the violation, some of the conditions to be met include the fact that the violation was inadvertent and that it does not indicate a lack of qualification on the part of the airline

5 In addition to promoting safety, the FAA's voluntary programs can become crucial given the limited number of FAA inspectors who oversee the airlines. As former airline mechanic Ted Ludwig claims in Chapter 6, because there are more aircraft being repaired than there are FAA inspectors, the agency relies on airlines to follow their own maintenance programs that the FAA have reviewed and certified.

6 A CMO specializes in the certification, surveillance, and inspection of major air carriers.

7 In addition the CMO inspectors allowed Northwest Airlines to continue to use the VDRP to report AD non-compliances. The report states: "this action directly conflicted with FAA and industry guidance that does not permit voluntary disclosures in anticipation of or during FAA inspection" (USDOT 2009: 6; see also FAA 2009: 4).

8 Although other aircraft have made use of composite materials, they have utilized these materials in smaller airplane structures, such as the Boeing 777 tail and horizontal stabilizer (Norris et al. 2005: 51). Composites are also now part of the cutting edge of engine development, and General Electric advances the use of composites in jet engines "in ways never before experienced" (ibid.: 56).

9 The development of the Boeing 747 can be seen as an example of this problem. In his reflections on the 747 project Joe Sutter, the head engineer, recalls several issues with management, especially regarding delays. For example, management wanted the 747 to fly on December 17, 1968, the 65th anniversary of the Wright Brothers' flight at Kitty Hawk. But Sutter knew that time frame was not possible (Sutter 2006: 167). Although management was unhappy about the delayed first flight, at least the flight was postponed until it was ready in February of 1969.

10 The report details some of the weaknesses of the designee program, including the "FAA's inconsistent monitoring of its designee programs and oversight of its designees" (US GAO 2004: 3).

References

Dal Bó, E. 2006. Regulatory capture: a review. *Oxford Review of Economic Policy*. Vol. 22, No. 2, 203–225.

Downer, J. 2010. Trust and technology: the social foundations of aviation regulation. *British Journal of Sociology*, Vol. 61, No. 1, 83–106.

Federal Aviation Administration (FAA). 2011a. *Delegated Engineering Representative*. Available at: www.faa.gov/other_visit/aviation_industry/designees_delegations/designee_types/der/ [accessed November 12, 2011].

Federal Aviation Administration (FAA). 2011c. Restrictions on operators employing former flight standards service aviation safety inspectors. *Federal Register*. Vol. 76, No. 162, 52231–52237.

Federal Aviation Administration. 2009. Advisory Circular 00-58B: Voluntary Disclosure Reporting Program. April 29.

Federal Aviation Administration (FAA). 2006. FAA approves new child safety device: government gives parents more options for safe air travel with children. Press Release, September 6. Available at: www.faa.gov/news/press_releases/news_story.cfm?newsId=7381 [accessed July 4, 2010].

Federal Aviation Administration (FAA). 1998. *Economic Analysis of Investment and Regulatory Decisions*. Office of Aviation Policy and Plans. Revised Guide. Form DOT F 1700.7 (8-72). January.

Federal Aviation Administration (FAA). 1996. *FAA Historical Chronology 1926-1996*. Available as a pdf at: www.faa.gov/about/media/b-chron.pdf [accessed March 8, 2012].

Kane, R.M. 2007. *Air Transportation*. 15th ed. Dubuque, IA: Kendall/Hunt.

Kant, I. 1993. *Grounding for the Metaphysic of Morals*. Translated by J.W. Ellington. 3rd ed. Indianapolis: Hackett.

Lowy, J. 2009. FAA will no longer call airlines "customers." *USA Today*. Available at: www.usatoday.com/travel/flights/2009-09-18-faa-customers-ap_N.htm [accessed October 30, 2011].

Mill, J.S. 2001. *Utilitarianism*. 2nd ed. Indianapolis: Hackett.

Mills, R.W. 2011. Collaborating with industry to ensure regulatory oversight: the use of voluntary safety reporting programs by the Federal Aviation Administration. Dissertation.

Available as a pdf at: http://etd.ohiolink.edu/view.cgi/Mills%20Russell%20William.pdf?kent13021027113 [accessed September 30, 2011].

National Transportation Safety Board (NTSB). 2004. Analysis of diversion to automobile in regard to the disposition of Safety Recommendation A-95-51. Office of Research and Engineering, Safety Studies and Statistical Analysis Division, August 3. Available as a pdf at www.ntsb.gov/aviation/sr_a-95-51_diversion_analysis.pdf [accessed August 9, 2010].

Norris, G., Thomas G., Wagner M., and Smith, C.F. 2005. *Boeing 787 Dreamliner: Flying Redefined*. Perth, Australia: Aerospace Technical Publications International.

Ostas, D.T. 2004. Cooperate, comply, or evade? A corporate executive's social responsibilities with regard to law. *American Business Law Journal*, Vol. 41, No. 4, 559–594.

Potoski, M., and Prakash, A. 2004. The regulation dilemma: cooperation and conflict in environmental governance. *Public Administration Review*, Vol. 64, No. 2, 152–163.

Saenger, A. 2003. Customer service connection defined. *Federal Air Surgeon's Medical Bulletin*. Vol. 41, No. 3, 1, 13. Available as a pdf at: www.faa.gov/library/reports/medical/fasmb/media/F2003_03.pdf [accessed March 7, 2012].

Shapiro, S.A., and Steinzor, R. 2007-08. Capture, accountability, and regulatory metrics. *Texas Law Review*, Vol. 86, 1741–1785.

Stigler, G.J. 1971. The theory of economic regulation. *Bell Journal of Economics and Management Science*. Vol. 2, No. 1, 3–21.

Stimpson, E.W., Babbitt, J.R., McCabe, W.O., Sparrow, M.K., and Vogt, C.W. 2008. *Managing Risks in Civil Aviation: A Review of the FAA's Approach to Safety*. Report of the Independent Review Team, September 2. Available as a pdf at: www.aci-na.org/static/entransit/irt_faa_safety_9-08.pdf [accessed: July 6, 2010].

Sutter, J. [with J. Spenser] 2006. *747: Creating the World's First Jumbo Jet and Other Adventures from a Life in Aviation*. New York: HarperCollins.

US Department of the Interior. Office of Inspector General (DOI OIG). 2010. Investigative Report: Island Operating Company et. al. Case No. PI-GA-09-0102-I. March 31. Available as a pdf at: https://abcnews.go.com/images/Politics/MMS_inspector_general_report_pdf.pdf [accessed July 11, 2018].

US Department of Transportation (DOT). Office of Inspector General. 2009. Report on FAA oversight of Airworthiness Directive of Compliance at Northwest Airlines. OSC File No. DI-08-2971. December.

US Department of Transportation (DOT). Office of Inspector General. 2008. Review of FAA's Safety Oversight of Airlines and use of Regulatory Partnership Programs. Report No. AV-2008-057. June 30. Available as a pdf at: www.oig.dot.gov/sites/dot/files/pdfdocs/SWA_Report_Issued.pdf [accessed March 5, 2012].

US Government Accountability Office (GAO). 2010. *Aviation Safety: Improved Data Quality and Analysis Capabilities are Needed as FAA Plans a Risk-Based Approach to Safety Oversight*. Report to Congressional Requesters. GAO-10-414. May.

US Government Accountability Office. 2004. *Aviation Safety: FAA Needs to Strengthen the Management of its Designee Programs*. Report to the Ranking Democratic Member, Subcommittee on Aviation, Committee on Transportation and Infrastructure, House of Representatives, GAO-05-40. Available as a pdf at: www.gao.gov/new.items/d0540.pdf [accessed November 12, 2011].

US Office of Special Counsel. 2010. *Department of Transportation Report Substantiates Allegations of FAA Oversight Failures*. [Online, July 22]. Available at: www.osc.gov/documents/press/2010/pr10_14du.pdf [accessed August 23, 2010].

White House. 1994. Regulatory planning and review. Executive Order 12866. June. Available at: http://govinfo.library.unt.edu/npr/library/direct/orders/2646.html [accessed March 7, 2012].

10

ETHICS IN AVIATION FROM THE PERSPECTIVE OF A FLIGHT ATTENDANT

Gail L. Bigelow

Flight attendants assigned to flight duty are responsible for performing or assisting in the performance of all safety, passenger service, and cabin preparation duties. The application of these duties requires flight attendants to be responsible for handling passenger carry-on items, as required in order to secure the cabin for take-off and landing. Flight attendants are responsible for fulfilling certain FAA requirements and company and operational requirements. Thus, the purpose of this chapter is for the reader to gain an overall understanding of the ethical role played by flight attendants, especially in relation to the FAA, passengers, labor relations, safety, and security.

The responsibilities of flight attendants

The flight attendant profession is a largely unsupervised job. Most of the time a flight attendant spends at work is on the aircraft or at a layover destination. Although there is a chain of command (namely, Captain, First Officer, First Flight Attendant, etc.), flight attendants are basically responsible for themselves. The way they conduct themselves, their level of integrity and personal ethics are highly visible to the traveling public, and reflect not only on themselves, but also on their carrier and all the other flight attendants in the industry. Since the inception of the flight attendant role in the 1930s, the position has been one that has sparked curiosity and watchful attention; it is no wonder that carriers hold flight attendants to the highest standards of performance.

It probably does not occur to most flight attendants that a good part of their day involves practicing ethics. But from the time we check in until the end of our debriefing, we are faced with choices that amount in no small measure to ethical decisions. Take, for example, checking in. At one carrier, flight attendants are required to check in one hour prior to departure. Failure to do so results in points assessed for being late. Accumulating 12 points can result in termination, so there is strong incentive to avoid attendance-related issues. Flight attendants at one carrier check in by scanning their ID cards at a computer. If for some reason the scanner does not work, they can call crew scheduling and ask them to be shown as checked in. Consider the case of a flight attendant running late, who has not yet arrived at the airport. Why not call crew scheduling and tell them the computer scanner is not working and then ask to be checked in? What does it hurt? The carrier knows that the flight attendant plans to cover the trip, and one can thereby avoid the attendance points. Aside from the potential disciplinary action a flight attendant might incur for such an infraction, the deceit itself is damaging. When one chooses to walk a fraudulent path and escapes detection, the risk is that path will become a thoroughfare to additional ongoing and unethical choices. It is not surprising, therefore, that one instance of intentional dishonesty, a display of ethical failing, can result in termination of employment.

Demands on the flight crew

Company policy might require that flight attendants conduct a pre-flight briefing covering their safety positions and their responsibilities in the event of an emergency. Flight attendants, together with pilots, might be required to conduct a pre-flight briefing which covers such issues as projected flight time, en route weather and emergency communication procedures, just to name a few. These crewmembers have an ethical duty to each other, certainly, but also to their customers, to conduct these briefings. But today's world of aviation is all about saving time, being on time, and giving time to the customer. Crewmembers are pulled in many different directions by the demands placed upon them, even as resources disappear. One makes choices, sometimes improper ones, in an effort to save time.

Amid what will end up being a 13-hour duty day on the last day of a 3-day trip, a flight attendant notices that something in the cabin has broken—nothing that would endanger anyone or affect the integrity of the flight. The attendant knows that if he or she writes it up, as one is supposed to do, and then turns it into the flight deck, upon arrival at the next stop (one of say, 5 or 6 stops that day), the flight deck will be required to call out a mechanic to examine the broken item and either repair it or place it on the minimum equipment list. This issue will result in a delay of at least minutes, but probably longer. Either way the flight attendant is ready to go home. One makes a choice—one decides to write it up on the way to the home domicile, so the flight will not incur a delay en route. In some cases, other crew members encourage this sort of unethical action. Even pilots have been known to say: "Are you sure you want to tell me about that right now?" Training supervisors have been known to give a wink and a nod when they tell flight attendants that they must write up any cabin discrepancy as soon as they become aware of it. The unspoken message to flight attendants is: consider your actions and how they affect the carrier's on-time performance.

Multiple choices are made each day. It is another long flight, perhaps transcontinental and almost a full load. However, the last row of seats is unoccupied. Do the flight attendants look for customers who could use additional room and offer them the row, or do they block the seats so that they and their fellow crewmembers can sit in these rows after the service is finished?

Speaking of the service, the Inflight Service Guide requires three beverage carts, but after the flight attendants pass through twice, the passengers seem to be satiated. Does the crew do the third cart because it is a service requirement or skip it? It depends on their sense of obligation. Recycle? What a hassle, but the carrier requires it—do it or not? Pour beverages or hand out cans? The carrier's policy is strict on this: pour—it saves money. But if you think about it, most people really would prefer a full can, so why not just skip a step and give them what they want at the outset?

All of these actions, or inactions, require flight attendants to perform a gut check of their personal and professional ethics. If they work with the same crew each flight, it is possible that the crew has, without realizing it, made a group gut check and has become automated in their approach to their jobs and customer service. Such automation may or may not include unethical decisions.

By the nature of bidding for flying on a monthly basis, rearranging trips by trading with other flight attendants, and reassignment by crew scheduling, it is not uncommon to fly with several different flight attendants during a multiple day trip. Each time the group dynamic changes, the gut check of one's ethics is performed, whether one realizes it or not. "I usually just hand out cans, how about you?" becomes a test of the newcomer's sense of responsibility.

Pertinent FAA regulations

The Federal Aviation Regulations (FARs) are detailed and strict regarding the ability of a certificate holder (airline company) to operate an aircraft under Part 121 (revenue flying). Some of those regulations, and the resulting ethical responsibilities of flight attendants and customers alike, are discussed in this section. The first and foremost FAR requires that flight attendants be on board the aircraft in the first place. It specifically stipulates that aircraft having a maximum seating capacity of more than 9 but less than 51 passengers have 1 flight attendant. As the seating capacity rises, so does the required number of flight attendants for the certificate holder to operate that aircraft. Certificate holders must "prove" that the number of flight attendants assigned is able to evacuate each aircraft within 90 seconds (depending on the number of emergency exits) before being allowed to use that aircraft for revenue flying.

The FAA requires that all flight attendants be registered and that, while on duty, they have accessible to them their registration card and flight attendant manual. There are multiple federal regulations about which the flight attendants are obliged to educate the public and do their best to enforce while on board. The flight attendant manual is a compilation of almost all the procedures for both emergency and standard operations. There are extensive instructions for medical emergencies and on the operation of all emergency equipment.

Manual revisions, including emergency interim bulletins (new procedures or FAA directives that are issued between substantial manual revisions), are a reading requirement and need to be inserted into the flight attendant manual within a certain number of days of receipt. Failure to do so or error in insertions/deletions—such as missing pages, duplicate pages, or out-of-date material—is the personal responsibility of the flight attendant. This task is conducted largely on one's own time, either at home or at a layover point. Does the flight attendant read each word of the revision, scan for obvious changes, or just add the pages where they belong, snap the manual shut, and stick it back in the flight bag? It depends on his or her sense of obligation to know and understand what is in his or her manual.

The FAA randomly chooses domestic and international flights for inspection. Inspectors arrive at the aircraft without advance notification. Flight attendants are required to produce both their registration cards and manuals for inspection. A violation of a FAR pertaining to a flight attendant, whether it is a manual violation or a violation of a regulation which the flight attendant is directly responsible for enforcing, can result in a personal fine of up to $10,000 against the flight attendant. It would appear that the prospect of a $10,000 financial hit would be the motivator that would keep flight attendants in line regarding the regulations; but there is more to it, as we will discuss below.

Flight attendants and safety

Flight attendants, as a rule, are caretakers who have a strong sense of personal and professional ethics. It is out of the ordinary to come across a flight attendant who has little or no regard for their customers and fellow crewmembers, or who is ethically or professionally lazy. Much of what is involved in a flight attendant's day revolves around concern for the safety and integrity of the flight. Although customer service is secondary to safety, flight attendants are, for the most part, conscientious about all of their inflight duties and do not take kindly to fellow flight attendants who may not be as dedicated to upholding the tradition of the capable, caring, nurturing flight attendant.

That smiling flight attendant who hands you a cup of coffee may also be the person who saves your life. Flight attendants train thoroughly to be able to fight fires, resuscitate passengers, evacuate aircraft in the worst conditions, and operate state-of-the-art equipment which assists in accomplishing these tasks. In addition, since 9/11 flight attendants are scrupulously trained to observe each and every passenger who boards a flight and evaluate them—are they a potential threat to the safety of the flight? In training, flight attendants are offered sophisticated courses in self-defense, and are taught how to disable a passenger who may be significantly larger than themselves, handcuff them, and turn them over to the police upon landing. In short, while on duty, flight attendants must act as doctors, nurses, psychologists, teachers, parents, and police officers. They must be able to tenderly comfort a crying child traveling between divorced parents in one moment, aggressively fight a fire a moment later, and then resuscitate a passenger in the very next moment. Obviously, it would be a bizarre flight in which all of these examples occurred at the same time; but it could happen, and flight attendants are trained to be alert and responsive when what "could" happen does in fact arise.

FARs require that flight attendants monitor all carry-on baggage, and that they strictly enforce the stowing of that baggage on board the aircraft. In the post-9/11 rush to patrol items carried onto an aircraft, the Transportation Security Administration (TSA) temporarily accomplished that for which flight attendants had lobbied for years—more baggage checked and less on board. When, in 2008, oil prices surged, airlines began charging for checked luggage. This new procedure abruptly ended the customer's momentary flirtation with checking their luggage. Space in an aircraft cabin is once again at a premium, and competition for overhead bin space on a full flight, particularly to a vacation destination, has become fierce. What does not fit in the overhead either must be stowed beneath a seat or checked. Passengers can be very creative in finding areas in the cabin to tuck their personal belongings. Flight attendants, often worn out by arguing with customers regarding carry-on luggage and its stowage, might be tempted to turn a blind eye to a bag stowed at a bulkhead or behind a row of seats, but for their ethical duty to the safety of their fellow crew members and their customers. Improperly stowed luggage, in an accident or rapid decompression, can become a deadly projectile. The textbox recounts an instance of a flight attendant experiencing a passenger who engaged in questionable decision making.

HEDGEHOG ON A PLANE

During a nighttime domestic flight, a passenger in the first row of coach (economy) seating rang the call button and told the flight attendant that something had scurried over his leg. The attendant looked around but did not see anything. About 15 minutes later the same passenger rang the call button and reiterated: "Something is definitely on this plane." Because most of the passengers were sleeping, the flight attendant got out a flashlight and found a hedgehog sitting in the middle of the aisle near the first-class cabin. She decided not to use the PA system, and instead asked the first-class cabin if anyone had lost a critter. One woman immediately woke up and said it was her pet hedgehog. The passenger had brought the hedgehog on board by carrying it in her purse and had not informed anyone of the animal. Because no one complained about the matter, the flight attendant decided not to report it. However, this type of situation could become unsafe, especially if a passenger smuggles on board something more dangerous.

Passenger responsibilities

It is important to note that airline passengers have shared ethical responsibilities. For example, flight attendants are required to speak directly to passengers seated in exit rows to ascertain whether, in the event of an emergency, they would be willing and able to open that exit. Determination of this ability includes: the need for personal assistive devices; an understanding of English that allows them to follow the commands of the flight crew; and whether or not the individual passenger is indeed willing to assist in an emergency situation. Exit row seating is typically more comfortable than other rows because the seat pitch (distance between the edge of one seat to the back of the seat in front of it) is longer, thereby allowing significantly more legroom. For this reason, many savvy travelers, particularly business women and men, request these seats. These travelers carry an ethical obligation to not only respond honestly to the questioning flight attendant, but also to read the information in the card printed specifically for those rows, to understand that information, and actually to look at the window exit they would be required to operate should an emergency occur. Often the flight attendant will address a row of heads bent down to their reading material, and the flight attendant will need to be insistent in questioning these passengers. When a customer requests exit row seating, a cooperative effort between the customer and the flight crew automatically arises. The customer is, in effect, promising that, in the event of an emergency, he or she will open that exit and begin passenger egress through that hatch. By choosing those seats, customers accept an ethical obligation toward their fellow passengers and the flight crew.

Seating considerations and related decent conduct are not limited to where and how adult passengers are accommodated. Once a child has reached its second birthday, the FAA requires that they occupy a seat of their own and not be held on the parent's lap. It is not atypical for a parent to attempt to avoid the price of a ticket for a child who has just passed that point by lying when questioned about the child's age. This seating issue is supposed to be determined prior to boarding the aircraft. Because children do come in various sizes and weights at two years and a few months, it is not difficult to understand a flight attendant's reluctance to challenge a parent. Nevertheless, it must be done.

Flight attendants have long held the opinion that any child, whether two months or two years, should be in their own restraint system. The G forces at play during a rapid (emergency) descent are substantial and, if combined with turbulent air, it could become impossible for a parent, even in the most protective moment, to hold on to a child. Severe injury or even death is not beyond the realm of possibility. Customers informed of these facts are not often enthusiastic in their response to the flight attendant. While the flight attendant may make a professional ethical decision on whether or not to educate the passenger as to the reasons for the rule, it really is not his or her decision to make regarding whether or not to enforce it.

The FAA is exacting in its regulations surrounding the type of child restraint systems (CRSs) used on board an aircraft. Such seats must bear certain labels to be approved for use, and booster seats, vest-type, harness-type, or lap-held CRSs are strictly prohibited. Here again, cooperation between the customer and the flight attendant is essential. Parents, especially those new to their role or unused to traveling with small children, can be very spirited in their defense of their chosen CRS. Often, they cannot understand why having their infant strapped to their body is not the safest way to fly. The boarding process, almost never a calm or unhurried affair these days, is not the best time to come between a parent and their child. Unfortunately, the mandated and ethical responsibility requires a showdown. Faced with the oft-heard declaration that "the flight attendant on the last flight let us use it," one can understand why a flight attendant might adopt a slightly forbidding countenance as he or she insists that the federal regulation be enforced.

In addition to the above-mentioned issues, post 9/11 regulations dictate that passengers may not congregate around the flight deck door while waiting to use the lavatory. Do they cooperate, or does their sense of privilege override their sense of ethics? For example, upon observing the economy-class passengers deplaning before him and his fellow first-class cabin passengers, one man opened an emergency hatch and slid down the chute! Of course, he was promptly arrested.

Labor relations

Since the industry was deregulated in 1978, airline companies have co-existed in what amounts to a dog-eat-dog environment. Constantly spying on one another to determine competitors' ticket prices, boarding procedures, inflight service, prices for buy on board items, seat pitch, aisle width, charges for checked baggage or premium seats in the cabin and who knows what else, airline management is demanding in their expectations of all of their employees. Their "frontline" employees—those who spend the most time with the customers—are at the eye of the storm. One would think that given the impact flight attendants have on creating loyal customers, companies would be, if not solicitous, at least mindful of the need to keep the flight attendants happy. Yet flight attendants to this day need to fight just to keep what they have gained during their 70-plus year history. Breaking new ground is almost unheard of.

Prior to the FAA-mandated crew rest regulations, flight attendants at unorganized carriers were required to work ridiculously long hours. Without an authority holding them in check, companies could, and did, keep flight attendants on duty until they were simply too exhausted to continue. After years of lobbying on behalf of all flight attendants, the largest flight attendant union in the world, the Association of Flight Attendants-Communication Workers of America (AFA-CWA), succeeded in getting crew rest provisions passed, despite vigorous anti-regulation pressure from the airline consortiums.

Trade unionism is strong in the airline industry. Even with carrier efforts and some successes to outsource labor, the majority of flight attendants remain committed to the union model. Most of today's major carriers have organized flight attendant work forces. One exception is Delta Airlines. The effort by Delta management to defeat organization of its flight attendants is legendary. In both the 2001 and 2008 organizing drives at Delta, management was accused of using heavy-handed tactics to crush the proposal to organize, including a 2008 bulletin posted in the flight attendant crew room advising attendants that they should just rip up and toss away their ballots. By this suppression of voter turnout, Delta kept the flight attendants from unionizing since not voting at all counted as a "No." AFA-CWA organizers can attest to the fact that Delta flight attendants whispered to them that they had cast their vote. A free and fair election would not require that employees whisper that they have voted.

The plight of the flight attendant at unorganized carriers remains the worst, but their organized colleagues in today's aviation world are not on Easy Street either. Unprotected by the Occupational Safety and Health Administration (OSHA), yet required to work long hours in an unhealthy cabin environment—with dry and unpredictable cabin air quality, exposed to extreme doses of radiation, jet fuels, and oils, and an array of germs and communicable diseases—flight attendants are today's canaries of the skies.[1] Without a voice to move their issues forward, the employees are dependent on the benevolence, indeed the ethical behavior, of their management.

The Railway Labor Act (passed in 1926 with later amendments) manages flight attendant labor relations. Contracts do not expire; instead they become amendable. Thus, a management team bent on rejecting improvements in a contract can drag out negotiations for years. The "what the market will bear" mentality of most airline management teams practically insures that where contract negotiations do take place, they will be long and acrimonious. Company attempts to capture back improvements in pension plans, sick leave days, family leave days, vacation days, insurance benefits, and hours of service rules ignore their ethical responsibility to recognize adequately by compensation the importance of those frontline employees, and their impact on customer loyalty.

At more progressive carriers, usually those with a long history of organized labor, a more cooperative tone exists. Rather than go head to head on every issue, management and labor are learning to team up to fight the competition. While management is still intent on getting back as much as it can in dollar-based benefits and rules, it is more willing to give way on areas that do not immediately affect the airline's bottom line. While this does not go far enough to keep flight attendant compensation packages on a par with the cost of living, it can help provide flexibility in quality of life, which offers rewards of its own.

Post-9/11 issues

There is no question that 9/11 caused a profound collision between the present working world and that previously experienced by flight attendants. Overnight the flight attendant's primary job changed, dramatically and forever. Behind the friendly greeting at the cabin door or at row six or mid-cabin during boarding is a flight attendant's careful analysis. Is this passenger a threat to the safety of this flight? Is he or she wearing the clothes, reading the book, or exhibiting the behavior that causes this flight attendant concern? But, except for this passenger's ethnicity, would the flight attendant be concerned? Should that passenger be questioned, challenged, or not? A passenger's extended time in a lavatory no longer raises just the concern regarding the possibility of fighting an inflight fire due to an illicit smoke and discarded cigarette, but is he or she building a bomb? An unclaimed bag no longer invokes a search to return the item to a thankful passenger, but rather a careful examination of the bag to determine whether it might be a camouflaged explosive. A call from the flight deck no longer simply concerns a pilot ordering yet another cup of coffee, but a request for the flight attendant to guard the flight deck door and the overall integrity of the flight deck as the pilot exits to use the lavatory. The call to the flight deck for the captain or first officer to come out and convince the troublemaker in seat 20D to stop is no longer possible. The flight crew is locked into its tiny space, inaccessible during normal operations, and in complete lockdown if any serious trouble arises in the cabin.

Flight attendants can no longer afford not to be suspicious. While 9/11 may be a distant memory to many, every day a flight attendant reports to work it is in the back of one's mind. As one flight attendant phrased it: "It is difficult to be vigilant and gregarious at the same time." Increasingly often, flight attendants are required to do a gut check of their personal and professional ethics as their responsibility for the integrity, the very safety, of the cabin experience rests on their shoulders alone.

Final thoughts

In this increasingly challenging airline environment it is clear that a sense of personal and professional ethics is desirable for all participants. It is so easy to look around and witness inappropriate behavior. Whether it is airline management's honesty in its labor relations, labor's ethical responsibility to one another and to the customer, or the customer's recognition of and compliance with the rules of the road, so to speak, it is becoming more imperative than ever that everyone act in an ethical manner and treat each other with respect.

Note

1 In the early days of coal mining, canaries were sent into the shafts with the men. A canary dying indicated that toxic fumes were present and served as an early warning system for the men to immediately leave the mine.

11

VALUJET 592 AND CORPORATE RESPONSIBILITY

Kenny Frank

On May 11, 1996 a ValuJet DC-9-32, registration number N904VJ, crashed into the Florida Everglades exactly ten minutes after its departure from Miami International Airport. The loss of 110 innocent lives sparked one of the most dedicated investigations ever performed by the NTSB, and it also became one of the highest media-covered disasters in US aviation history. On May 13, only two days after the accident, the public first heard about the oxygen generators which had been loaded in the forward cargo compartment of the aircraft destined for Atlanta. The NTSB investigation later concluded that the probable cause of the crash of Flight 592 was the ignition of one or more of these oxygen canisters. The report speculated that SabreTech, ValuJet's contract maintenance provider, improperly labeled and stored these generators as empty, but in reality they were actually full. Much of the public and the media following the accident felt relief in knowing the cause of this terrible disaster. However, some believed that this investigation was cut short. Many questions remained unanswered regarding ValuJet's questionable rapid expansion and maintenance procedures in the months prior to the accident. Some of these issues even sparked the FAA to begin an investigation of ValuJet. This chapter will develop both theories of the NTSB and those who believe something else brought down the DC-9 on that sunny Florida afternoon.

The accident investigation and the NTSB report

The crash of Flight 592 sparked a major recovery effort by the NTSB, the FBI, local authorities, and volunteers. For months they slowly searched the Florida Everglades, battling snakes, alligators, extreme heat, and blood-borne diseases in an attempt to recover all available wreckage, remains, and any information supporting the investigation. Due to their efforts, a total of 90 percent of the aircraft was recovered, but nothing larger than the size of the main landing gear. The first interest of the NTSB centered on the 50–60 oxygen generators that were loaded in the forward cargo bin. SabreTech removed the canisters from other ValuJet aircraft months before the accident due to their pending expiration, and the airline requested that the canisters be shipped back to ValuJet's maintenance facility in Atlanta upon cleaning out SabreTech's facilities.

These types of canisters are commonly installed on passenger aircraft to provide a source of oxygen in the case of an emergency or loss of cabin pressure:

> Typically, an oxygen canister contains a sodium chlorate pellet or cylinder and an igniter. The igniter can be triggered by friction or impact. It generates enough heat to start the sodium chlorate reaction, and then the heat of the reaction sustains itself. The sodium chlorate does not burn—its decomposition just happens to give off lots of heat and lots of oxygen. (HowStuffWorks 2000)

The canisters used on the DC-9 contained spring-loaded igniters secured by a safety device commonly known as a retaining pin. One end of the pin is inserted into the canister to prevent ignition, and the other end has a long cord which attaches to the oxygen masks located in the overhead compartments. In an emergency a passenger will pull down on the mask and, as a result, the pin is pulled out of the canister. With the pin removed, the spring-loaded igniter will strike the canister, which in turn creates a chemical reaction that produces oxygen. The amount of heat that these generators produce is what makes them extremely dangerous and considered as hazardous materials.

On November 6 and 7, 1996 the NTSB conducted a series of five tests at the FAA's Fire Facility to test their theory on the likelihood that an oxygen canister started the fire that caused the crash of Flight 592. The first 2 tests each involved 1 box with 28 generators, and the next 3 tests contained 5 boxes with 24 generators in each box. All tests were done in a controlled environment, and ranged from sitting them on a metal floor to being arranged in the cargo bin similar to the accident aircraft. All retaining pins were manually removed from the generators, which require 1–4 lb of pressure. The first test was unable to produce a fire, only minor smoke generation. The second test resulted in a fire, and within 10 minutes produced heat up to 2,000°F. In the third test the five boxes were placed on the floor, but the test was unable to produce a fire, only minor smoke. The fourth test performed was similar to the third, and fire was produced with temperatures reaching almost 3,000°F in 16 minutes. In the fifth and final test, 2 boxes were placed atop a main gear tire pressurized to 50 psi; the other 3 boxes were placed around the tire, and luggage was stacked around them. In 11 minutes 30 seconds the temperatures reached 3,200°F, and in 16 minutes the main tire ruptured.

By the end of November 1996, seven months after the accident, the NTSB determined that ValuJet's contract maintenance provider, SabreTech, was at fault for mislabeling the generators as empty when in fact they may have possibly been full. The evidence included the fact that recovered areas of the cargo compartment provided traces of a fire and that the Cockpit Voice Recorder (CVR) also indicated the presence of smoke and fire prior to the crash. The NTSB concluded that the smoke and fire crippled the aircraft and incapacitated the flight crew, resulting in an uncontrollable descent into the Everglades (NTSB 1997: 53–54).

Another possible cause

In the months and years following the accident, several other theories regarding the crash of Flight 592 began to surface. One major counterargument to the NTSB report is that this tragedy was not caused by ignited oxygen generators, but rather by an electrical failure within the aircraft due to poor maintenance—notably not by the outside vendor SabreTech, but by ValuJet itself. The accident compelled several anonymous sources to come forward and share their experiences and information on aircraft 904VJ as well as the condition of the entire ValuJet DC-9 Fleet.

With airline operations beginning in October 1993, ValuJet grew rapidly and became one of the fastest-growing airlines to make a profit in aviation history. Their fleet was composed mostly of DC-9s many of which were nearly 30 years old. Because these aircraft cost a fraction of new ones, the fleet would give ValuJet a better chance of turning a profit in its early years. By using a business model to operate less-expensive aircraft, which included utilizing the aircraft as often as possible, the airline was able to gross $21 million in 1994, the same year it went public. This huge profit allowed ValuJet the opportunity to place orders on 50 brand new MD-85 jets to expand their future fleet (Morley 2015).

A common phrase heard in the aviation industry is: "A plane not in the air is a plane not making money." While this concept of utilizing an aircraft as much as possible has proved profitable for many airlines, with it comes great responsibility. As an aircraft increases its hours on its flight time, engines, and overall structure, it is equally important that it gets time on the ground for routine maintenance inspections to ensure its safety and airworthiness. As passenger loads and destinations increased for ValuJet in its early years, the DC-9s became the workhorses of the airline, with each aircraft working many flight legs each day. At the time of the accident, aircraft 904VJ was approximately 27 years old and had logged 68,395 hours (Aviation Safety Network 2006). It was originally purchased in 1969 by Delta Air Lines, and later sold to ValuJet in 1993.

While it is not uncommon for commercial aircraft to be used for many years and sold from airline to airline, it is important to remember that technological advancements in the aviation industry develop with each new aircraft that comes off the assembly line. Many materials that were once used to produce aircraft have since been discarded and replaced with newer, safer, more cost-effective ones. In ValuJet's case, aircraft 904VJ and most of the DC-9s in the fleet were wired with a material known as Poly Vinyl Chloride (PVC), which was first produced in 1950 and used for wiring only until 1979. Today this wire does not meet the specifications of the FARs; nor is it installed on any passenger aircraft. It was discontinued due to its major disadvantages. In terms of its negative characteristics, the wire is "heavy, thick, chafing, aging, flammable, [and it] burns readily, produces smoke, gasses, and hydrochloric acid" (Types of Wire 2006).

Going beyond the wire that was installed on the DC-9, ValuJet's N904VJ was known to many that worked on the aircraft as a mechanical disaster. Anonymous sources claimed that mechanics were constantly monitoring this aircraft for its continuous electrical problems. Interviews of the ValuJet mechanics by the NTSB revealed startling facts about the aircraft. In the six months prior to the accident, N904VJ "received six write-ups for partial electrical power losses" (Aviation Today 1998). It is also important to note that a major electrical failure was reported on the flight deck right before they reported smoke. A ValuJet mechanic also told CNN (1996) that, "hours before the ValuJet plane crashed, mechanics found an electrical problem in the cockpit area." The problem centered on a hydraulic pump that continually tripped a circuit breaker behind the pilot's seat, and it was discovered before the plane departed Atlanta for Miami on the flight prior to 592. Some reports show that the problem was not fixed, and it was noted that it would be fixed in the near future. After the crash investigators discovered that the circuit breaker area contained soot deposits (CNN 1996). According to the NTSB, the circuit breaker had been repaired in Atlanta. While en route from Atlanta to Miami, the public-address system stopped functioning but, once on the ground, it began to work again. Upon arrival at the gate in Miami, a SabreTech mechanic met the aircraft at the request of the captain to inspect the PA system. While working in the electronics bay under the cockpit, he found that the PA amplifier was loose in its mount, so he repaired it and made a log book entry.

After review of the CVR, the NTSB report determined that the aircraft pushed back from the gate at 13:42. The crew proceeded to run through their usual check lists and soon began to taxi. At 13:43 the electrical panel was set. After an Air Traffic Control (ATC) hold, the aircraft finally began its takeoff at 14:03, 20 minutes after it left the gate. At 14:07 the crew turned on the radar to review the weather. At 14:10 a noise was heard, and four seconds later the electrical bus went out (AirDisaster.com). In that one long minute the aircraft lost all its electrical power, and encountered smoke and fire in both the flight deck and the cabin. The CVR stopped recording 50 seconds before the aircraft impacted the ground due to the electrical failure (CNN 1996).

On May 14, 1996 an anonymous caller spoke with FAA inspector Jim Cole. The caller stated that he was maintenance personnel at ValuJet, and he had concerns about the maintenance that was being performed by the company. In his opinion the airline hid many things from the FAA, including the recycling of parts, shortage of wire, and the condition of several aircraft. He stated that "aircraft 901-940, and 930-940 needed to be 're-wired,' specifically number 904 [N904VJ]." The caller also said that "904 was delayed 40 minutes on its flight from Atlanta to Miami due to electrical problems." As a result, the mechanic bypassed two circuit breakers behind the captain's seat and said that the 904's wiring was notoriously bad

(Airliners.net 2002). He also referred to the fact that personnel did not like to work on 904 because any time they needed to move wires under the cockpit, the wires would break and short out. These problems also included PA amplifiers, fuel flow transmitters, and, on the day of the crash, wiring and circuit breakers shorting out. If the circuits really had been jumped, it would be impossible to shut off certain circuits individually. The witness alone had replaced six black box transmitters in the previous 30 days. He finally stated that many of the maintenance problems were not logged, and that the filing system used by ValuJet prevented the FAA from seeing the actual cause of the accident. (See the textbox for the potential connection with Swissair 111.)

SWISSAIR 111: A SIMILAR CAUSE?

The wiring in aircraft 904VJ is closely related to Kapton, which was determined to be a main factor in the Swissair 111 crash on September 2, 1998. Within one hour of its flight from New York to Geneva, Switzerland, a fire began in an area directly above the flight deck. When laboratory tests were performed on the Kapton wire, investigators discovered that once the wire shorted and caught fire, the insulation within the wire reacted like a stick of dynamite. After a four-year investigation, the Transportation Safety Board of Canada (TSB 2013) concluded that the fire probably started due to wire arcing that "ignited the flammable cover material on nearby metallized polyethylene terephthalate (MPET) covering on the thermal acoustic insulation blankets." The Kapton wires surrounding the unit ignited, and the fire quickly became uncontainable. While the flight crew did everything they could to try to extinguish the fire, severe smoke inhalation quickly overtook them. The aircraft was lost in the Atlantic Ocean just off the coast of Nova Scotia, and all 229 on board were killed. If the alternative theory regarding the cause of the ValuJet 592 crash is correct, then perhaps the Swissair tragedy could have been avoided.

Ethical considerations

In the case of the anonymous sources, many pieces of information support their theory. It is possible that as ValuJet quickly grew, the airline was forced to make rash decisions in relation to their expansion. This accident very well could have started from the improper monitoring of maintenance by ValuJet and its unethical methods with respect to human lives. It is possible that everything was normal on the aircraft since no problems were reported by any of the crew members during taxi or departure. At 14:10:03 the pilots heard a chirp followed by a beep over the PA/interphone channel. Four seconds after the noises were heard, they began to lose all electrical power. Even the pilots themselves stated at 14:10:15: "We got some electrical problem" (NTSB 1997: 170–171). The NTSB claimed that the noises heard by the pilots were those of the main tire exploding.

The possibility remains that the noises could have been due to a wire arc or some other type of electrical explosion. If in fact one or more of the circuit breakers were jumped prior to the flight, this type of scenario could have caused an extreme electrical failure. It could have taken place in the electrical bay located directly under the floor of the cockpit, reacting with the jumped circuits and cracked bundled wires which started a chain reaction. If the power put through the wire was too strong, it would have no circuit breaker to protect the wire and, as a result, this wire would most likely short and split in half. If the circuit was jumped, there would be no way to turn it off, and the live wire could have reacted with the flammable insulation and other wires that were installed on the aircraft. As the fire grew, it would have spread thick smoke created by the burning wires into the cabin and cockpit areas and begun to eat away at the cargo bin and control cables. The CVR soon shut off, and the DC-9 went into an uncontrolled descent toward the Everglades.

This theory greatly differs from that of the NTSB. More importantly, it now moves the blame from SabreTech completely on to ValuJet.

The NTSB report also contains some questionable issues. One is the claim that the sounds on the CVR are those of the main tire exploding. Aircraft tires have been known to be extremely durable and hard to puncture, which means that the fire would have needed to reach extreme temperatures. When this tire was recovered, it contained several rips and tears, including an X-shape tear, indicating that the tire exploded outwardly. It is quite possible that all of the damage to the tire occurred on impact. But if it did in fact rupture due to the fire, this supports both the oxygen generator and wire theories.

Of the 28 pieces of oxygen generator recovered, 9 had evidence of indentations on their percussion caps, showing they had been ignited. Some speculated that these occurred during the flight, but they could have been depleted when mechanics removed them from the other aircraft. In the NTSB laboratory tests, all of the oxygen generators were full. However, regarding the accident itself, it remains undetermined if all, some, or even any of the generators were full. Additionally, in the lab tests all pins were manually pulled to start the ignition of the generators. No tests were performed to dislodge the pins by shaking or rattling the canisters in the same way that they would have been in the forward cargo hold.

If this other theory is accurate, the ethical issue behind this accident directly relates to ValuJet's corporate responsibility rather than SabreTech, who became the scapegoat. During the NTSB investigation they interviewed a ValuJet Senior Vice President who served the company from June 1994 until he retired in February 1996, three months prior to the accident. He had been responsible for both flight operations and maintenance. In his interview he told Safety Board investigators:

> when he joined ValuJet, the air carrier was operating 13 airplanes. He said that although he had expected the company to be running smoothly, when he began his new job he found a number of discrepancies, included maintenance records that were "not in great shape" and "lots of sloppiness due to rapid growth." (NTSB 1997: 58–59)

Being a fast-growing airline, ValuJet's management may have tried to cut costs in any way possible; this includes not buying parts that were necessary for the safety of their fleet.

ValuJet would have known that their 30-year-old airplanes had dangerous wiring, and a more recent incident supports this point. After re-emerging in 1997 as AirTran Airways, one of the previous ValuJet DC-9s had an inflight fire on August 8, 2000. "The cabin crew reported dense smoke and could barely see each other. . . . They also saw smoke and sparks coming from the area in front of the flight attendant's jump seat" (King 2000: 1). No one was hurt, and the plane landed safely. The fire was caused by a bad relay and wire arcing; five other bad relays were discovered in other DC-9 aircraft.

The crash of ValuJet 592 was both devastating and eye-opening to many in the aviation industry. The rapid growth of ValuJet came to a halt, and their reputation was so damaged that they were forced to merge with a much smaller regional airline, AirWays Corporation (the holding company for AirTran), to change the now tarnished name they once wore with pride. While ValuJet had basked in the glory of success, they soon realized their own demise after the 1996 crash. With airline expansion there arises great responsibility regarding air safety. Many times, an airline can be torn between ensuring that its aircraft are safe and putting them in the air to make a profit. In ValuJet's case they wanted to be successful; but, in the crash of 592, safety was lost in the balance.

In the end it is irrelevant whether the true cause of the accident was due to the oxygen canisters or faulty wiring, for, ultimately, the true responsibility falls on ValuJet's management. They had an ethical responsibility to ensure that their success depended not only on themselves but also on any person or company affected by their decisions. On that sunny Florida afternoon ethical responsibility was left by the way side; and, as a result, ValuJet may never be remembered for their success, but only for their failure.

While the next two chapters deal with more recent aviation accidents, the 1996 ValuJet crash continues to serve as a reminder that airline management needs to value safety as the number one priority. While the purpose of a business is to make a profit, if management cuts corners, such as in the case of ValuJet 592, then both profitability and public safety are compromised.

References

AirDisaster.com. Cockpit voice recorder: ValuJet, 1997–2006. Available at: www.airdisaster.com/cvr/vj592tr.htm [accessed: April 4, 2006].

Airliners.net. 2002. Request to re-open Flight 592 investigation. Available at: www.airliners.net/forum/viewtopic.php?t=156689 [accessed: July 11, 2018].

Aviation Safety Network. 2006. Accident description. Available at: http://aviation-safety.net/database/record.php?id=19960511-0 [accessed: April 4, 2006].

Aviation Today. 1998. Expanded inspections needed to assess wiring woes, experts say. Available at: www.aviationtoday.com/reports/wiring7.htm [accessed: April 4, 2006].

CNN. 1996. Divers going into crater caused by DC-9 crash. Available at: www.cnn.com/US/9605/18/valujet/index.html [accessed: April 4, 2006].

HowStuffWorks. 2000. How does an oxygen canister on an airplane work? How can heat generate oxygen? Available at: http://science.howstuffworks.com/transport/flight/modern/question258.htm [accessed: August 13, 2010].

King, J. 2000. Another AirTran/ValuJet DC-9 in-flight fire. Available at: www.flight592.com/Flight592discussion-current/_disc10/0000002c.htm [accessed: April 4, 2006].

Morley, D. 2015. Throwback Thursday in aviation history: ValuJet. *Airline Geeks.com*. Available at: https://airlinegeeks.com/2015/11/05/tbt-throwback-thursday-in-aviation-history-valujet/ [accessed March 8, 2018].

National Transportation Safety Board. 1997. *In-Flight Fire and Impact with Terrain: Valujet Airlines Flight 592, DC-9-32, N904VJ, Everglades, Near Miami, Florida, May 11, 1996.* Aircraft Accident Report NTSB/AAR-97/06. PB97-910406. Washington DC: NTSB.

Transportation Safety Board of Canada (TSB). 2013. *Swissair 111 Investigation Report: Executive Summary*. Available at: www.tsb.gc.ca/eng/medias-media/fiches-facts/A98H0003/sum_a98h0003.asp [accessed February 11, 2017].

Types of Aircraft Wire. 2006. Available at: www.cadmus.ca/wiretypes.htm [accessed: April 4, 2006].

12

THE SOCIOTECHNICAL COCKPIT

Ethical dimensions of piloting, planes, and programming

Eric B. Kennedy

At 35,000 feet above the Atlantic Ocean the captain of an Airbus A330-203 had just handed off pilot flying duties to one of his co-pilots. The June 1, 2009 flight had thus far been quite normal, with some light turbulence after departure from Rio de Janeiro, Brazil, typical storms brewing around the inter-tropical convergence zone, and slightly warmer than usual conditions preventing the heavily laden aircraft from climbing to its planned altitude of 37,000 feet. The pilot's decision to retire for a rest was similarly typical, as a requirement given the 13-hour "duty time" planned for the flight. Within 15 minutes of the captain's departure from the cockpit, however, the aircraft would plummet into the choppy ocean below.

The 2009 crash of Air France Flight 447, operating from Rio de Janeiro to Paris, captivated public attention. Part of this allure could be found within the drama of the recovery process—the tense wait until wreckage was first discovered, the continuous media coverage, the visceral image of the tail being retrieved, and the fear and grief conjured in the public's imagination. Yet, the intrigue also centered on the perceived impossibility of the crash: How could a perfectly functional aircraft, flying through typical weather for the region, so quickly kill 228 people?

Early in the discussions that followed among pilots, the media, and casual observers, the issue of automation entered the forefront. How much control and trust should we put into computers that fly aircraft? What kinds of interfaces should—and should not—exist between pilots and their control of the aircraft? And, in perhaps the most distracting line of questioning, is there something inherently less safe (or preferable) about Airbus aircraft relative to Boeing designs?

Broadly speaking these questions could not be more important. Indeed, the incredible safety of civilian air travel in today's world is thanks to the explicit analyzing, testing, debating, and re-examining of emerging and existing technologies alike. Yet, three pieces of context are essential when considering these kinds of ethical issues in aviation:

- Although these questions are often viewed through a technocentric lens, the intersection of technology and aviation safety is intrinsically a *sociotechnical* space, with human, social, economic, and technoscientific dimensions.
- While questions proximate to the cause of accidents are popular conversation topics, they represent only a narrow swath of the broad and important ethical issues that arise in the cockpit.
- Although hindsight makes criticism of the system remarkably easy, particular practices in aviation—such as transparent investigation of failures, iterative learning, and ongoing correction—also offer a fertile set of lessons for other fields that grapple with similarly weighty ethical questions.

The study of aviation by practitioners, industry, academics, and outsiders alike has yielded a rich body of literature that considers issues with major ethical questions. Accordingly, this chapter does not set out to deliver an exhaustive treatment of any of these issues, or offer conclusive guidance on their possible resolutions. Rather, it intends to provide a high-level sketch of some of the vexing and complex problems found in aircraft cockpits, work that has been done on these challenges, and topics fertile for future examination. In addition it aims to highlight two sets of ethical questions that arise in considering these issues—questions related to the specifics of particular sociotechnical systems and their implementations, as well as a series of more open and thought—provoking quandaries about the general deferral in aviation towards pragmatic solutions, technological changes, and systems improvements versus explicitly ethical deliberations.

We begin by exploring the emergence of the "glass cockpit" and the ethical questions that arise with emerging and changing sociotechnical systems. While Air France 447 is often a focal point of such conversations, we argue that its common interpretation (loosely characterized by the question "can pilots still fly?") fails to do justice to the complexity of such accidents and ethics. We then turn towards a host of broader ethical questions in the aircraft cockpit that often fail to garner the discussion they deserve, including three issues worthy of significantly more consideration in today's world of aviation: the tradeoffs inherent within security, inclusivity, and reliability. Finally, we take a broader view of aviation ethics to examine the systems used to address failure and improve outcomes. We argue that these institutions and processes function effectively as collective metacognitive practice, and offer both an explanation for some of aviation's successes and an opportunity for improving ethical practice in other fields.

The glass cockpit: or, "can pilots still fly?"

As with nearly all aviation accidents, the cause of the crash of Air France 447 was not a singular event, but rather a combination of factors, misunderstandings, and cascading circumstances. According to the official incident report by the Bureau d'Enquêtes et d'Analyses (BEA 2012: 200), a series of six events resulted in the accident:

- Temporary inconsistency between the airspeed measurements, likely following the obstruction of the Pitot probes by ice crystals that, in particular, caused the autopilot disconnection and the reconfiguration to alternate law,
- Inappropriate control inputs that destabilized the flight path,
- The lack of any link by the crew between the loss of indicated speeds called out and the appropriate procedure,
- The late identification by the PNF [pilot not flying] of the deviation from the flight path and the insufficient correction applied by the [pilot flying] PF,
- The crew not identifying the approach to stall, their lack of immediate response, and the exit from the flight envelope,
- The crew's failure to diagnose the stall situation and consequently a lack of inputs that would have made it possible to recover from it.

These six events tell the rough story of what happened in the cockpit of AF 447 as it passed through the inter-tropical convergence zone (ICTZ).[1] A failure that was initially mechanical—the blocking of the Pitot tubes because of the formation of ice crystals—rapidly incorporated human factors like disorientation, confusion, and differing actions among the crew members.

Yet even before the acute portion of the accident, human and social dimensions were significant. The BEA report, for instance, highlights an earlier conversation between the PF and the captain around possible responses to the ICTZ (such as climbing to a different altitude or adjusting course). According to the report, however, the captain "neither expressed nor explained his position clearly" to a PF whose "anxiety was noticeable" about the conditions ahead (BEA 2012: 168). In the moments that followed, it was unclear if the captain made appropriate decisions about when to take his relief break, or whether sufficient thought had been given to the selection and transition to a new designated relief captain (BEA 2012: 170).

The notion of human factors as being significant to aviation accidents is hardly novel. In both civil and military contexts there is significant incentive to reduce error rates and costly failures. While the number of accidents attributable only to mechanical causes has decreased drastically since the advent of aviation, such improvements have yet to be made to a similar degree with respect to human error (Shappell and Wiegmann 1996). As a result, for instance, the National Transportation Safety Board (NTSB) established a Human Performance Division in the early 1980s to broaden their investigations from a more technocentric approach (Sumwalt 2014).

In NTSB investigations the investigators consider three concentric elements of human performance: the individuals involved (whose failures are more proximate to a particular accident); the organizational influence/corporate culture; and the regulatory influence (the latter two of which are more upstream causes of failures). As such, ethical questions around the responsible performance of duties, the establishment and improvement of appropriate training protocols, and the development of reliable practices at each level are essential. The evolution of pragmatic responses to these challenges can be readily seen, such as: the emergence of and adherence to checklists for nearly every cockpit function (Gawande 2007); extensive requirements around training hours both in simulators and in aircraft; and an emphasis on the importance of so-called "safety culture" among airlines, operators, and maintainers (Sumwalt 2009).

It is also useful, both for pragmatic purposes and ethical considerations, to contemplate a slightly more granular taxonomy of possible failure points for these human factors. One method is through the Human Factors Analysis and Classification System (HFCS), which was developed and tested in for the US military to examine the causes of aviation accidents (Wiegmann and Shappell 2001). This taxonomy considers four levels of failure points: unsafe acts by individual actors; the preconditions that led to those errors; the role of supervision; and higher-level influences from the overarching organization. For individuals, failure comes in two flavors: errors, whether in a mistaken decision, inaccurate perception, or skill gap; and known violations, whether routine or exceptional. These can arise because of two kinds of preconditions: either substandard conditions, such as an adverse mental state, psychological state, or other limitations; or substandard practices, like mismanaging resources or personal readiness. Supervisors can also play a role in errors, whether through planning inappropriate actions, inadequate supervision, failure to correct problems, or violations in their supervisory duties. Finally, much like the final concentric stage of the NTSB model, organizational climate, processes, or management can also lead to failures. Not only did Wiegmann and Shappell find that such a framework provided sufficient categorization to capture the many different civilian aviation incidents that the NTSB investigates, but the framework has also been applied in other contexts, such as aircraft maintenance (Schmidt et al. n.d.).

These sorts of frameworks provide at least two levels of ethical questions well worth pondering—and which will remain laden in the case studies explored through the remainder of this chapter. At a pragmatic level, it behooves us to think clearly about the sorts of ethical frameworks that undergird our solutions to each of these potential failure points. Should interventions be designed that lead to the greatest reduction in the most costly failures (whether calculated by hull-losses, fatalities, or something else) regardless of their impacts on the pilots? Or, are there different principles—be they neo-Kantian, procession, and so on—that should guide interventions? What do we make of trade-offs? Just how much are we willing to spend on each such interventions? And, how should we deal with the reality—like we will discuss mid-way through this chapter—that ethically sound interventions in one location (such as hard-boxing the cockpit from passengers) may well open up a suite of negative possibilities in other contexts (for example, the 2015 Germanwings accident)?

However, there is also a second, more meta-reflective question that arises. In many of these interventions, an ethical standpoint is not necessarily considered in day-to-day operations. Although the NTSB has included many new forms of disciplinary expertise since its inception, including particularly on the human factors side, its reports remain pragmatically oriented. For regulators, airlines, and passengers alike ethical questions are likely viewed as rather tangential to their principal goal: safe operations, where everyone walks away to fly another day. But from an ethical standpoint does this represent a problematic dearth of reflexivity and moral consideration? Would the presence of an ethicist in NTSB investigative activities and cockpit design operations enhance a report? Or, does the cockpit represent a place where ethical considerations are—at least, by and large—sufficiently agreed upon tacitly in which a utilitarian standpoint provides sufficient functional guidance in design and decision-making? And, if so, at what stage should more explicit conversations begin to arise regarding crew welfare, inclusivity, and previously unconsidered trade-offs?

Pragmatic approaches to safe cockpit practices: the advent of crew resource management

To grapple with this second suite of questions, we need to consider crew resource management (CRM). We will do this in two ways. First, here, we will briefly explore the ways in which pragmatic considerations have provided incredibly fruitful ways forward in enhancing aviation safety. Then, later in the chapter, we will consider an intersecting issue—the inclusivity of cockpit design and practices—that complicates this narrative of success, and at the very least suggests that the failure to ask particular questions (namely, who is included and who is excluded by de facto norms) has limited the potential improvements attained.

On December 28, 1978 United Airlines Flight 173 was completing a flight from New York's John F. Kennedy Airport to Portland, Oregon via Denver, Colorado. As the aircraft descended through 8,000 ft., the first officer requested that the landing gear be lowered. According to the captain, however, the gear seemed "to go down more rapidly" in an unusual manner, with a different "thump" than normal, and without the normal gear door lights. The first officer also remarked that the aircraft "yawed to the right" upon the landing gear being lowered (NTSB 1979: 2, 3).

Over the next hour, the flight crew focused on troubleshooting the landing gear problem. The aircraft held, circling, at 5,000 ft., while the crew worked various potentially troubled systems, dealt with other traffic in the area, and considered possible outcomes like a crash landing. At 18:13 local time, however, the flight engineer reported a remarkably different problem: that two of their four engines had flamed out. Within 20 seconds, the first officer declared a mayday with all four engines flaming out. Within 2 minutes the aircraft crashed into a wooded area only about 6 miles short of the airport where the crew had intended to land.

The findings of United 173 would prove to be foundational in the aviation industry. According to the NTSB report, the primary cause of the crash was not the problems with the landing gear. Rather, it was the failure of the flight crew to monitor the fuel status of the aircraft until its ultimate exhaustion over an hour after the planned landing time (NTSB 1979: 24). In the words of the NTSB:

> The Safety Board believes that this accident exemplifies a recurring problem—a breakdown in cockpit management and teamwork during a situation involving malfunctions of aircraft systems in flight. To combat this problem, responsibilities must be divided among members of the flight crew while a malfunction is being resolved. In this case, apparently *no* one was specifically delegated to the responsibility of monitoring fuel state. (NTSB 1979: 26, emphasis original)

The crash of United 173 would lead to the 1981 implementation of Crew Resource Management (CRM) at United Airlines, an advent that other major airlines would adopt within the decade. A pragmatically oriented solution, CRM training offers behavioral strategies "as error countermeasures that are employed to avoid error, to trap errors committed, and to mitigate the consequences of error" (Helmreich et al. 1999). CRM offered a different perspective on accidents from the prevailing view. Although it is clear that accidents can have both technical (for example, the failure of a mechanism or system) and human causes (for instance, failure to perform up to the expected skill level), CRM suggests that failures by pilots cannot be adequately explained through a single action or non-action.[2] Rather, they must be situated in the context of larger systems—be they human systems, like the failure of Flight 173's crew to collectively monitor the fuel status, or sociotechnical interfaces, like the NTSB's recommendations following the accident for clearer and more easily read gauges.

Cockpit technologies: can pilots still fly?

One relatively common point of media discussion following the crash of Air France 447 was the comparative differences in the design of Boeing and Airbus aircraft. *The Telegraph*, for instance, ran a leader reading: "With the report into the tragedy of Air France 447 due next month, Airbus's 'brilliant' aircraft design may have contributed to one of the world's worst aviation disasters and the deaths of all 228 onboard" (Ross and Tweedie 2012). For many, the design of the Airbus itself was complicit not only in the crash of the A330 into the Atlantic, but also in the broader eroding of pilot skills around the globe.

At a basic level, Airbus and Boeing each espouse a different philosophical approach when it comes to the relationship between the pilot and the aircraft. Airbus aircraft are designed with a "Flight Protection Envelope" at the forefront. This envelope provides so-called "hard protections" against pilot maneuvers deemed unsafe by the flight computers, thereby overriding the pilot (Young et al.

2007). A pilot who aggressively moves the stick to the right, for instance, will find the aircraft responds only with a safe bank, rather than an unsafe angle (Traufetter 2009). By contrast, Boeing aircraft generally provide "soft protection" where automation provides an aid to pilots, who ultimately retain comparatively more control over the suite of possible actions (Young et al. 2007).

Several other differences exist between the technologies present in each cockpit. Airbus aircraft, for instance, are controlled with side sticks located beside the pilots' seats (loosely resembling a computer joystick), whereas Boeing aircraft generally have physically linked yokes more evocative of traditional aircraft cockpits. The side sticks provide a host of advantages, including weight (and corresponding fuel) savings, fewer moving components, less physical exertion required for its use, and the ability to let go of the stick and allow the turn or other maneuver to be continued (see Tarnowski 2003 for further discussion). Yet, it was precisely this separation of control between the pilots—when one adjusts the stick, the other stick remains stationary—that led to the Air France crash. As the PF pulled the stick back, forcing the plane into a stall, the other crew in the cockpit did not recognize the erroneous input being entered.

The Airbus cockpit also features a series of differing flight laws under which the flight computer can operate. This too would prove contributory in the Air France crash. When the Pitot tubes iced over, the computer lost its airspeed data. This not only resulted in the disconnection of the autopilot, but also switched the computer into one of its "alternate laws." In the alternate law fewer computer controls—the hard protections referred to earlier—are exerted over the pilot's actions. Simply put, when the computer is operating in alternate law, it becomes much easier for a pilot to stall the airplane, precisely what happened when the side stick was held back.

These comparisons lead relatively naturally to the question of whether the Airbus or Boeing design philosophies are safer in practice. Yet the evidence is relatively equal and inconclusive. One analysis (Young et al. 2007) suggests that Airbus aircraft do indeed experience a larger absolute number of automation-related crashes than Boeing aircraft do, at a roughly 2:1 ratio (note that the study did not account for the relative differences in use levels of either brand). Yet, Boeing aircraft were just under twice as susceptible over the same time period to pilot-related accidents, suggesting a counterpoint of sorts.

Boeing's own numbers (2014: 19) suggest a relatively neutral breakdown as well. Compare the Airbus A318/319/320/321 series of aircraft against Boeing's 737 (300–900) series, for instance. Between their introduction and 2013, the Airbus fleet had a hull loss rate of 0.24 per million departures, with 0.14 involving fatalities. By contrast, Boeing's 737 series had 0.81 hull loses per million departures, with 0.37 of those involving fatalities. Yet, these numbers also are poorly controlled: While the A318/319/320/321 series is broadly comparable to the 737 series, for instance, sometimes the A321 is used for a role that is more similar to a Boeing 757. Nor do these statistics account for operator differences in types flown between North America, Europe, and elsewhere (though the report does involve a parallel breakdown by geography that is worth reading). And, it conceals one of the most important variables involved: the age of the particular models. The hull loss rate of the more modern 737-600/700/800/900 is just half of the hull loss rate of the previous 300/400/500 fleet.

The statistics above did not include the very early 737-100/200, which had a hull loss rate over five times higher than the current versions. Likewise, the earlier A310's hull loss rate is nearly ten times the current Airbus series considered above. Airbus versus Boeing safety comparisons—poorly evidenced to begin with—disappear when compared to the within-manufacturer variation achieved by iterative improvement of both computer and pilot-first approaches.

Despite their potentially limited impact on real-world outcomes, these contrasting design philosophies and other orientations towards technology present among manufacturers, airlines, and operators a set of interesting and contentious ethical issues in their wake. If per the Boeing data Airbus aircraft do indeed see an ever so slightly lower hull loss accident rate versus Boeing equipment, what are we to make of the fact that those accidents are comprised of a higher fraction of automation-related incidents (per Young et al. 2007)? Should we see accidents caused by automation failures (and resulting pilot failures to remedy the situation) ethically different than those that were initially triggered by a pilot? If, at least hypothetically, fewer accidents resulted from more strongly automated systems, is a decrease in the overall fatality rate to be tempered by the knowledge that some of the new victims would not have lost their lives under a less automated regime? And, finally, what if some of those victims of automation are actually victims of more malicious activities, such as computer hacking?[3]

These questions do not have straightforward answers. Moreover, while airlines have been at the vanguard of many of these debates, the ethical questions of automation are reaching far beyond the cockpit and into each of our lives. Parallel questions to those above exist in an increasing number of spheres, including everything from the vulnerability of so-called "smart cities" and "smart" or "connected" homes to outside hacking (or even malicious manufacturer algorithm manipulation) to the ethics of self-driving cars. Indeed, for those interested in how some of these ethical considerations are being grappled with in comparable technological systems, consider the significant research and thinking being done with respect to the ethics of autonomous cars (for example, Goodall 2014; Hevelke and Nida-Rümelin 2014; Lin 2013; Dresner and Stone 2007) and autonomous weapons systems (see for instance, ICRC 2014; Sparrow 2009; Arkin 2008; Lin et al. 2008).

To this point, we have focused on ethics relating to the sociotechnical interfaces in an increasingly glass cockpit. While iterative improvement from hard lessons learned in previous crashes has led to significantly better CRM and human factors research, the ever-increasing potential for automation of flying duties raises new questions and concerns. Yet, these technologically focused questions are only the tip of the iceberg of ethical questions that concern both pilots and passengers alike. Before the end of this chapter, therefore, we turn a brief eye towards two other aspects: a broader set of ethical questions that must be considered as part of this sociotechnical interface; and the value of creating systems and institutions of reflection and iterative improvement.

The importance of broadening ethical considerations

As many aircraft accidents over the past decade have demonstrated, many factors other than flight control play essential roles in safe aviating. Tragic accidents have occurred because of hijackers and suicide alike, performance-based pressures, the actions of ground personnel, and many other factors. It behooves us, therefore, briefly to consider some of the other issues that ought to be front and center in any such reflections on ethics and the commercial cockpit. We will examine four factors that warrant such attention: the tradeoffs present in security decisions; the inclusivity of airlines and cockpits; conflicting pressures placed on pilots and airlines; and the people involved in making the aviation system work.

When increasing security increases other vulnerabilities

On Tuesday, March 24, 2015, the tradeoffs present in cockpit security became all too vivid. An A320 operating for Germanwings (a Lufthansa subsidiary) took off from Barcelona (Spain) with 144 passengers, four flight attendants, and two pilots. The plane reached its planned 38,000 feet 27 minutes after takeoff. With everything being normal on this short flight back to Düsseldorf, the captain left the cockpit, leaving the co-pilot in charge. Only 30 seconds after the captain had shut the cockpit door, the autopilot was set from its intended 38,000 feet to a mere 100 feet—far too low for flying over the continent, let alone the mountainous terrain below.[4] The audio recording recovered from the black box tells a chilling story, with sounds of (presumably) the captain knocking on the door, asking to be let back in, and eventually launching violent blows against the hardened, locked surface (BEA 2015).

In a cruel irony, the very system that prevented the captain from returning to the cockpit had been implemented for the purposes of keeping aircraft safe. Hardened cockpit doors—first via a closed door, then increased locking mechanisms, then significantly more robust door construction to defend against bullets and other attacks—arose in response to the threat posed by a malicious few intending to hijack aircraft. The door system is quite complex, and includes an escape hatch, several cameras allowing pilots to see who is attempting to enter, and, ultimately, cockpit-only control over who is able to enter. Even the keypad system, designed to allow the cabin crew to access the cockpit in the case of flight crew incapacitation, ultimately provides anyone in the cockpit a 15-second window in which to override the access (BEA 2015: 14–18).

Significantly more attention has been paid toward the mental health of pilots since the accident. The British Civil Aviation Authority (CAA) acknowledged that it had grounded 350 pilots over the previous five years for mental health-related reasons (Matthew 2015). Mental health is one of the many challenging tradeoffs brought to light by the Germanwings crash. To what extent, for instance, does instituting protective measure for pilots with mental health challenges (for example, grounding or license suspensions) risk decreasing the diagnosis or reporting of such issues, or increasing stigmatization by pilots flying together? How can future vulnerabilities be anticipated in light of security measures like hardened cockpits? What other technological improvements—such as increasing automation or even increased real-time reporting of location and flight data (as many have called for in the wake of the disappearance of Malaysia Airlines Flight 370 in 2014)—risk opening up unanticipated vulnerabilities and risks? And, much like questions of automation, how should we deal with the ethics of a few new accidents that might occur because of technologies, even if others are prevented?

Cockpit inclusivity

Particularly after the terrorist attacks of September 11, 2001 in the US, the connection between race and flying became evident for passengers. For many this is tied to the "widespread increase" in discriminatory behaviors faced by passengers who belong to a visible minority (see, for example, Chandrasekhar 2003). Questions of race, gender, and other discrimination, however, are also prominent for pilots as well, a topic that will be addressed more fully in Part IV.

One of the key features of effective CRM, as discussed earlier, is the ability of anyone in the cockpit to raise concerns and speak effectively. Yet, significant literature has demonstrated that pilots of different status (for instance, captains vs. lower status) tend to initiate communications at different frequencies while serving as PF (Milanovich et al. 1998). This is particularly problematic if coupled with other status factors such as race or gender.

While taking great pains to emphasize the progress that has been made by African American professionals, an in-depth ethnography with pilots of color indicated that there remain reasons to be concerned. These pilots reported racial comments being made by training personnel and other pilots, and even a case of a passenger leaving the plane because it was being "flown by a black person" (Evans and Feagin 2012: 656). This led, in the experience of the interviewees, to a state where any pilot of color had to suppress their emotions to avoid being perceived as an "angry black man" (Evans and Feagin 2012: 657). These pilots faced scrutiny for actions based not on their decisions but on their race, or feeling as though they could get written up for behavior that other pilots would not have to worry about (Evans and Feagin 2012: 658–659). In 2014 a female WestJet pilot was the subject of a note left behind on the aircraft by a passenger, stating that "the cockpit of an airliner is *no* place for a woman" (Sathiyanathan 2014, emphasis original). The passenger had apparently asked flight attendants before takeoff whether the female pilot, with 17 years of experience, had completed sufficient flight hours.

As discussed more fully in Part IV, these gendered and racialized roles have a long history. Historical studies, like one of British Airways' long journeys toward a more equal airline, can offer insight into both the degree of the challenge (for example, it took 68 years before BA hired its first female pilot) and some of its origins: as examples, the construction of heroic masculine norms for pilots; the carryover of wartime symbolism; and workplace and recruitment practices (Mills 1998). More contemporary ethnographic studies shed light on the current extent of the problem (for example, in 2007, only 524 of the 10,735 pilots in the UK were women), the role of management in changing these practices, and the way that cultural sexism can be manifest in the cockpit (Neal-Smith and Cockburn 2009; Davey and Davidson 2000).

As much as there is a risk of seeing aircraft accidents as simply technical in nature, it is easy for one to view the factors of inclusivity as social questions. Yet, these too ought to be considered as sociotechnical ethical issues. Take, for instance, the matter of the physical cockpit layout and design bias. As women tend to typically have smaller limbs, torso lengths, and lower upper body strength, certain physical designs can restrict how well different bodies are able to perform certain tasks. While design specifications and contracting processes have led to slightly better performance in military cockpits than civilian cockpits (Weber 1997), problems remain in both. A study of one type of military helicopter, for instance, found that seat design led to challenges in some smaller females attaining an adequate over-the-nose field of view and sufficient reach to all critical controls (Gordon 2002: 6, 9).

Taken together, these findings compel a serious amount of reflection: For whom should cockpits be designed? How should we respond when some are excluded? And, what are the practices—both technologically and socially—that lead to inclusion or exclusion and can remedy the challenges faced today?

Faster, further, and safer

As air travel becomes increasingly commonplace, societal expectations for its performance also rise. Passengers expect on-time performance, even during adverse weather conditions. Airlines are incentivized to cut costs whenever possible, both for the sake of profit and because customers demand ever lower fares. Airports also feel the pressure to be open as often as possible, perform in as environmentally friendly a manner as possible (for instance, new continual decent profiles for approach), and squeeze in as many landings as possible.

> For the pilots of American Airlines Flight 1420, on June 1, 1999, the day was a relatively typical sequence. They departed just before noon from Chicago for Salt Lake City, Utah. Their second flight of the day took them onwards to Dallas/Fort Worth, with a 39-minute airborne hold on the arrival because of adverse weather. The final flight of the day, delayed by

> over two hours, would take their McDonnell Douglas DC-9 onwards to Little Rock, Arkansas. Weather warnings suggested that there would be severe thunderstorms along the route and including at Little Rock upon arrival. Just after takeoff, the airline flight dispatcher sent the flight crew an ACARS message indicating that the weather around Little Rock might be a factor during the arrival. The dispatcher suggested that the flight crew expedite the arrival to beat the thunderstorms if possible, and the flight crew acknowledged this message. (NTSB 2001: 2)

The cockpit voice recorder captured much of the conversation, including the captain's desire to "get over there quick" and expedite the approach, and "get down as soon as we can" (NTSB 2001: 2, 185). Although the mood on the flight deck was light at times, there was also significant discussion about the crosswind speeds being very close to the limits—though with significant uncertainty about what the actual limit was, and how it varied on a wet runway. As the approach continued, there was confusion about the location of the airport (which was lost in the clouds), the leg of the approach they were on, whether the landing could be maintained, and, eventually, about being way off course (NTSB 2001: 212). Ultimately the plane would overrun the end of the runway, killing 11 of its passengers and injuring several more. The NTSB attributed the crash to the failure to discontinue the approach when thunderstorms became too severe for landing, and the failure to ensure spoilers had extended after touchdown.

Transportation safety boards take significant pains to ensure that their findings are in the service of improving aviation safety and reducing future accident potential, rather than an exercise in placing blame. Yet, in reading between the lines of the official findings, questions arose concerning why the pilots chose to continue an approach that they knew was abutting limits and dangerous given the weather conditions. How then should airlines, aircraft designers, and pilots balance competing priorities of safety, of on time/as planned performance, of environmental aims (like continual descent, fuel minimization, or reduced engine running time)?

One additional brief example can be found in the downing of Malaysia Airlines Flight 17 over Ukraine in 2014. Best evidence suggests that the Boeing 777 was shot down by "a large number of high energy objects from outside the aircraft" (DSB 2014: 30), likely in an action related to ground conflict below. After the aircraft was lost, the practice of flying over combat zones received significant attention from the International Civil Aviation Organization (ICAO 2014) and others. So, when should a route be moved away from a potential threat? Many major carriers joined Malaysia Airlines in flying the route, which was not affected by a Notice to Airmen at that altitude or advised against by Ukraine or ICAO (Freed 2014). The more direct route offered favorable fuel burn for the airline, as well as overflight fees for Ukraine—the country responsible for determining whether the route was safe to fly in the first place (Halsey 2014).

HELIOS AIRWAYS FLIGHT 522: AVIATION AS A SOCIOTECHNICAL SYSTEM

Early on the morning of August 14, 2005, two Hellenic Air Force F-16 fighters intercepted a Boeing 737 as it flew in circles around the KEA VOR holding pattern near Athens International Airport in Greece. One of the F-16 pilots positioned his fighter close to the cockpit, observing someone slumped over the controls. Further back in the cabin, oxygen masks were dangling, and three passengers were spotted wearing the masks. Only a minute before the left engine flamed out from lack of fuel, and the plane began descending, the F-16 pilot saw someone enter the cockpit without an oxygen mask; 13 minutes later, the aircraft crashed into hilly terrain northwest of the airport where it was supposed to land.

Helios Airways Flight 522 crashed for a relatively straightforward reason: The flight crew failed to recognize that the pressurization system was in manual rather than automatic mode at takeoff. When alarms sounded later, they also failed to interpret the problem correctly, continuing onwards to the intended cruising altitude, where the air was much too thin to afford survival. Yet, it was actually much earlier—during unscheduled maintenance—that the switch had been put into the incorrect mode (AAIASB 2006). What occurs on the ground during maintenance, as in the case of Helios 522, can prove fatal in the air. Thus, it is important to keep in mind the broader sociotechnical implications of this accident.

Concluding thoughts: ethical learning and creating reflexive systems

We began this chapter by highlighting that ethical questions on the flight deck related to technology are in fact sociotechnical questions. There is a corresponding lesson at a broader level, however: not only are aircraft highly sociotechnical, but they are also part of massive systems. For instance, what kind of cargo is packed and how it is secured can prove costly—as the directive against transport of large volumes of lithium ion batteries suggests (Pasztor 2015)—and has become all too obvious, as when the shifting cargo load on board National Airlines Flight 102 (a 747 departing Bagram, Afghanistan) caused a fatal stall just after takeoff (Hradecky 2013). Furthermore, there is concern about the low level of wages for regional and entry pilots, and potential safety implications that could hypothetically result (Schaal 2013).

Ethical issues on the flight deck may at first appear to be largely technical in nature. Indeed technology—whether increasing automation or differing design philosophies—is central to both many of the accidents that have occurred in the past and the many ethical questions that arise as we go forward. Yet they are also highly sociotechnical ethical questions that become rich for exploration only when we consider the intersection of humans and technology, and when we can situate these sociotechnical interfaces in their larger systems context. But, given the litany of ethical issues that arise and are often resolved through pragmatic fixes rather than explicit ethical reflections, what lessons can we take from the world of aviation?

One of the striking things when studying aviation accidents is just how low the rates of hull losses and fatalities are, especially when compared to the early days of aviation. Flying remains one of the safest modes of transportation available, alongside other public options like buses and trains. There have been many contributors to this incredible reduction in accident rates, including everything from improved understanding of the fundamentals of aircraft materials to technological advancements in air traffic control.

It is not an accident that this chapter has heavily referenced investigative reports by transportation safety boards in the US, France, and elsewhere. These documents offer thorough accounts of the particular events that happened, possible causal pathways, and a probable set of causes behind the accident. They also, however, offer a unique locus of reflection around the more distal causes and conditions behind these acute events, and possible pathways for iteration and improvement in technological, human, and institutional systems. The reports certainly are not without their faults—allegations of mistaken causes, questionable interests, and ineffective implementation of recommendations abound; but, in many ways they offer a unique example of the role of reflective, empirically driven institutions in helping deal with life and death challenges in an evolving field.

As such it is not terribly surprising that they have many potential parallels with good ethical inquiry. The reports involve experts and perspectives from as many vantage points as possible. Rather than being fixed on finding particular problems (for instance, human vs. technological causes), the investigations involve a process of data-driven inquiry towards a complex reality. They function with deeply incomplete information, but simultaneously iterate themselves on how better information might be gathered in the future, before, during, and after accidents. Finally, the reports rely on legitimacy garnered through the transparent publication of findings and obtain credibility through attempting to interpret value-laden data and theories in as objective a manner possible.

It is fitting, then, to conclude the chapter by returning to the two layers of ethical questions that were raised earlier. Many of the quandaries we have considered here involve very real, tangible decisions about how particular technologies and systems should be shaped. How much power should be given to autonomous systems? Do systems that disconnect the pilot from the experience of flying offer benefits that outweigh costs, or vice versa? And how ought we to construct our sociotechnical systems so as to appropriately address tradeoffs, be they in inclusivity, security, reliable performance, or the many different components important to the system itself?

We also, however, flagged the potential for a more "meta" layer of ethical questioning. What should we make of processes, like accident investigation, that explicitly avoid dealing with moral questions like those of blame or culpability, but focus instead on solutions and fixes? Is it appropriate, admirable, or questionable to have the core functions of investigation and improvement be processes that do not necessarily engage explicitly with these ethical questions? How should we interpret this lack of explicit ethical conversation when the safety improvements have been so great in its occasional absence?

Ethical issues are deeply intertwined with the flight deck. They are inherent to sociotechnical systems because of both their sociotechnical and systems natures. It is through both their examination—and the process of thinking about how, when, where, and why we think about ethics—that we can not only engage with ethical issues in aviation, but also apply the many lessons learned in the skies to other emerging systems and technologies.

Notes

1 Several authors and organizations have written significantly fuller accounts of the accident. The BEA report (2012) is a particularly thorough—and official—account, while Palmer (2013) provides additional narration and analysis. A more accessible introduction can be found in *Popular Mechanics* (Wise 2011), providing a layer of commentary over the black box transcripts, or more recently in *Vanity Fair* (Langewiesche 2014).

2 Note that there is the occasional accident that is viewed as being primarily technological. The post-landing fire on China Airlines Flight 120 in 2007, for instance, was attributed to a leak-inducing puncture caused by a missing washer on the retracting slats post landing. Yet, even in this case, the Japan Transportation Safety Board (JTSB 2009) placed blame on the manufacturer, airline, and maintenance personnel for creation of maintenance protocols and failure to report the difficulty of inspections in that area.

3 The question of whether modern aircraft are vulnerable to hackers is an open and contentious one. Some, such as security researcher Chris Roberts, have claimed to be able to access vital functions on board flying aircraft (D.N. 2015). United Airlines and other "industry experts" have disputed the degree to which he actually had access. In addition to the ethics of particular vulnerabilities opened by electronic systems, there are a whole host of questions about such endeavors to hack the systems. Should they be publicly announced (like Roberts did via Twitter), and should airlines embrace those that disclose vulnerabilities or respond critically to such attempts?

4 Note that the actual altitude programmed into the autopilot varied significantly and is visualized here in BBC (2015).

References

Air Accident Investigation and Aviation Safety Board (AAIASB). 2006. *Aircraft Accident Report: Helios Airways Flight HCY 522.* Available as a pdf at: www.aaiu.ie/sites/default/files/Hellenic%20Republic%20Accident%20Helios%20Airways%20B737-31S%20HCY522%20Grammatiko%20Hellas%202005-085-14.pdf [accessed September 1, 2016].

Arkin, R. C. 2008 (Mar). Governing lethal behavior: Embedding ethics in a hybrid deliberative/reactive robot architecture part I: Motivation and philosophy. In *Human-Robot Interaction (HRI): 2008 3rd ACM/IEEE International Conference*: 121–128.

BBC. 2015 (May 6). Germanwings crash: Co-pilot Lubitz "practised rapid descent." Retrieved from: www.bbc.com/news/world-europe-32604552 [accessed September 1, 2016].

Boeing. 2014. *Statistical Summary of Jet Airplane Accidents*. Available as a pdf at: www.boeing.com/resources/boeingdotcom/company/about_bca/pdf/statsum.pdf [accessed September 1, 2016].

Bureau d'Enquêtes et d'Analyses (BEA). 2015. *Preliminary Report: Accident on 24 March 2015 (Germanwings)*. Available as a pdf at: www.bea.aero/docspa/2015/d-px150324.en/pdf/d-px150324.en.pdf [accessed September 1, 2016].

Bureau d'Enquêtes et d'Analyses (BEA). 2012. *Final Report: Accident on 1 June 2009 (Air France 447)*. Available as a pdf at: www.bea.aero/docspa/2009/f-cp090601.en/pdf/f-cp090601.en.pdf [accessed September 1, 2016].

Chandrasekhar, C. A. 2003. Flying while brown: Federal civil rights remedies to post-9/11 airline racial profiling of South Asians. *Asian American Law Journal*, 10: 215. Available as a pdf at: http://scholarship.law.berkeley.edu/cgi/viewcontent.cgi?article=1014&context=aalj [accessed September 1, 2016].

D. N. 2015 (May 21). Hacking aircraft: Singing like a bird. *Economist*. Retrieved from: www.economist.com/blogs/gulliver/2015/05/hacking-aircraft [accessed September 1, 2016].

Davey, C. L., and Davidson, M. J. 2000. The right of passage? The experiences of female pilots in commercial aviation. *Feminism & Psychology*. 10 (2): 195–225.

Dresner, K. M., and Stone, P. 2007 (Jan). Sharing the road: Autonomous vehicles meet human drivers. *International Joint Conference on Artificial Intelligence*. 7: 1263–1268.

Dutch Safety Board (DSB). 2014. Preliminary report: Crash involving Malaysia Airlines Boeing 777-200 Flight MH 17. Available as a pdf at: www.onderzoeksraad.nl/uploads/phase-docs/701/b3923acad0ceprem-rapport-mh-17-en-interactief.pdf [accessed September 1, 2016].

Evans, L., and Feagin, J. R. 2012 (Feb 28). Middle-class African American pilots: The continuing significance of racism. *American Behavioral Scientist*. 56 (5): 650–665.

Freed, J. 2014 (Jul 21). Ukraine responsible for aircraft safety: IATA. *Sydney Morning Herald*. Retrieved from: www.smh.com.au/business/aviation/ukraine-responsible-for-airspace-safety-iata-20140720-zuzmp.html#ixzz388H87JsB [accessed September 1, 2016].

Gawande, A. 2007 (Dec 10). The checklist. *New Yorker*. Retrieved from: www.newyorker.com/magazine/2007/12/10/the-checklist [accessed September 1, 2016].

Goodall, N. J. 2014. Machine ethics and automated vehicles. In G. Meyer and S. Beiker (eds), *Road Vehicle Automation*. Cham: Springer: 93–102.

Gordon, C. C. 2002 (Jun 12). Memorandum. *Consortium for Science, Policy & Outcomes*. Available as a pdf at: http://cspo.org/wp-content/uploads/2014/08/Female-Pilot-Accommodation-in-the-UH-60M.pdf [accessed September 1, 2016].

Halsey III, A. 2014 (Jul 23). Ukraine gets $200 million a year for allowing overflights. *Washington Post*. Retrieved from: www.washingtonpost.com/blogs/worldviews/wp/2014/07/23/ukraine-gets-200-million-a-year-for-allowing-overflights/ [accessed September 1, 2016].

Helmreich, R. L., Merritt, A. C., and Wilhelm, J. A. 1999. The evolution of crew resource management training in commercial aviation. *International Journal of Aviation Psychology*. 9 (1): 19–32.

Hevelke, A., and Nida-Rümelin, J. 2014. Responsibility for crashes of autonomous vehicles: An ethical analysis. *Science and Engineering Ethics*, 21 (3): 619–630.

Hradecky, S. 2013. Crash: National Air Cargo B744 at Bagram on Apr 29th, 2013, lost height shortly after takeoff following load shift and stall. *Aviation Herald*. Retrieved from: http://avherald.com/h?article=46183bb4&opt=0 [accessed September 1, 2016].

International Civil Aviation Organization (ICAO). 2014. Joint statement on risks to civil aviation arising from conflict zones. Retrieved from: www.icao.int/Newsroom/Pages/Joint-Statement-on-Risks-to-Civil-Aviation-Arising-from-Conflict-Zones.aspx [accessed September 1, 2016].

International Committee of the Red Cross (ICRC). 2014 (Mar). *Autonomous Weapons Systems: Technical, Military, Legal and Humanitarian Aspects*. Available as a pdf at: www.icrc.org/en/download/file/1707/4221-002-autonomous-weapons-systems-full-report.pdf [accessed September 1, 2016].

Japan Transport Safety Board (JTSB). 2009 (Aug 28). *Aircraft Accident Investigation Report: China Airlines B18616*. Available as a pdf at: www.mlit.go.jp/jtsb/eng-air_report/B18616.pdf [accessed September 1, 2016].

Langewiesche, W. 2014 (Oct). The human factor. *Vanity Fair*. Retrieved from: www.vanityfair.com/news/business/2014/10/air-france-flight-447-crash [accessed September 1, 2016].

Lin, P. 2013 (Oct 8). The ethics of autonomous cars. *The Atlantic*. Retrieved from: www.theatlantic.com/technology/archive/2013/10/the-ethics-of-autonomous-cars/280360/ [accessed September 1, 2016].

Lin, P., Bekey, G., and Abney, K. 2008 (Dec 20). *Autonomous Military Robotics: Risk, Ethics, and Design*. Ethics and Emerging Sciences Group at California Polytechnic State University. Available as a pdf at: http://ethics.calpoly.edu/onr_report.pdf [accessed September 1, 2016].

Matthew, S. 2015 (Jul 27). 350 British pilots grounded in the past five years because of mental illness. *Daily Mail*. Retrieved from: www.dailymail.co.uk/news/article-3176861/Review-British-pilots-mental-health-wake-Germanwings-disaster-finds-350-grounded-past-five-years-276-returned-skies.html [accessed September 1, 2016].

Milanovich, D. M., Driskell, J. E., Stout, R. J., and Salas, E. 1998. Status and cockpit dynamics: A review and empirical study. *Group Dynamics: Theory, Research, and Practice*. 2 (3): 155.

Mills, A. J. 1998. Cockpits, hangars, boys, and galleys: Corporate masculinities and the development of British Airways. *Gender, Work & Organization*. 5 (3): 172–188.

National Transportation Safety Board (NTSB). 2001. *Aircraft Accident Report. Runway Overrun during Landing: American Airlines Flight 1420*. PB2001-910402. NTSB/AAR-01/02. DCA99MA060. Available as a pdf at: www.ntsb.gov/investigations/AccidentReports/Reports/AAR0102.pdf [accessed September 1, 2016].

National Transportation Safety Board (NTSB). 1979. *Aircraft Accident Report: United Airlines N8082U*. NTSB-AAR-79-7. Available as a pdf at: http://libraryonline.erau.edu/online-full-text/ntsb/aircraft-accident-reports/AAR79-07.pdf [accessed September 1, 2016].

Neal-Smith, S., and Cockburn, T. 2009. Cultural sexism in the UK airline industry. *Gender in Management: An International Journal*. 24 (1): 32–45.

Palmer, B. 2013. *Understanding Air France 447*. Retrieved from: www.amazon.com/Understanding-Air-France-Bill-Palmer-ebook/dp/B00E5W9YZG [accessed September 1, 2016].

Pasztor A. 2015 (Jul 18). Boeing warns carriers about flying bulk shipments of lithium batteries. *Wall Street Journal*. Retrieved from: www.wsj.com/articles/boeing-warns-carriers-about-flying-bulk-shipments-of-lithium-batteries-1437256427 [accessed September 1, 2016].

Ross, N., and Tweedie, N. 2012 (Apr 28). Air France 447: "Damn it, we are going to crash." *Telegraph*. Retrieved from: www.telegraph.co.uk/technology/9231855/Air-France-Flight-447-Damn-it-were-going-to-crash.html [accessed September 1, 2016].

Sathiyanathan, L. 2014 (Mar 4). "No place for a woman": WestJet pilot slams sexist napkin note. *CBC News*. Retrieved from: www.cbc.ca/newsblogs/yourcommunity/2014/03/

no-place-for-a-woman-westjet-pilot-slams-sexist-napkin-note.html [accessed September 1, 2016].

Schaal, D. 2013. The US airline pilots who barely make minimum wage. *Skift.com*. Retrieved from: http://skift.com/2013/08/28/the-u-s-airline-pilots-who-barely-make-minimum-wage/ [accessed September 1, 2016].

Schmidt J., Lawson, D. amd Figlock, R. n.d. Human Factors Analysis & Classification System: Maintenance Extension (HFACS-ME) review of select NTSB maintenance mishaps: An update. Federal Aviation Administration. Available as a pdf at: www.faa.gov/about/initiatives/maintenance_hf/library/documents/media/hfacs/ntsb_hfacs-me_updated_study_report.pdf [accessed September 1, 2016].

Shappell, S., and Wiegmann, D. 1996. US Naval aviation mishaps 1977–92: Differences between single- and dual-piloted aircraft. *Aviation, Space, and Environmental Medicine*. 67: 65–69.

Sparrow, R. 2009. Building a better WarBot: Ethical issues in the design of unmanned systems for military applications. *Science and Engineering Ethics*. 15 (2): 169–187.

Sumwalt III, R. 2014 (Jan 12). The role of human factors in accident/crash investigation: The legacy and future. NTSB. Retrieved from: www.ntsb.gov/news/speeches/rsumwalt/Documents/Sumwalt_140112.pdf [accessed September 1, 2016].

Sumwalt III, R. 2009 (Sep 2). Human factors: It's not just all about humans, you know. NTSB. Available as a pdf at: www.ntsb.gov/news/speeches/rsumwalt/Documents/FAA-Maint-HF-Conference-4.pdf [accessed September 1, 2016].

Tarnowski, E. 2003. Cockpit automation philosophy. NATO Research and Technology Organization. Available as a pdf at: www.dtic.mil/get-tr-doc/pdf?AD=ADA422301 [accessed September 1, 2016].

Traufetter, G. 2009 (Jul 31). Will increasing automation make jets less safe? *Der Spiegel*. Retrieved from: www.spiegel.de/international/world/the-computer-vs-the-captain-will-increasing-automation-make-jets-less-safe-a-639298-3.html [accessed September 1, 2016].

Weber, R. N. 1997. Manufacturing gender in commercial and military cockpit design. *Science, Technology & Human Values*. 22 (2): 235–253.

Wiegmann, D., and Shappell, S. 2001. *A Human Error Analysis of Commercial Aviation Accidents Using the Human Factors Analysis and Classification System (HFACS)*. Federal Aviation Administration. Available as a pdf at: www.faa.gov/data_research/research/med_humanfacs/oamtechreports/2000s/media/0103.pdf [accessed September 1, 2016].

Wise, J. 2011 (Dec 6). What really happened aboard Air France 447. *Popular Mechanics*. Retrieved from: www.popularmechanics.com/flight/a3115/what-really-happened-aboard-air-france-447-6611877 [accessed September 1, 2016].

Young, M. S., Stanton, N. A., and Harris, D. 2007. Driving automation: Learning from aviation about design philosophies. *International Journal of Vehicle Design*. 45 (3): 323–338.

13

MALAYSIA AIRLINES FLIGHT 370

Ethical considerations

Richard L. Wilson

Malaysia Airlines Flight 370 (MH370) disappeared on Saturday, March 8, 2014 while en route from Kuala Lumpur International Airport to Beijing Capital International Airport in the People's Republic of China. The last message from the aircraft was received less than one hour after takeoff. The Boeing 777-200ER was carrying 12 Malaysian crew members and 227 passengers from 15 nations. The July 2015 discovery of debris including a wing section, or flaperon, has provided a possibility of where the aircraft may be located (Ng and Adamson 2015). However, as of August 2018, very little flight debris has been found, and no crash site has been determined.

With aviation accidents the recovered physical evidence aids in identifying the probable causes. Numerous aviation disasters—such as TWA Flight 800, United Airlines Flight 232, and Air France Flight 4590—provided opportunities to investigate flight debris and accident sites, which in turn led to the identification of the technical problems associated with each of these flights. The difficulty with identifying any technical issues involved in MH370 is primarily due to the lack of any substantial amount of flight debris and a crash site. Because the technical problems with the flight have not been identified, an ethical assessment of this tragedy may seem difficult to determine as well.

Despite these problems, this chapter will provide both a technical and ethical analysis based on the possible reasons for the disappearance of MH370, including pilot actions, inflight disruptions, passenger and crew involvement, dangerous cargo, power interruption, unresponsive crew, or a hypoxia event. In addition, there are also ethical issues related to search efforts and the sharing of information about the flight by the Malaysian and other governmental authorities. Identifying the ethical problems requires adopting a method. This chapter will analyze the ethical dimensions involved in MH370 by addressing what is known as stakeholder theory (Wilson 2013, 2014b). A stakeholder analysis involves identifying ethical and social problems related to the stakeholders affected by the details of the case. The chapter will first offer a background of MH370, which includes a brief discussion of Boeing 777 aircraft features. This section will be followed by the time line of the flight, knowledge of what occurred after the aircraft disappeared, a discussion of the possibilities of what could have gone wrong, and an examination of how stakeholder analysis applies to this accident. Finally, the chapter will provide speculative conclusions based on what may have happened to MH370.

Background on the Boeing 777

The accident aircraft, a Boeing 777, is a long-range, twin-engine airliner developed by Boeing for a variety of reasons, including the replacement of older, wide-body aircraft as well as an increased flight range. With a flight range of 5,235 to 9,380 nautical miles, it was designed to bridge the capacity difference between 747s and 767s (Eden 2008: 106).

The Boeing 777 is the world's largest twin-engine jet with a seating capacity of 314 to 451 passengers. It has numerous unique features, which include the largest diameter turbofan engines of any aircraft, six wheels on each main landing gear, fully circular fuselage cross-section (Birtles 1998: 52), and a blade-shaped tail cone (Norris and Wagner 1996: 89). The 777 was developed in consultation with eight major airlines. It was Boeing's first fly-by-wire (FBW) airliner, meaning that it has a computer-mediated control system. It was also the first entirely computer-aided design of a commercial aircraft.

FBW systems were designed to replace the need for the manual flight control system, an innovation first introduced by Airbus. FBW allows pilots to input commands through electrical signals, which adds precision and safety to aircraft operation. The overall flight system monitors pilot commands to help guarantee that an aircraft remains within the flight protection envelop (North 2000). The FBW system also provides for maximum performance within safety margins for the aircraft (Airbus.com 2015). The movements of flight controls are converted to electronic signals transmitted by wires, which is why they are called fly-by-wire systems. Flight control computers determine how to move the actuators at each control surface to provide an ordered response in an aircraft. The FBW system also allows automatic signals sent by the aircraft's computers to perform functions without the pilot's input, including systems that automatically help stabilize the aircraft and that can help prevent unsafe operation of the aircraft outside of its performance envelope (Crane 1997: 224).

As discussed in the previous chapter, an important feature of the Boeing 777 airliners in comparison with Airbus is that in the 777 the two pilots can override the computerized flight-control system to permit the aircraft to be flown beyond its usual flight-control envelope during emergencies. This feature is an innovation over Airbus, where pilots have a more difficult time overriding the FBW system. Airbus's strategy began with the A320 and has been continued on the development of subsequent Airbus airliners (Briere and Traverse 1993; North 2000).

What makes the disappearance of MH370 more puzzling from a technical point of view is that the 777 does not have a history of any major defects or failures, although, as of December 2017, it has been involved in 18 accidents and incidents, including six hull loss accidents and three hijackings, with a total of 540 fatalities (Aviation Safety Network 2018). Adding to the mystery of the disappearance of MH370, the scheduled flight path was 2,700 miles, which is well within the range of a Boeing 777.

Timeline of events

The scheduled flight route of the aircraft was nearly due north, traveling over the Malay Peninsula and across the South China Sea, eventually taking the aircraft across Vietnam, Cambodia, Thailand, and Laos, where it would finally enter Chinese airspace until landing in Beijing (Cawthorne 2014: 2). The flight took off from Kuala Lumpur International Airport at 00:41 a.m. local time on Saturday, March 8, 2014. It was expected to land at Beijing Capital International Airport at 6:30 a.m. local time. The flight was expected to last approximately 5 hours and 45 minutes. After the flight took off, 20 minutes passed before it reached its assigned altitude of 35,000 feet and was traveling at 547 mph.

An important feature of aircraft since the 1980s is that planes are equipped with Aircraft Communications Addressing and Reporting Systems (ACARS) (Carlsson 2002). The ACARS system provides an automatic datalink that transmits vital information about aircraft to satellites and air traffic controllers (ATCs) at regular intervals. This allows an aircraft to be tracked even without voice communication. ACARS last communicated with MH370 at 1:07 a.m. Malaysian time (MYT). The last time a voice communication was heard from the aircraft was at 1:19 a.m., approximately 38 minutes after takeoff (Kohli 2014: 8). However, the aircraft continued to be tracked, and appeared to be flying until 8:11 a.m. MH370 had onboard an Inmarsat Classic Aero Terminal, allowing the aircraft to be tracked by an Inmarsat satellite until the last transmission (Quest 2016: 72–73 and 80–83). The following timeline is adapted from Soucie (2015):

12:47 a.m.	Flight MH370 departed
1:17 a.m.	MH370 last made voice contact with air traffic control at 01:19 a.m. MYT (17:19 UTC, March 7) when it was over the South China Sea, which was less than an hour after takeoff. The aircraft disappeared from air traffic controllers' radar screens at 01:21 a.m.
1:07 a.m.	Last ACARS contact
1:19 a.m.	Last Malaysian air traffic control voice contact
1:21 a.m.	MH370's transponder failed to respond to secondary radar
1:22 a.m.	MH370 disappears from Thai military radar
1:22–1:28 a.m.	The aircraft changed course from Beijing toward Penang, Malaysia, where there is a Malaysia Airlines System (MAS) maintenance base. The crew failed to respond to multiple radio contact attempts

1:30 a.m.	Civilian radar loses contact with the plane
1:37 a.m.	The expected ACARS transmission did not occur. The ACARS signal was supposed to transmit every half hour, which meant it should have sent a signal at 1:37 a.m.
2:15 a.m.	Malaysian military radar detected what was believed to be the plane. The flight was hundreds of miles off the course, about 200 miles northwest of Penang.
2:22 a.m.	The Royal Malaysian Air Force detected an unidentified flight, believed to be MH370, on its radar for the last time. The plane had by now swerved far off course and was over the Malaysian coastal area of Penang, heading in the direction of the Malacca Strait. (The deviation in MH370 flight patterns appears in Figure 13.1.)
7:13 a.m.	Malaysia Airlines tried to make a voice call to MH370, but there was no voice response from the aircraft.
8:11 a.m.	The sixth and final full Inmarsat satellite signal is sent from the aircraft.

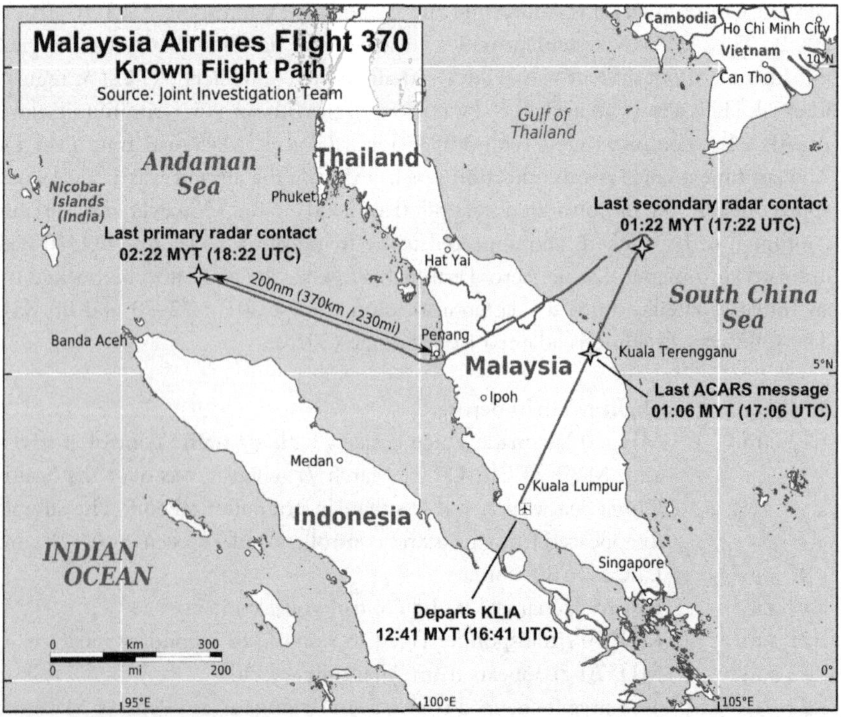

FIGURE 13.1 Malaysia Airlines Flight 370: known flight path (Attribution: Andrew Heneen)

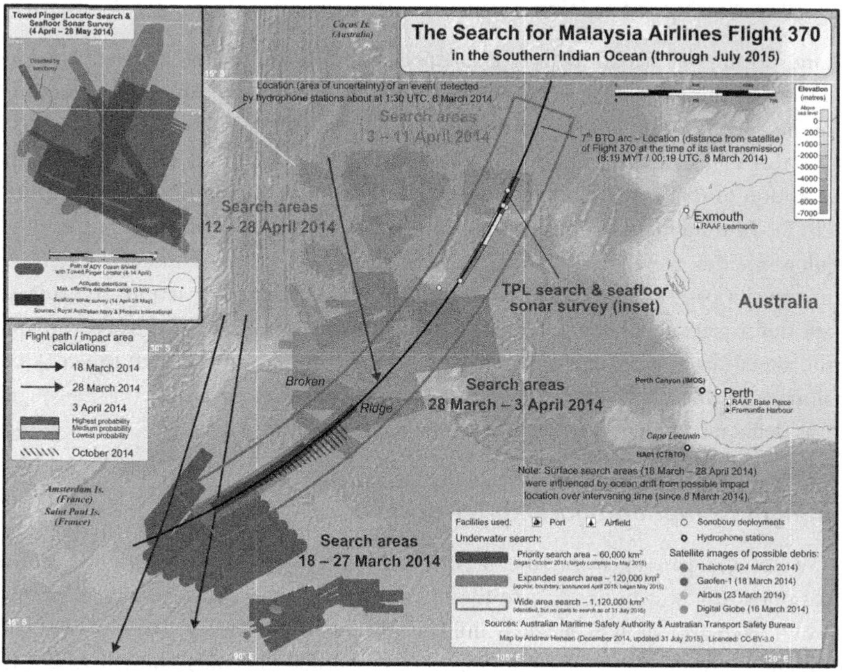

FIGURE 13.2 The search for Malaysia Airlines Flight 370 (Attribution: Andrew Heneen)

On March 24, 2014, the Malaysian government reached the conclusion that the final location determined by the Inmarsat satellite communication was far from any possible landing sites, concluding that "flight MH370 ended in the southern Indian Ocean" (Yang et al. 2014). Based upon the evidence related to its flight pattern, Figure 13.2 shows where Flight MH370 could have ended.

Possible causes of the disappearance of MH370

One way to answer the question of what happened to Flight MH370 is to construct a framework of possible events and scenarios that could have led to its disappearance. The goal of this analysis is to work with the most plausible scenarios, including both technical and ethical possibilities. The July 2015 discovery of a 777 flaperon on Réunion Island in the Indian Ocean, and subsequent discoveries of possible debris from Mozambique and the island nation of Mauritius, gives a rough indication of where the aircraft might be located. If MH370's black boxes are ever recovered, they could provide valuable insight into what went wrong during the flight. But since the black boxes remain missing, any inquiry into the disappearance must be speculative in nature. The following scenarios are some of the possibilities that could have led to the disappearance of the aircraft (see also Choisser 2014).

Pilot actions/pilot suicide

One possible explanation for the disappearance of the aircraft is related to the actions of the pilots (Tapaleao 2014). One of the pilots could have taken control of the aircraft, or it could be an example of pilot suicide (McCoy 2014). The 2015 crash of Germanwings Flight 9525 due to the first officer's actions has brought attention to the issue of airline disasters based on pilot suicides (New York Times 2015). Although these types of episode are somewhat rare, there is a history of pilot suicides which have been a factor in several aircraft disasters. Extensive investigations of both MH370 pilots have concluded that pilot actions and suicide are not a primary likelihood for the disappearance of the flight. However, with the alteration in the flight path and the lack of communication with the cockpit, along with the discovery of aircraft debris, the possibility remains that the flight was deliberately altered from its assigned route.

Inflight disruption/hijacking/terrorism

Two of the passengers on the passenger list for MH370 were carrying stolen passports. Investigations of these two men, as well as the entire passenger list and crew, have also shown that it is unlikely that they, or anyone else onboard the aircraft, were the cause of the disappearance of MH370 (Fuller and Perlez 2014). Another possibility is that the aircraft could either have been hijacked or terrorists could have tried to take control of the aircraft. What makes these scenarios problematic is the degree of technological knowledge required for someone to be able to take control of the aircraft and be able to operate it. The Boeing 777 is highly sophisticated, and thereby requires substantial knowledge to fly it effectively. Turning off the ACARS system also requires a certain amount of technical ability. These facts make the above scenarios less viable. However, as before, the discovery of debris shows some evidence supporting the claim that the flight path of the aircraft was deliberately diverted.

Dangerous cargo

Another possibility is that smuggled or illegal cargo was onboard the aircraft. If dangerous cargo were aboard MH370, it could have ignited or exploded, causing the loss of cabin pressure and subsequent loss of control of the aircraft. The difficulty with this interpretation is once again the alteration of the flight path. If everyone onboard the aircraft was accidentally incapacitated, there would be no one to take control.

Electrical problems

In another scenario an electrical fire could have caused a power outage onboard the aircraft. Any kind of electrical fire could have led to an outage that disabled the aircraft and incapacitated the pilots and crew. These events could have led to the loss of control of the aircraft. What also makes this interpretation of the disappearance problematic is, if the aircraft was not intentionally diverted, how did it move so far out of its intended flight path? One possibility is that once the problem arose, the pilot and crew could have attempted to regain control and failed.

Hypoxia event

Mechanical failure could have led to the rapid decompression of the aircraft. In the event of cabin decompression, the pilots could have lost control of the 777, due to being victims of hypoxia. Hypoxia is a physical condition that is related to low oxygen levels in the blood stream and tissues. This condition could have led to the loss of situational awareness by the flight crew, thus leading to problematic decision-making, including the inability to react appropriately to the emergency. This scenario also seems plausible due to the alteration in the flight path of the aircraft. The Inmarsat satellite data shows that the aircraft made a complete turnaround, which could have occurred before the pilots lost control of the aircraft.

This possibility is comparable to another aircraft disaster, Helios Airways Flight 522, where the pilots appear to have suffered from the effects of hypoxia, and a flight attendant may have attempted to land the aircraft but failed (Kohli 2014: 160–161). Aircraft cabin depressurization can occur rapidly, leading to a loss of control. However, in this case the aircraft would have continued on its flight path. Could an aircraft on autopilot alter course before the pilots lost consciousness? It does seem possible given the October 1999 Learjet 35 accident that took the life of the US golfer Payne Stewart (see textbox).

THE 1999 LEARJET 35 ACCIDENT: A PARALLEL WITH MH 370?

An example of an aircraft on autopilot deviating from its heading occurred in the South Dakota crash of a chartered SunJet Aviation Learjet 35. On October 25, 1999, this aircraft, with the US professional golfer Payne Stewart onboard, was scheduled to fly from Orlando, Florida to Dallas, Texas. While climbing to its assigned altitude, the cabin lost pressure and the crew and passengers were incapacitated, most likely due to hypoxia. According to the NTSB report, after losing contact with the flight crew, the plane deviated from its assigned course and failed to level at its assigned altitude (FL 390) (NTSB 2000: 27). In footnote 43 the NTSB states: "Because the airplane's ground track (and presumably its heading) was maintained for nearly the remainder

continued . . .

of the flight, it is likely that this right turn was initiated by human input to the autopilot heading select knob. However, the National Transportation Safety Board was unable to ascertain whether this input was the result of an intentional act." The possibility remains that an aircraft, even on autopilot, may deviate from its assigned course even though the cause is unclear.

Overview of the ethical analysis

Ethical issues relate to how people act and how they ought to behave towards one another. When events involving human actions are performed, an ethical analysis aids in determining who is responsible for the positive and/or negative consequences that those actions produced. Flight MH370 has two sets of problems: how the case has been treated and what we can learn through an ethical analysis. The difficulty with taking an ethical approach to MH370 is that the aircraft has not been located. Therefore, our discussion can only examine the probable reasons for the disappearance of the flight. While one may utilize any number of ethical approaches to MH370, our analysis will focus primarily on the stakeholders affected by its disappearance.

Stakeholder theory arose in the early 1980s, and one of its early proponents is R. Edward Freeman. He contends that the dominant 20th-century model of shareholder theory is no longer viable. Previously the dominant belief was that management should center its attention on the desires of the shareholders and create value for them (Freeman 2007: 6). However, this focus can put shareholders at odds with the needs of the customers, employees, suppliers, etc. Because of the potential for conflict, Freeman advocates for stakeholder theory, which takes into consideration the relationships among the different groups who each have a stake in the business activities. Rather than center decisions around maximizing profits for the shareholders, the executive needs to manage and shape the relationships among the different groups and then create value based on those relationships (Freeman 2007: 9). Of course, a business needs to make a profit; but if it solely focuses on money-making for its shareholders, other important values might be disregarded, such as safety.

Stakeholder theory revolves around two sets of people: primary and secondary stakeholders. Although the definition of each group varies, one way to distinguish them is as follows. Regarding the primary stakeholders, Freeman includes any groups that are necessary for the business to operate. Some examples are the employees, customers, suppliers, etc. Freeman describes the secondary stakeholders as any group that can indirectly affect a business, such as the media, competitors, advocacy groups, etc. According to Freeman (2007: 12), the executives must take each group into consideration in thinking about how to create value for the business.

The goal of stakeholder theory here is to analyze what is at issue for each of the stakeholders affected by—either directly or indirectly—the disappearance of flight MH370. Due to the investigations conducted into the possible intentions and actions of the pilot in command and the first officer, the crew, and the two passengers carrying stolen passports, there is no strong evidence—other than the change of the flight path of the aircraft—that any of these scenarios could have caused the disappearance of MH370. An exception to this would be if there was evidence that the transponder of the aircraft was intentionally turned off. Since there is no indication of crew or passenger involvement in the disappearance, we must consider other possible reasons for this accident. Some authors support the claim that an electrical fire, a power outage, and/or operator hypoxia could have caused the disappearance of MH370.

Background: UPS Flight 6

In order to provide a framework for the following stakeholder analysis, we will begin by examining possible technical problems aboard MH370 based upon the UPS Airlines Flight 6 accident on September 3, 2010 which killed its two pilots. The Boeing 747-400 crashed as the result of an inflight fire while on a scheduled flight from Dubai International Airport (United Arab Emirates) to Cologne Bonn Airport in Germany. Although there were no known hazardous materials being carried on UPS Flight 6, there were at least three shipments of lithium ion batteries that met Class 9 hazardous material criteria. Twenty minutes into the flight the crew informed Bahrain Area East ATC that there was an indication of an onboard fire on the forward main deck. Cargo in the main cargo area had ignited, and less than three minutes after the first warning to the crew, the fire resulted in severe damage to the flight control systems. The crew advised Bahrain East ATC that the cockpit was "full of smoke" and that they "could not see the radios" (Kohli 2014: 133).

Due to the UPS Airlines Flight 6 disaster, investigators made several technical recommendations for improving safety features of Boeing aircraft. One recommendation is for improved cargo container fire suppression methods. A second involves improvements to crew oxygen and vision equipment during a smoke event. Airlines should pay closer attention to the limitations of the current oxygen systems in aircraft given the possibility of an onboard electrical fire due to hazardous materials. Thirdly, one needs to reclassify lithium-based batteries as a class higher than ICAO Class 9 hazardous material. A fourth recommendation concerns the aircraft's fuselage. This recommendation relates to the integrity of the aircraft's hull and to the possibility of cracking of the fuselage skin (Cawthorne 2014: 95).

Stakeholder analysis of MH370

Based on the technical problems with Flight 6, we will now provide a stakeholder analysis of MH370. If Malaysia Airlines management had focused on its stakeholders more fully, while this may not have prevented the disappearance, it would have helped the various groups affected deal with this tragedy more clearly and honestly. The following account of the MH370 case will include both primary and secondary stakeholders (see Wilson 2013, 2014a, and 2014b). At the center of this analysis is what Freeman calls "the Firm," in this case Malaysia Airlines management, which is responsible for creating value for each of its stakeholders.

The primary stakeholders

For the purposes of this chapter, the primary groups for Malaysia Airlines management to consider are the flight crew, the passengers, and the Malaysian government. This list is not meant to be exhaustive, but rather provides suggestions for future interactions between airline management and its primary stakeholders.

Pilots and crewmembers

A key issue involves airline management's relationship with the crew members, especially the pilots. Given that the disappearance of MH370 may have been similar to UPS 6 (or even a hypoxia event such as the Payne Stewart accident), Malaysia Airlines should ensure that pilots and crew have adequate training to combat extreme circumstances such as the sudden loss of electricity and oxygen during flight. For the pilots and crew to function effectively in emergency situations, the proper safety equipment must be installed in the aircraft, such as fire suppression devices and oxygen tanks. Additional safety equipment would give pilots and crew extended time to be able to handle emergency situations involving fires, smoke, and lack of oxygen. Given the possibility of an onboard electrical fire due to the combustion of hazardous materials, the airline should reclassify dangerous cargo, including labeling batteries as hazardous. This could help avoid any negative consequences related to dangerous cargo. Not having the appropriate emergency equipment onboard an aircraft means that there is a limited opportunity for the pilots to remain in control.

Airline management should also ensure that the pilots and crew have training in areas that can influence the safety of the aircraft. Their training should include not only the above-mentioned safety issues, but also dangers relating to terrorists, hijackers, and other important factors such as pilot suicides.

The passengers

Passengers, along with the crew members, are most directly affected by the disappearance of MH370. Passengers have the intention of traveling safely, and they expect to arrive safe and sound at their intended destination. Malaysia Airlines has a duty to ensure that the expectations of passengers are met. If the disappearance of the aircraft is related to a safety problem, people have a right to be made aware of the issue. Management needs to be forthcoming about any possible system failures to help the public feel safer when flying on commercial aircraft.

Malaysian government

From the moment that irregularities were associated with MH370, questions arose about communication issues from both the airline and the Malaysian government. Public communication from Malaysian officials regarding the loss of the flight was from the beginning riddled with confusion (Hodal 2014). The government and the airline released imprecise, incomplete, and sometimes inaccurate information, with civilian officials sometimes contradicting military leaders about the details of the case (Straits Times 2014). Malaysian officials were criticized for the persistent release of contradictory material about the flight, specifically regarding the last point and time of contact with the aircraft (Denyer 2014). The Malaysian government, along with Malaysia Airlines management, failed to communicate accurate information about the flight to both the relatives and the public, who could in turn view this problem as failing to exercise basic virtues: honesty and truthfulness.

When an airline disaster occurs within a certain jurisdiction, that country's government must be more directly involved in addressing the issues related to the disappearance of the aircraft. This is especially true of MH370, and the government needs to make the issues known to the public by showing an intention to have aircraft operate safely within their authority. This could be accomplished by instituting the continual assessment and reassessment of regulations for the emergency equipment in aircraft, and by replacing equipment as needed. They must assess flight disasters and attempt to enhance future outcomes by recognizing the need for improved safety equipment. The government must also provide better awareness of potential threats such as pilot suicide, terrorist acts, and hijacking.

The secondary stakeholders

As mentioned previously, the secondary stakeholders are those who are indirectly affected by the disappearance of MH370. As with the above-mentioned groups, the following are not meant to provide an exhaustive list of secondary stakeholders. But, at a minimum, this analysis should show the need for airline management to determine value for anyone who was impacted by this accident.

Family members

The family members of the crew and passengers have a right to know the potential causes of accidents and the fate of their loved ones. Although the relatives of the people onboard flight MH370 were not harmed physically, they would feel mental anguish due to the loss of a loved one (and possible financial loss when, for example, a spouse is lost). With MH370 the family members also have the uncertainty of knowing what actually happened to their loved ones. Therefore, these families need accurate information communicated to them. Counseling could also be provided to help family members cope with their grief.

Some have severely criticized Malaysia Airlines management for how they treated the MH370 disappearance since they confused family members by presenting inconsistent information about the disappearance of the flight and the status of the investigation. This problem relates to virtue ethics and the lack of appropriate character traits of Malaysia Airlines management, a problem that can lead to distrust. Relatives of those affected by the disappearance of the aircraft most likely feel that they have a right to accurate information about all aspects of the flight.

Future customers

Unless the airline and the government institute the safety measures enumerated above, potential passengers could lose faith in traveling with Malaysia Airlines. Future customers will have the expectation of safely reaching their destination. From the perspective of these potential passengers, Malaysia Airlines must be committed to carrying the best safety equipment and establishing best practices, including instituting training that will help guarantee passenger wellbeing.

Stockholders

There is also the issue of the stockholders in the airline. Those who have invested in Malaysia Airlines saw a financial loss from a drop in the stock price due to bad publicity. Potential liability issues and a decrease in overall business can have a harmful impact upon the future ability of Malaysia Airlines to operate safely and efficiently.

International Air Safety Administration (IASA)

It is crucial that the IASA examine regulations related to aircraft operating in a wide variety of air spaces. This is crucial because, as stated previously, the route for MH370 would have traversed several countries, such as Vietnam, Cambodia, and Thailand. After the flight first shifted from its assigned route, this diversion also indicated the aircraft was crossing different airspaces without authorization. Air traffic must be monitored and regulated in such a way as to guarantee aircraft safety, particularly when an aircraft operates in a variety of airspaces.

Conclusions

When and if the wreckage of Flight MH370 is located, it will be possible to provide an updated analysis of the technical and ethical issues with MH370. Until that time arises, our analysis will have to rest on the most plausible possibilities for the disappearance of the flight. As seen from Figure 13.2, the tracking of the aircraft when it disappeared showed an extreme deviation from its original flight plan. Since the time one of the aircraft's flaperons washed ashore on Réunion Island in July of 2015 and additional debris was discovered in Mozambique in

2015 and 2016, along with suspected debris on Mauritius in 2016, these pieces give us an indication of a possible resting site for the aircraft (see Ng and Adamson 2015; Scruton et al. 2017). This is an extremely large area and it seems safe to assume that it may take a great deal of time to find the aircraft. Despite the lack of definitive answers, this chapter will draw tentative conclusions based on both possible technical issues and stakeholder theory.

Conclusions based on possible technical problems

Because the aircraft has not been found at the time of writing this chapter, the technical conclusions are primarily based upon the UPS Flight 6 accident. A fire event could have occurred onboard Flight MH370. If fire and smoke incapacitated the pilots and crew, there is a clear need for improved safety equipment. In case of an electrical fire event, pilots and crew need the time and equipment to regain control of the aircraft. In addition, we should focus on installing improved safety equipment and reclassifying dangerous cargo. International regulations for tracking and monitoring aircraft must be strengthened. Area radar control must provide swift action, and contact with the aircraft must be maintained.

Conclusions based on the stakeholders

Malaysia Airlines management provided misleading and confusing information to several stakeholders, especially the families, the public, and foreign governments. This is a clear-cut ethical issue of truthfulness and calls into question the management's credibility. Official airline personnel must communicate in a more organized fashion with its stakeholders, such as relatives and friends of those onboard aircraft, when there is an accident or major incident.

The possibility remains that individuals may have played a primary role in the flight's disappearance. Some American intelligence reports claimed that the aircraft may have been subject to interference from someone in the cockpit who took control of the aircraft (Perez 2015). This suggests that pilot, crew, or even passengers may have interfered in the operation of the flight. Given the first officer suicide on Germanwings Flight 9525, one conclusion is that airline management should provide more rigorous screening of its crew, including psychological screening. No one should be left alone in an aircraft cockpit.

However, until the aircraft and possibly the black boxes are found, the final conclusions of this chapter are at best speculative. The issues enumerated above—including crew training, improved fire suppression methods, increasing emergency oxygen, more developed tracking systems, and improved communication (in this case from Malaysia Airlines and the Malaysian government)—are all crucial conclusions we can draw. No matter what the cause of the MH370 disappearance may be, this chapter has tried to show that an ethical analysis, such as stakeholder theory, can help provide important insight into the appropriate ways to communicate with the people affected by this tragic loss.

References

Airbus.com. 2015. Fly-by-wire. Available at: www.airbus.com/innovation/proven-concepts/in-design/fly-by-wire/ [accessed July 6, 2015].

Aviation Safety Network. 2018. Boeing 777 statistics. *Flight Safety Foundation*. Available at: https://aviation-safety.net/database/types/Boeing-777/statistics [accessed March 5, 2018].

Birtles, P. 1998. *Boeing 777: Jetliner for a New Century*. St. Paul, MN: Motorbooks International.

Briere D. and Traverse, P. 1993. Airbus A320/A330/A340 Electrical flight controls: A family of fault-tolerant systems. *Proceedings of the 23rd International Symposium on Fault-Tolerant Computing* (FTCS-23). Toulouse, France: 616–623.

Carlsson, B. October 2002. GLOBALink/VHF: The future is now. *The Global Link*. No. 23. Available as a pdf at: http://web.archive.org/web/20060211160041/http://www.arinc.com/news/newsletters/gl_10_02.pdf [accessed February 25, 2016].

Cawthorne, N. 2014. *Flight MH 370: The Mystery*. London: John Blake.

Choisser, J.P. 2014. *Malaysia Flight MH 370 Lost in the Dark: In Defense of the Pilots: An Engineer's Perspective*. CreateSpace Independent Publishing Platform.

Crane, D. 1997. *Dictionary of Aeronautical Terms*, third edition. Newcastle, WA: Aviation Supplies & Academics.

Denyer, S. 2014 (Mar 12). Contradictory statements from Malaysia over missing airliner perplex, infuriate. *Washington Post*. Available at: www.washingtonpost.com/news/worldviews/wp/2014/03/12/contradictory-statements-from-malaysia-over-missing-airliner-perplex-infuriate/ [accessed March 14, 2014].

Eden, P., ed. 2008. *Civil Aircraft Today: The World's Most Successful Commercial Aircraft*. London: Amber Books.

Freeman, R.E. 2007 (Jan). Managing for stakeholders. Available as a pdf at: file:///C:/Users/EA/Downloads/SSRN-id1186402.pdf [accessed February 17, 2018].

Fuller, T. and Perlez, J. 2014 (Mar 11). Stolen passports on plane not seen as terror link. *New York Times*. Available at: www.nytimes.com/2014/03/12/world/asia/malaysia-flight.html?_r=0 [accessed March 15, 2014].

Hodal, K. 2014 (Mar 14). Flight MH370: a week of false leads and confusion in hunt for missing plane. *The Guardian*. Available at: www.theguardian.com/world/2014/mar/14/flight-mh370-false-leads-confusion-hunt-missing-plane [accessed March 25, 2014].

Kohli, S. 2014. *Into Oblivion: Understanding MH370*. Edited by Pranav Kohli. CreateSpace Independent Publishing Platform.

McCoy, T. 2014 (Mar 11). The surprising frequency of pilot suicides. *Washington Post*.

National Transportation Safety Board (NTSB). 2000. *Crash of Sunjet Aviation, Learjet Model 35, N47BA*. NTSB Number AAB-00-01. NTIS Number: PB2003-910402. Available as a pdf at: www.ntsb.gov/investigations/AccidentReports/Reports/AAB0001.pdf [accessed August 26, 2016].

New York Times. 2015 (Mar 24). Where the Germanwings plane crashed. Available at: www.nytimes.com/interactive/2015/03/24/world/europe/germanwings-plane-crash-map.html?ref=liveblog [accessed March 25, 2015].

Ng, E. and Adamson, T. 2015 (Aug 2). Malaysia seeks help in finding more possible MH370 debris. *Indian Express*. Available at: http://indianexpress.com/article/world/world-others/malaysia-seeks-help-in-finding-more-possible-mh370-debris-2/ [accessed August 8, 2015].

Norris, G. and Wagner, M. 1996. *Boeing* 777. St. Paul, MN: Motorbooks International.

North, D. 2000 (Aug 28). Finding common ground in envelope protection systems. *Aviation Week & Space Technology*: 66–68.

Perez, E. 2015 (Aug 2). US intelligence assessment focuses on cockpit activities of MH370. *CNN.com*. Available at: www.cnn.com/2015/07/30/politics/mh370-cockpit-activities-u-s-intelligence/ [accessed February 23, 2016].

Quest, R. 2016. *The Vanishing of Flight MH 370*. New York: Berkeley Books.

Scruton, P., Phipps, C., and Levett, C. 2017 (Jan 17). Missing Flight MH370: a visual guide to the debris found so far. *The Guardian*. Available at: www.theguardian.com/world/ng-interactive/2017/jan/17/missing-flight-mh370-a-visual-guide-to-the-parts-and-debris-found-so-far [accessed March 9, 2018].

Soucie, D. 2015. *Malaysia Airlines Flight 370: Why It Disappeared—and Why It's Only a Matter of Time before It Happens Again*. New York: Skyhorse Publishing.

Straits Times. 2014 (Mar 12). Missing Malaysia Airlines plane: air force chief denies tracking jet to Strait of Malacca. Available at: www.straitstimes.com/asia/se-asia/missing-malaysia-airlines-plane-air-force-chief-denies-tracking-jet-to-strait-of [accessed September 28, 2014].

Tapaleao, V. 2014 (Mar 27). Flight MH370: Terrorism expert backs theory of pilot suicide flight. *New Zealand Herald*. Available at: www.nzherald.co.nz/nz/news/article.cfm?c_id=1&objectid=11227090 [accessed March 27, 2015].

Wilson, R.L. 2014a. Interdisciplinary anticipatory ethical stakeholder analysis. Association for Interdisciplinary Studies Conference. Unpublished presentation. Michigan State University, East Lansing. October 15–19.

Wilson, R.L. 2014b. Anticipatory ethical stakeholder analysis. Society for Ethics Across the Curriculum. 15th annual conference. Unpublished presentation. Arizona State University, Scottsdale, October 2–4.

Wilson, R.L. 2013. Event based ethical stakeholder analysis. Association of Practical and Professional Ethics. 22nd Annual meeting. Unpublished presentation. San Antonio, Texas, February 28–March 1.

Yang, J.L., Wan, W., and Halsey III, A. 2014 (Mar 24). Malaysian prime minister says Flight MH370 ended in the southern Indian Ocean. *Washington Post*. Available at: www.washingtonpost.com/world/asia_pacific/malaysian-prime-minister-says-flight-mh370-ended-in-the-southern-indian-ocean/2014/03/24/cb28ffc6-b370-11e3-8020-b2d790b3c9e1_story.html [accessed February 23, 2016].

PART IV
Diversity in aviation

Race and gender issues have accompanied the advent of the United States, and have continued in the field of aviation. From the early days of flight doors were often closed to women and minorities. Even in today's world, of all the Airline Transport-rated pilots only 4.2 percent are women, 2.7 percent African-American, and 5 percent Hispanic or Latino. In addition to problems of diversity, age discrimination against pilots was a topic of debate with the mandatory retirement age of 60 for Part 121 pilots enforced until 2007, when it was changed to 65. However, this new rule only applies to Part 121, leaving the question open for Part 91 and Part 135 operators. The insights of the following four chapters reveal the depth and complexity of diversity issues from the early days of aviation through today.

This part of the book addresses not only contemporary issues but also the early days of aviation. Chapter 14, which provides an overview of the history of discrimination against black pilots, allows the reader to readily see why racial discrimination continues to play a role in US aviation. Flint Whitlock first examines early black aviators in order to show the ways in which they were able to overcome racial barriers. Some of the key areas of focus are Bessie Coleman's path to gaining a pilot's license and acceptance, the Tuskegee Airmen of World War II, and the fight Marlon Green faced on his way to becoming a pilot for Continental Airlines. His chapter ends with a brief overview of the Organization of Black Aerospace Professionals (OBAP).

Chapter 15 then turns to contemporary issues by examining the gender and racial barriers that confront pilots today. Some problems include a lack of role models, a lack of exposure to aviation careers, cost, prejudice, etc. As James E. Sulton III points out, people have different learning styles, and by employing a variety of learning techniques, anyone should be able to become a pilot. However, with the lack of mentors and role models, especially in terms of flight training, the possibility of successful motivation for women and minorities is greatly minimized. Thus, Sulton shows us not only the reasons why the majority of pilots continue to be white as well as male, he also indicates some of the solutions to the barriers that continue to confront the aviation industry.

After examining barriers for pilots, Chapter 16 addresses the ways in which the industry can improve diversity opportunities in aircraft maintenance. Mirroring the problems associated with flight training, maintenance is another field that tends to dissuade women and minorities from entering the workforce. Similar problems, such as the lack of mentors and role models, plague the maintenance field. In addition, maintenance is not usually considered a desirable career choice. Because of these issues, Paul Foster addresses several ways in which diversity can be improved, beginning early on with the education of children and instilling in them an interest in the subject, especially by promoting respect, understanding, and professionalism.

Finally, Chapter 17 investigates the historical and political background behind the Age 60 Rule for Part 121 pilots, along with why it finally changed to 65. Michael Oksner first provides the historical context for the Rule in order to show that it was never really an issue of pilot health. Instead the situation was based on factors such as economics rather than scientific evidence. He analyzes different studies in order to uncover flaws in the rationale for the Rule. As Oksner contends, throughout the entire Age 60 Rule era the FAA never collected meaningful data even when mandated by Congress. In examining ICAO and their move to a 65 retirement age, Oksner shows how the tide changed in the US, especially with FAA Administrator Marion Blakey announcing in 2007 that the time had come to change this rule.

14

RACIAL DISCRIMINATION AGAINST PILOTS

An historical perspective

Flint Whitlock

Perhaps the ethical issue that has had the most lasting impact in the United States concerns racism. Blatant racial discrimination in the domestic airline industry was rampant for many decades, and continues to arise today. Although the first commercial flight with a black pilot took place between Tampa and St. Petersburg, Florida in 1914, the first African-American flight crew member did not take his seat in the cockpit of a regularly scheduled major airliner until 50 years later. From the very beginning, aviation was a "white man's game." Persons of color were discouraged from pursuing a career in aviation for a variety of reasons, perhaps not the least of which was white America's perception of blacks as being incapable of learning the highly complex skills necessary to fly. Even after this perception was proven to be groundless, the airlines were still afraid that bigoted white passengers would boycott any carrier that employed black pilots.

This chapter provides an historical analysis of racial discrimination against US pilots of African descent. It will be followed by Chapter 15, which provides the key reasons why both gender and racial discrimination continue to exist and, more importantly, some of the ways in which discrimination can be overcome. In examining the historical context of several prominent African-American aviators, this chapter also reveals those pioneers who dispelled the prejudicial attitudes of white America. It covers the early years of aviation through the 1976 formation of the Organization of Black Airline Pilots (OBAP)—now known as the Organization of Black Aerospace Professionals; and from this analysis one can readily discern why the early struggles of Bessie Coleman and other important aviators is far from over.

Bessie Coleman and other early African-American pilots

One of the first persons of African descent to prove that blacks could fly just as well as whites was a woman, Bessie Coleman. She became the first black American to be issued an international pilot's license. Born into poverty in 1892 in Atlanta, Texas, the ambitious Coleman graduated from high school and attended one semester at the Colored Agricultural and Normal University (now Langston University) in Langston, Oklahoma. A lack of finances forced her to drop out of college, and she moved to Chicago in 1915. While working as a beautician, she saw a newsreel on aviation and began to dream about a career in the skies. Her brother, who had been a soldier with the American Expeditionary Force in France during World War I, told her stories about French women learning to fly. Due to her brother's prodding, Bessie decided that anything French women could do, she could do better.

She applied to numerous flight schools across the US, but because of her race and gender no one was willing to admit her. She found a better-paying job, taught herself French, and began saving money for a move to France. After several sponsors agreed to fund her dream, the gutsy young woman sailed for Paris in November 1920. Within seven months she had completed the ten-month course at the École d'Aviation des Frérès Caudron (Caudron Brothers' Flying School) in Le Crotoy, the Somme, where she quickly learned to do loops, barrel rolls, and spins in a Nieuport aircraft over the cratered battlefields that had been the scene of years of ferocious fighting. She received her license on June 15, 1921—the only woman out of the 60 candidates to graduate from the Fédération Aéronautique Internationale during a six-month period.

After returning to the US, Coleman hoped to open the first African-American flight school and make her living in aviation, but the racial barriers were too high. Becoming a barnstormer—an aerial daredevil and stunt flier—seemed to offer her a way to fly and make money. She flew in her first air show in Garden City, New York, in 1922; her beauty, style, and grace helped her become a celebrity. She not only broke down gender barriers but also racial ones, refusing to perform in Texas unless black and white spectators were allowed to enter the air show grounds through the same admission gate; the stands themselves remained segregated, however. Her career was tragically cut short on April 30, 1926, when her mechanic, testing a JN4D "Jenny" from the front seat, lost control, causing the plane to flip and throwing Coleman from the plane to her death. But fortunately her legacy did not die with her. Because of her example several Bessie Coleman Aero Clubs were established, and they helped hundreds of African-Americans learn to fly (Hardesty 2008).[1]

Some of the other early black aviators to gain fame include Hubert Julian, born in either 1897 or 1900 and nicknamed "the Black Eagle" (Nugent 1971). Two years after Charles A. Lindbergh's historic 1927 transatlantic flight, Julian duplicated the feat. In 1931 he then became the first black pilot to make a transcontinental flight across the US. He later trained with the famed Tuskegee Airmen during World War II but was discharged from the service before graduation (Hardesty 2008).

Charles Alfred "Chief" Anderson, born in Bryn Mawr, Pennsylvania in 1907, became one of the most important persons in the history of black aviation.

Anderson fell in love with flying as a youngster, and at 20 tried to sign up for flying lessons but found that no flight school would take a black student. Undiscouraged, he sought out individual pilots willing to give him private lessons. In August 1929 he soloed and received his private pilot's license. Shortly thereafter Anderson made the acquaintance of Ernst Bühl, a German Air Force pilot in World War I who had come to the US and operated an airport near Philadelphia. Seeing Anderson's prowess in the cockpit, Bühl gave the black pilot further instruction and helped him earn his commercial and air-transport pilot's license in 1932—the first African-American to receive such a license. In 1933 and 1934, to promote the idea of flight to the black population, Anderson made several long-distance flights around the country and, with Dr. Albert Forsythe (a black physician from Atlantic City, New Jersey, whom he had taught to fly), made a trans-continental flight in 1934. The two also set many other firsts, such as piloting the first flight to land on the island of Nassau, Bahamas, and being the first black pilots to make an international flight (Atlantic City to Montreal, Canada).

World War II and the Tuskegee Airmen

In 1939 the federal government created flight schools at colleges around the country to train pilots in the event the US might be drawn into the growing world conflict. However, as none of these flight schools were established at black colleges, a student at Howard University in Washington, DC sued the government, and Anderson was subsequently hired to head Howard's civilian pilot-training program. These programs, such as the one at Howard, paved the way to help overcoming the blatant racial discrimination prevalent in the US.

Perhaps the most important flight-training program began in 1940 at Tuskegee Institute, a black college in Tuskegee, Alabama, with Anderson chosen to head the program. In April 1941 Eleanor Roosevelt, the wife of President Franklin D. Roosevelt, visited the institute. Impressed with the array of planes lining nearby Moton Field, she asked Anderson if Negroes were really capable of flying such machines. He answered in the affirmative and invited the First Lady to go for a ride with him in a Piper Cub. Horrified, the Secret Service agents tried to talk her out of it, but to no avail. When the Secret Service called President Roosevelt in a panic to inform him of her impending venture, he replied: "If she wants to do it, there's nothing we can do to stop her." The ride went off without a hitch, and shortly after Mrs. Roosevelt's return to Washington, the decision was made in September 1941 to form the first Black Air Corps at Tuskegee. The Army Air Corps also established a technical training school for African-Americans at Chanute Field, in central Illinois (Hardesty 2008).

One of the biggest boosts to the aviation dreams of African-American pilots came during World War II. Although racial segregation was strictly enforced in the US military until 1948, World War II did much to establish the competency of black pilots. The all-black 99th Pursuit Squadron, famously known as the "Tuskegee Airmen," compiled one of the most stellar combat records of any unit within the US Air Force. Captain Benjamin O. Davis—a 1936 graduate of the US Military Academy at West Point, New York, and the first black Army Air Corps officer to solo—was one of the original 13 cadets in the Tuskegee program, and earned his pilot's wings in March 1942. The 99th Pursuit Squadron became one of the most storied units in US military history. The unit was deployed to North Africa in April 1943, where its aircraft supported the Allies in the Mediterranean Theater of Operations. The squadron compiled an outstanding record in just four months, and Davis was promoted to colonel and assigned to the all-black 332nd Fighter Group, flying P-47 "Thunderbolts" and later the P-51 "Mustangs" in escort of American bombers hitting Nazi targets from Italy to Berlin. Under the umbrella of 332nd Fighter Group were the 99th, 100th, 301st, and 302nd Fighter Squadrons. The group took part in the invasions of Sicily and Italy, with the pilots of the 332nd flying 15,533 sorties and shooting down 111 enemy aircraft, including 3 of the super-fast German ME-262 jet fighters. The group also accounted for 150 German aircraft on the ground, 950 railcars, trucks, and other motor vehicles, and one German Navy destroyer. Members of the group earned 150 Distinguished Flying Crosses, along with 744 Air Medals, 14 Bronze Stars, and 8 Purple Hearts. However, 66 Tuskegee Airmen lost their lives during the war, and another 32 were downed and taken prisoner (Hardesty 2008).[2]

The Tuskegee Airmen, with pilots such as Davis and Daniel "Chappie" James, proved to the world that black warriors could be just as good—if not better—than any white fighters, including those of Hitler's "Master Race." While it is impossible to prove that the Tuskegee Airmen were directly responsible for desegregating America's armed forces, one could argue that their magnificent contributions to the war effort certainly aided the cause of integration.

Discrimination in commercial aviation after WWII

Despite the advances made by the Tuskegee Airmen, a racially divided US was not yet ready for black pilots to fly mostly white passengers traveling on commercial airliners. Instead they were forced to pick up "odd jobs" if they wanted to pursue their love of flying. For example, August H. "Augie" Martin, who had trained pilots in World War II, could only find work flying for low-paying, low-prestige, non-scheduled airlines. In 1949 he finally became a captain, but not with a US-based airline. El Al, the newly formed national airline of the new Jewish state of Israel, hired him. He later worked as a test pilot for Lockheed. Finally, in 1955 Martin became a pilot at Seaboard and Western Air (later Seaboard World), a cargo carrier that Flying Tigers eventually acquired. Martin was killed in 1968 when his plane crashed while on a mercy flight delivering emergency medical supplies to war-torn Nigeria (AvStop.com n.d.).

On February 5, 1957 Perry H. Young, Jr. became the first black American pilot generally credited with carrying passengers on a commercial flight. At the controls of a 12-passenger New York Airways helicopter, Young, who had been a flight instructor at Tuskegee during the war, made the nine-minute trip from LaGuardia

Airport to Idlewild. Like Wilbur and Orville Wright's flight in 1903, Young's journey was short, but its historic implications were enormous (LeDuff 1998). Nor were airlines eager to integrate the passenger cabin. It was not until 1958 that the New England regional carrier Mohawk Airlines became the first airline to hire a black flight attendant—Ruth Carol Taylor (Jet 1997).

In 1957 Marlon D. Green, a black Air Force pilot from Arkansas, resigned his military commission to pursue his dream of becoming a commercial airline pilot; but despite over 3,000 hours' experience in multi-engine aircraft, he found the doors to the cockpit barred to him. He spent months knocking on the doors of every airline in the country to no avail. One airline executive even told Green to his face that no matter how qualified he was an airline would never hire him because of his race. Only one airline—Denver-based Continental—even gave Green a flight test, and that was only because he had purposely failed to submit a photograph of himself and had not checked "Negro" on the application form in the space marked "Race." Although he had more than twice the number of hours as the other applicants (all white), Continental hired all of them except for Green.

Convinced that he had been the victim of racial discrimination, Green chose to pursue the matter through the courts, but the Denver District Court dismissed his complaint against the airline. The Colorado Supreme Court also sided with the carrier, citing case law that said a state could not impose regulations on a carrier engaged in interstate commerce. A unanimous decision by the US Supreme Court in 1963, however, overturned the Colorado judgment and ruled that Continental had discriminated against Green solely because he was black, and ordered the airline, which received federal funds for carrying mail, to admit Green to its next training class. The airline dragged its feet, but finally hired Green in 1965 during the turbulent years of the Civil Rights Movement. Green flew for Continental for 15 years and retired in 1979. His landmark victory in the US Supreme Court also opened the door for other African-American pilots (Whitlock 2009).

It should also be noted that in 1963 Trans World Airlines (TWA) offered a job to William DeShazor, a black, former US Marine Corps pilot, but he committed suicide in February 1964, before he completed his commercial pilot training (Poughkeepsie Journal 1964). Also, David Harris, a black pilot who was hired by American Airlines on December 3, 1964, after Green's Supreme Court victory, actually flew before Green took to the air for Continental (Whitlock 2009).

The Organization of Black Airline Pilots

Gradually the major US airlines began allowing black pilots into the cockpit. In addition to Green, Continental Airlines also hired Elra "Doc" Ward in 1968, followed by Ron Jennings. In 1976 Ben Thomas, a black pilot flying for Eastern, saw the success of two groups—Black Wings and Tuskegee Airmen, Inc.—and decided that black airline pilots needed an organization of their own. As a result, he created the Organization of Black Airline Pilots (OBAP). At that time US passenger airlines and freight carriers employed only 80 black pilots. The mission of OBAP was to prepare young blacks and other minorities for a career in aviation, and to ensure that the airlines engaged in fair employment practices. OBAP also worked with the Smithsonian's National Air and Space Museum in Washington, DC to create the "Black Wings" exhibit, which opened in 1982. The exhibit, which continues to grow and evolve, showcases the often overlooked contributions of blacks to the field of aviation.[3]

Ten years after OBAP's founding, nearly 400 African-American pilots were flying commercially. By 2006 that number had jumped to 674, including 14 female pilots. Of course, the percentage of black pilots is still minuscule compared to the more than 71,000 pilots working for US carriers. As OBAP (2010) has pointed out:

> The struggle to expand African-American pilot presence in the faces of unfair hiring/retention practices continues to be an uphill effort and promises to become increasingly difficult as the generation of black pilots (hired in the 60s) has already begun to reach retirement age. Additionally, the military, which serves as a traditional source of airline pilots, especially black pilots, is rapidly being downsized.

While changes in the law and hiring practices over the past half-century have eliminated most of the blatant, overt manifestations of discrimination in the airline industry that once kept women and minorities out of the cockpits and passenger cabins, other forms of bias continue to plague the field. They may be subtle and less visible, but they are no less pernicious to the cause of equality and civil rights. In the next chapter James E. Sulton, III will address some of the ongoing problems along with possible solutions for overcoming racial and gender barriers in flight training.

Notes

1 Further information on Bessie Coleman can be found at www.BessieColeman.com.
2 For more information see www.acepilots.com.
3 For more information visit the Black Wings website at: https://airandspace.si.edu/explore-and-learn/topics/blackwings/index.cfm.

References

AvStop.com. n.d. Black airline pilots: August Martin. Available at: http://avstop.com/history/blackairlines/augustmartin.htm [accessed January 21, 2018].

Hardesty, V. 2008. *Black Wings: Courageous Stories of African Americans in Aviation and Space History*. New York: Collins.

Jet. 1997 (May 12). First black flight attendant is still fighting racism (Ruth Carol Taylor). *Jet*. Available at: www.highbeam.com/doc/1G1-19391567.html [accessed July 2, 2018].

LeDuff, C. 1998 (Nov 19). Perry H. Young Jr., 79, pioneering pilot, dies. 1998. *New York Times*. Available at: www.nytimes.com/1998/11/19/nyregion/perry-h-young-jr-79-pioneering-pilot-dies.html [accessed July 2, 2018].

Nugent, J.P. 1971. *The Black Eagle*. New York: Stein & Day.

Organization of Black Aerospace Professionals (OBAP). 2010. Available at: https://obap.org/aboutus/aboutus-history.asp [accessed June 9, 2010].

Poughkeepsie Journal. 1964 (Feb 10). Local negro pilot found dead, suicide suspected.

Whitlock, F. 2009. *Turbulence before Takeoff: The Life and Times of Aviation Pioneer Marlon DeWitt Green*. Brule, WI: Cable Publishing.

15

GENDER AND RACIAL BARRIERS IN FLIGHT TRAINING

James E. Sulton, III

> The presence of white men in key aviation jobs . . . is the legacy of both explicit discrimination in hiring and an internal culture that from the beginning of commercial aviation gave heavy emphasis to the masculine nature of flying. (Hansen & Oster 1997: 114)

Flight training began with Orville and Wilbur Wright in 1910 at their flight school in Montgomery, Alabama. While other inventors of "flying machines" were challenging the control they exerted over the flying market, the Wright brothers needed pilots to demonstrate their aircraft. They developed a touring company that conducted flying exhibitions and lessons to promote the sale of their airplanes. The lessons required brief explanations of the operation of aircraft and were the beginning of aviation education. At the time, there were fewer than ten fully qualified aviators in the world—most of whom were white men (Ennels 2002).

The introduction of women represented the first sign of diversity in the industry. In 1910 Baroness Raymonde de la Roche (also Laroche) became the first woman in the world to earn a pilot's license after being trained in Chalones, France. As discussed in the previous chapter, in 1921 Bessie Coleman traveled to Paris, France in search of flight training and became the first African-American—male or female—to earn an international pilot's license. A female captain for a major US airline, however, was not seen until 1986 when American Airlines provided Beverley Bass with captain's stripes. Despite these accomplishments, the aviation industry remains overwhelming white and male.

Today flight training and aviation education are conducted at flight schools located at airports—known as Fixed Based Operators (FBOs)—institutions of higher education, and high schools. These organizations share a common interest in advancing the education of those who wish to participate in the industry.

For example, Florida's Embry–Riddle Aeronautical University (2017) describes its curriculum as being "the world's most comprehensive collection of academic programs focused in aviation, aerospace, business, engineering, and security." At Aviation High School in Long Island City, New York, "students complete both rigorous vocational and academic programs that provide excellent preparation for aviation-related careers as well as college" (Aviation High School 2017). Institutions of aviation education should be especially sensitive to the needs of women and minorities to ensure their success and to take advantage of this historically underutilized source of pilots.

In order for more women and minorities to successfully become pilots, the obstacles confronting them must be addressed and eliminated. The principal concerns are a lack of exposure to aviation careers and a dearth of role models within the industry (Sharp 1994). In 2016, of the 104,382 instructor pilots in the US, only 6,683 (6.4 percent) were women (Federal Aviation Administration 2017). Moreover, of the 157,894 Airline Transport-rated pilots, just 6,705 (4.2 percent) were women. As a result, there have been few role models to inspire women and minorities to pursue advanced flight ratings, which exacerbates the paucity of underrepresented groups in aviation.

Flight instruction of women and minorities

Having suitable flight instructors is regarded as the most critical element of the flight-training process (Eichenberger 1990; Federal Aviation Administration 2007b; Rodwell 2003). They should be "well qualified technically, possess the desired teaching skills and motivation, and demonstrate the ability to implement these [flight] skills" (Rodwell 2003: 239). The profession is primarily dominated by white male pilots with relatively little experience and whose long-term goals are centered on other facets of the industry, including the airlines and corporate aviation (Eichenberger 1990; Rodwell 2003; Federal Aviation Administration 2007a).

The Federal Aviation Administration (FAA) provides flight instructors with one of the greatest tasks in the industry: full responsibility for student flight training. Operating in this role, it is the responsibility of the flight instructor to train pilots in all areas of knowledge and skills needed to safely operate an aircraft. This training includes flying, judgment, and decision-making skills (Federal Aviation Administration 2007b). Because instructors forge a direct link between aspiring flight students and their careers as pilots, they are at the crux of the learning process and become natural role models. As role models flight instructors are at the highest level of importance because their behavior shapes that of their students. When they exhibit poor discipline and haphazard procedures, their students are likely to follow suit, which results in a poorly trained pilot who becomes a detriment to the aviation industry (Frazier 2001).

Women and minorities are underrepresented as flight instructors, pilots, and aviation career-oriented role models, resulting in a lack of their presence in flight-

TABLE 15.1 Flight instructors 2010–2016

Gender	*2010*	*2011*	*2012*	*2013*	*2014*	*2016*
Women	6,359	6,350	6,371	6,386	6,521	6,683
Men	90,114	91,059	91,957	92,456	94,472	97,699
Total	96,473	97,409	98,328	98,842	100,993	104,382

Source: Federal Aviation Administration (2017).

training programs. The FAA's 2016 data indicates that 6.4 percent of US Certified Flight Instructors (CFIs) were women (Federal Aviation Administration 2017). These figures have been historically low, and the trend continues today. Table 15.1 depicts the disproportionate representation of women flight instructors as compared with men for the years 2010–2016. In 2010, of 96,473 flight instructors in total, only 6,359 (6.6 percent) were women. By 2016, the number of women flight instructors decreased to 6.4 percent of the total.

Varying learning styles

The operation of a sophisticated aircraft relies on the performance of a team of professionals. Differences in gender and cognitive processes result in variations of learning styles. An understanding of these differences and preferences, with appropriate consideration of learning theories, will provide for effective teaching and learning in aviation (Bye & Henley 2003). As such, effective training efforts will consider the learning style and characteristics of the learner (Henley & Bye 2003).

Learning is a lifelong process requiring different methodological approaches as one continues to age. Adults tend to approach learning differently than children. Adults want and need to learn for self-fulfillment or societal advancement. This propels them through the learning process. Children, however, are often forced to learn in a structured educational setting by parents or by legal requirements. Therefore, children's desire for learning and educational needs differ from adults because of the immediate need for critical knowledge.

Because of this discrepancy in wants and needs, different learning characteristics become evident. These critical differences have resulted in two types of learning models: pedagogy and andragogy. Pedagogy is a didactic, traditional, or teacher-directed method of learning. This is the most widely used form of instruction for adolescent learners. While this approach is effective for educating youth, instruction of adult learners requires a different approach. Andragogy is "a set of core adult learning principles that apply to all adult learning situations," and has been the basis for instructional design of adult education programs for the past three decades (Knowles et al. 1998: 64).

There are four crucial assumptions about the characteristics of adult learners, as illustrated in Table 15.2. First, adult learners shift from being entirely dependent

TABLE 15.2 Pedagogical and andragogical assumptions

Critical Element	*Pedagogy*	*Andragogy*
Self-concept	Total dependency	Movement towards independence
Role of experience	Not essential to learning process	Valuable resource in discussion and problem solving
Readiness to learn	Learn what is required and expected	Learn what is needed
Orientation to learning	Subject-centered	Problem-centered

Source: Knowles (1980).

on the instructor for guidance to being self-directed. Second, they are exposed to many experiences that result in invaluable resources for teaching and learning. Third, they become concerned more with why learning is necessary and how it can be applied to their personal lives. Finally, adult learners see education as a necessary tool for achieving their full potential in life (Knowles et al. 1998).

Studies suggest that aviation students relate significantly more to the andragogical model than students of other disciplines (Brady et al. 2001). Furthermore, students involved in collegiate aviation programs differ from some other college students. The primary distinction is that aviation students are not actively looking for a career; rather, they have found one and are taking the necessary steps to accomplish their clearly identified and specific career goal. Thus, aviation educators should utilize adult education learning strategies and methodologies to provide effective instruction to aviation students (Brady et al. 2001).

In addition to utilizing adult education learning strategies, aviation educators should develop strategies that meet the needs of women and minority flight students because their learning styles differ from those of white males. Women flight students require more confidence in the airplane, more hours before they are ready for a solo flight, and usually display a stronger fear of maneuvers-stalls, spins, and unusual attitudes in an airplane. Although only a small percentage of women choose to pursue pilot career paths—fewer still who are of color—they share the same motivation: the challenge and excitement of flying (Sitler 1998).

Learning motivation

In aviation certification and ratings are required to become a pilot. As student pilots progress through the training and education process, they only pursue the learning requirements needed to achieve the necessary skills and ratings that will allow them to reach their ultimate goal of becoming an aviation professional. This results in the application of their learning to more immediate and relevant life circumstances that will satisfy a specific need or accomplish a goal such as becoming a career pilot (Knowles et al. 1998).

When considering the motivation of women and minorities in white male-dominated arenas, inter-group perceptions often are critical factors. Some minority students perceive success in primary and secondary school as "acting white," and may view academic success as being in opposition to their societal norms (Fordham 1988; Fordham & Ogbu 1986). A recent study describes harassment by minorities against others "who have the temerity to take their studies seriously. According to the poisonous logic of the harassers, any attempt at acquiring knowledge is a form of 'acting white'" (Austen-Smith & Fryer 2005: 551). This perception results in the purposeful academic under-achievement of minority male and female students in a wide array of educational paths, especially those paths that are highly technical.

To enhance the motivation of women and minorities in flight training, aviation educators and flight instructors must anticipate the challenges of participating in the white male-dominated realm of aviation: exposure to stereotypes, prejudice, discrimination, and self-perceptions of inferiority. Moreover, these challenges should be utilized in the context of learning to identify desired outcomes and establish task-based goals that include career objectives (Kaplan & Maehr 1999).

Women and minorities can be motivated by career goals and have strong sentiments of pride and resiliency. However, women and minorities are not receiving adequate counseling on viable career options (including aviation), resulting in a lack of motivation in their educational efforts beyond high school (Fleezanis 2001). Because many flight hours are required for pilot certification, women and minorities must identify aviation as a viable career option early in their education process in order to compete successfully for aviation jobs.

Historical and current barriers faced by women and minorities

As addressed more fully in the previous chapter, the end of World War II changed the roles of organizations aimed at enhancing diversity in aviation—such as the Women's Air Force Service Pilots (WASPS) and the Ninety-Nines (the International Organization of Women Pilots)—and marked a new era of discrimination in flight training. None of the women or minorities who flew in the war were hired by major airlines as pilots. Instead, minority men were hired into service jobs—including as ground handlers and skycaps (airport porters)—while women were forced to fill roles as flight attendants or aircraft cabin cleaners. The only successful agents of change were court battles and the struggles of the Civil Rights Movement, which slowly created more opportunities in the cockpit for women and minorities (Hansen & Oster 1997).

Discrimination against women and minorities in the cockpit continues today. Airlines and the military maintain employment policies that have explicitly or implicitly barred underrepresented groups from jobs as pilots (Hansen & Oster 1997). Moreover, despite their abilities women and minorities will likely face discrimination as they pursue their goals in this science-based industry (Shalala 2007). One such example is Lt. Colonel Beverly Armstrong's struggle for equality in the United States Air Force. She experienced this discrimination first-hand, and recently prevailed after nearly 15 years of litigation. "It's not just me. The Air Force and airlines continue to exclude us [African-American women] from these positions by attacking [our] skills and questioning [our] abilities," she said (Personal communication, November 23, 2007).

Discrimination in collegiate aviation programs

Racial prejudice and discrimination also continue to exclude women and minorities from educational opportunities and contribute to stereotyping. One of the consequences of this unfair treatment is its adverse effect on career opportunities and professional advancement:

> The power of stereotypes lies in their function of justifying social relations. For example, stereotypes about Whites being overly organized and Mexicans and Blacks being stupid, lazy, and violent, work to positively uphold beliefs about White superiority and the right to dominance, while serving as negative reminders of the naturalness and necessity for racial subordination. Stereotypes solidify racial dichotomy wherein White is good and non-White is bad. Racialized stereotypes remind students, faculty, staff, and administrators that people of color are the abnormal, the marginal, because they stand in contrast to Whites, who are naturally the norm, the center. (Allen & Solorzano 2001: 248)

Therefore, as women and minorities pursue college degrees, they are held to lower expectations, are rarely encouraged to participate in fields traditionally dominated by white men, and appropriate flight-training strategies are not applied.

Additional challenges arise when colleges and universities make cosmetic attempts to avoid discrimination and prejudice. These are most clearly exemplified by implementing diversity-education plans with many shortcomings, which often include celebrations that lead to little change within educational communities. Rather than focusing on superficial events, educators should design programs that, at a minimum, are progressive, helpful for all students, and promote social justice (Nieto 2002). When this is accomplished in collegiate aviation education programs, women and minorities likely will experience additional opportunities for successful pursuit of their career goals.

Another example of discrimination is what is known as the "good ole boy" network. This creates a glass ceiling for women and minorities pursuing a collegiate aviation education. "Women must constantly break through this network of predominantly white men and their attitudes toward women—that is, women do not belong in the cockpit, they cannot do the job as well as [white] men, etc." (Luedtke 1993: 70). The glass ceiling in aviation is an unfortunate metaphor for barriers that block women and other underrepresented groups from achievement. This barrier is ubiquitous in aviation (especially in pilot and management capacities) as white men continue to dominate the industry (Hansen & Oster 1997).

RACIAL DISCRIMINATION ALLEGATIONS AGAINST UNITED AIRLINES

An article on the internet forum *FlyerTalk* by Jackie Reddy (2016) provides a recent case in racial discrimination. Reddy writes that 18 black pilots who are members of the United Coalition for Diversity asked for a federal investigation into racial discrimination at United Airlines. The coalition claims that there is a secret, racist organization called "the Vault" that prevents African-American pilots from professional advancement within the airline. The coalition also claims that pilots have received racist threats and photos. A spokesperson for United countered the claims by stating that 5 out of 8 chief pilots are persons of color, including 3 African-American male pilots. According to the activist Gary Flowers (2012), following 2010 Equal Employment Opportunity Commission (EEOC) complaints, "United Airlines hastily hired three African Americans for management positions." Flowers also claims that "African Americans and people of color are woefully underrepresented in managerial ranks in proportion to their total numbers in the United workforce."

Prejudice within collegiate aviation programs has resulted in wasted talents and skills of women and minorities. This has hampered, and will continue to hamper, the ability of the US to compete on a global scale as the country fails to utilize a much-needed resource (Singer 2006). Moreover, historical exclusions of women and minorities from careers in aviation have resulted in current imbalances of their presence in the industry. This problem can be remedied by enlarging the pool of qualified candidates by including underrepresented groups through recruitment efforts (Hansen & Oster 1997; see also textbox).

Current barriers faced by African-American women

Barriers to the participation of African-American women in flight training include the high cost of training, absence of role models, lack of available aviation career information, and an alienating climate that makes them feel unwelcome in the industry. In order to enhance the presence of African-American women in flight-training programs, these barriers can be eliminated.

Cost

Many potential pilots are prevented from enrolling in or drop out of flight-training programs because of excessive costs. Scholarships are difficult to obtain, and many students worry that few or no jobs will be available upon course completion (Whitnah 1998). Furthermore, in a collegiate environment, flight training can add more than $7,500 annually to the regular cost of post-secondary education (Hansen & Oster 1997). With the 2016–2017 average annual tuition costs of four-year private and public college courses exceeding $33,480 and $9,650 respectively (College Board 2017), any additional flight-training costs would only further hinder the participation of African-American women. Should they opt for flight training at fixed-base operators (FBOs), their average expected costs would be $7,000 for a private pilot certification (AOPA 2018).

Lack of role models and mentors

In addition to the high costs of flight training, African-American female pilots are at a unique disadvantage due to a lack of role models and mentors who could provide needed industry support. African-American women have traditionally been underrepresented as flight instructors and have fewer opportunities to explore their interests in this exciting field. As a result, there have been few role models and mentors for underrepresented groups of African-American female pilots in succeeding generations (Sharp 1994). Role models and mentors have been "identified as an important component in the training process particularly for women and minority groups" (Turney 2004: 224). Furthermore, women and minorities continue to be underrepresented as faculty members in science-based programs, and are rarely seen as role models or mentors in these programs (Hamilton 2004; Black Issues in Higher Education 2004). Because one of the greatest benefits of having an appropriate representation of women and minorities on faculties is to provide mentors, motivators, and role models, underrepresented groups are again without a valuable resource (Louque 1994).

Lack of available information

Without access to role models and mentors, potential African-American women pilots are forced to search for information on their own. Airlines do not regularly publish this type of information (Hansen & Oster 1997). In fact, published information describing the hiring process, selection criteria, and roles of professional pilots is scarce and contributes to the unawareness of careers in aviation (Fessenden 2002). By contrast, white men often have family members or colleagues in the aviation industry, providing them with easy access to career information. Their resources provide crucial career information and professional guidance.

Unwelcoming environment

When African-American women gain access to the aviation industry despite the previously mentioned barriers, they are often subjected to an unwelcoming environment that is dominated by white men. The environment often includes attitudes, behavior, language, and actions that are meant to deter their participation. This has caused many to leave without accomplishing their goals, and has contributed to the further exclusion of underrepresented groups from careers as pilots (Turney 2000).

Conclusion

Aviation is an essential component of the current and future global marketplace. To maintain its competitive edge, the US must create additional opportunities for all its citizens to participate fully in the field of aviation. Eliminating the barriers women and minorities encounter in pursuing their careers as pilots is an important first step.

References

Aircraft Owners and Pilots Association (AOPA). 2018. Learn to fly. Available at: www.aopa.org/training-and-safety/learn-to-fly [accessed January 30, 2018].

Allen, W.R. & Solorzano, D. 2001. Affirmative Action, educational equity and campus racial climate: a case study of the University of Michigan Law School. *Berkeley La Raza Law Journal.* 12 (2): 238–320. Available at: http://scholarship.law.berkeley.edu/blrlj/vol12/iss2/6 [retrieved January 30, 2018].

Austen-Smith, D. & Fryer, Jr., R.G. 2005 (May 1). An economic analysis of "acting white". *Quarterly Journal of Economics.* 120 (2): 551–583.

Aviation High School. 2017. *School History.* Available at: www.aviationhs.net/school_history [accessed January 30, 2018].

Black Issues in Higher Education. 2004 (Feb 12). Noteworthy news: women, minorities rare on science, engineering faculties. *Black Issues in Higher Education.* 20 (26): 19.

Brady, T., Stolzer, A., Muller, B., & Schaum, D. 2001. A comparison of the learning styles of aviation and non-aviation college students. *Journal of Aviation/Aerospace Education &* Research. 11 (1). Retrieved from https://commons.erau.edu/jaaer/vol11/iss1/1 [accessed January 30, 2018].

Bye, J., & Henley, I. 2003. Learning theories and their application to aviation. In Henley, I. (ed.), *Aviation Education and Training.* Burlington, VT: Ashgate: 3–29.

College Board. 2017. Trends in college pricing: 2016. Available at: https://trends.collegeboard.org/sites/default/files/2016-trends-college-pricing-web_0.pdf [accessed January 30, 2018].

Embry–Riddle Aeronautical University. 2017. Admissions. Available at: http://daytonabeach.erau.edu/admissions/index.html [accessed January 30, 2018].

Eichenberger, J.A. 1990. *General of Aviation Law.* Blue Ridge Summit, PA: TAB Books.

Ennels, J. A. 2002. The "Wright Stuff" pilot training at America's first civilian flying school. *Air Power History.* 49: 22–32.

Federal Aviation Administration. 2017. US civil airmen statistics. Available at: www.faa.gov/data_research/aviation_data_statistics/civil_airmen_statistics/ [accessed January 12, 2018].

Federal Aviation Administration. 2007a. 2006 US civil airmen statistics. Retrieved from www.faa.gov/data_statistics/aviation_data_statistics/civil_airmen_statistics/2006/ [accessed November 20, 2007].

Federal Aviation Administration. 2007b. *Airplane Flying Handbook.* New York: Skyhorse Publishing [accessed November 20, 2007].

Fessenden, N. B. 2002. *Exploring the Experiences of Middle-Aged Women in Becoming Private Pilots: A Phenomenological Study.* Unpublished doctoral dissertation, Capella University, Minneapolis.

Fleezanis, N. 2001. *Baccalaureate Degree Persistence among Adult Learners: The Case of Female African-Americans*. Lansing: Michigan State University.

Flowers, G. 2012 (Jun 1). United Airlines' unfriendly skies for black pilots. *Roland Martin Reports*. Retrieved from: http://rolandmartinreports.com/blog/2012/06/gary-l-flowers-united-airlines-unfriendly-skies-for-black-pilots/ [accessed June 7, 2017].

Fordham, S. 1988 (Feb). Racelessness as a factor in black students' school success: pragmatic strategy or pyrrhic victory? *Harvard Educational Review*. 58 (1): 54–84.

Fordham, S. & Ogbu, J.U. 1986. Black students' school success: coping with the "burden of 'acting white' ". *Urban Review*. 18 (3): 176–206.

Frazier, D. 2001. *Controlling Pilot Error, Training & Instruction*. New York: McGraw-Hill.

Hamilton, K. 2004 (Mar 25). Faculty science positions continue to elude women of color. *Black Issues in Higher Education*. 21 (3): 36–39.

Hansen, J.S., & Oster, C.V. 1997. Taking flight education and training for aviation careers. Washington, DC: National Academy Press.

Henley, I.M.A., & Bye, J. 2003. Learning styles, multiple intelligences and personality types. In Henley, I. (ed.), *Aviation Education and Training*. Burlington, VT: Ashgate: 92–117.

Kaplan, A., & Maehr, M.L. 1999. Enhancing the motivation of African American students: an achievement goal theory perspective. *Journal of Negro Education*. 68: 23–42.

Knowles, M.S. 1980. *The Modern Practice of Adult Education: From Pedagogy to Andragogy*. New York: Cambridge Adult Education.

Knowles, M., Holton, E., and Swanson, R. 1998. *The Adult Learner*. Houston, TX: Gulf.

Louque, A. 1994. *The Participation of Minorities in Higher Education*. Doctoral dissertation, Pepperdine University.

Luedtke, J.R. 1993. *Maximizing Participation of Women in Collegiate Aviation Education* Doctoral dissertation, Oklahoma State University.

Nieto, S. 2002. *Language, Culture, and Teaching Critical Perspectives for a New Century*. Mahwah, NJ: Erlbaum.

Reddy, J. 2016 (Sept 16). Black pilots call for investigation into United Airlines discrimination. *Flyertalk*. Retrieved from: www.flyertalk.com/articles/32274.html [accessed June 7, 2017].

Rodwell, J.F. 2003. *Essentials of Aviation Management a Guide for Aviation Service Business*. Dubuque, IA: Kendall/Hunt.

Shalala, D. 2007. *Beyond Bias and Barriers: Fulfilling the Potential of Women in Academic Science and Engineering*. Retrieved from www7.nationalacademies.org/ocga/testimony/Beyond_Bias_and_Barriers. asp [accessed November 17, 2007].

Sharp, A.J. 1994. *Academic Achievement, Career Expectations and Self-Efficacy of African American Students in Airway Science*. Doctoral dissertation, University of Miami.

Singer, M. 2006. Beyond bias and barriers. *Science*. 314 (5801): 893–893.

Sitler, R.L. 1998. *The Cockpit Classroom: Women's Perceptions of Learning to Fly and Implications for Flight Curriculum and Instruction*. Unpublished doctoral dissertation, Kent State University, Ohio.

Turney, M.A. 2004. *Tapping Diverse Talent in Aviation: Culture, Gender, and Diversity*. Aldershot, UK: Ashgate.

Whitnah, D.R. 1998. *US Department of Transportation: A Reference History*. Westport, CT: Greenwood.

16

DIVERSITY RECRUITING IN AVIATION MAINTENANCE

Paul Foster

Imagine a typical day at a major airport in the US, such as Los Angeles International Airport. Normally, you will see minorities working in the following positions: black males can normally be found working as Skycaps; black females are generally working the security checkpoints; females in general dominate the ticket counters for most airlines. While Latino employees are generally working in the catering services, Latina workers can be found in housekeeping, cleaning aircraft for the next flight. Asians (females) are usually found working the food service areas in the airport terminal. As one gets closer to the aircraft, the ethnicity of the employees begins to change. One sees individuals in uniforms such as flight attendants, who are typically female. Some of the males working as flight attendants are often mistaken for pilots. Other individuals found in uniform are pilots, who are typically white and male. If you happen to look out the window onto the ramp area you will notice individuals working around the aircraft, and, again, the majority of these workers are white males, with a spattering of other ethnicities, and also a few females.

Hansen and Oster (1997) have noted that aviation occupations, although changing, do not mirror the diversity of the overall American work force. Although aviation employees as a group are not dramatically different in sex, race, and ethnic makeup from all employees, the representation of women and racial minorities varies substantially from occupation to occupation. Hansen and Oster have also noted that pilots and senior managers continue to be predominantly white and male; and aviation maintenance technicians are less likely to be white than are pilots and managers, but are still mostly men.

As a major training ground for aviation maintenance technicians, the US Air Force's aviation-related work force is noticeably more diverse than the civilian aviation work force. Minorities and women are better represented in military aviation specialties than they used to be; but, with the exception of minority male technicians, their presence in these jobs is small and still significantly lagging their representation in the overall population. One could argue that the aviation culture should represent the nation's culture and not continue to represent the twentieth-century male (white)-dominated culture. Encouraging minorities from all backgrounds to consider aviation careers will enable the industry to draw from a much larger pool of prospective students and, eventually, have a dramatic effect on diversity in the aviation work force.

Minorities in aviation maintenance

The findings from FAA research beginning in 1993 indicate that a critical shortage of aviation maintenance professionals will be upon us before most of the current high school students will be available for employment (FAA n.d. and FAA 1993). This shortage is the result of a large number of retirements, a large number of experienced mechanics leaving for higher-paying jobs or transferring to other occupations, and fewer entrants from the military. If not enough mechanics are available to make repairs, airlines will be forced to park planes that would normally fly.

Aviation maintenance-training programs need only look to the underrepresented population of minorities to find their future generations of students. Currently, this population is virtually untapped by aviation technical schools. The prospective students need to be made aware of the potential for excellent careers, which can be achieved by attending aviation technical programs. Given adequate preparation, women have proven to be excellent employees in various aviation occupations. The Federal Glass Ceiling Commission (US DOL 1995) found that 57 percent of the working population was comprised of women and/or minorities. One would think that the number of women in aviation would match that figure.

In a study that examined how gender stereotyping creates and maintains barriers for women who wish to enter male-dominated fields of employment, Ruble et al. (1984) concluded that gender stereotyping does exist, and that it operates in ways that limit the employment possibilities and advancement of women in the work force. Douglas (1991) also noted that, regardless of the color of their skin, women had to overcome several obstacles. First and foremost, they were not wanted in the workplace. The most difficult prejudice for them to overcome was not the fact that women's mechanical aptitude lacked social acceptance but, rather, the notion that women "belonged in the home."

Statistics compiled by the FAA suggest that women are making significant strides toward greater representation in the most popular aviation careers. However, these statistics also reveal that the representation of women among the ranks of certified mechanics remains very low. Estimates from 1995 to 2000 placed the number of active female aircraft mechanics at between 3,914 and 5,047. This represented slightly more than 1 percent of the total number of active mechanics in the work force. From 2001 to 2014, the number of active female aircraft mechanics was reported as 5,295–8,141 (FAA 2018). This represents a dramatic increase to 2.4 percent of the total number of active mechanics in the work force.

Declining enrollments by all populations in aviation technical programs have also been linked to the poor professional status of aviation maintenance careers. Aviation's use of the term "mechanic" does not project exciting thoughts among young minorities and their families. Phillips (2000) notes that although that perception may be changing, the word "mechanic" still has some negative connotations attached to it, and youth and their parents do not always view blue-collar employment as very desirable. Fiorino (2000) argues that the industry's problem (challenge) is to remove the stigma of the aviation mechanic. He recommends that the industry move away from the grease monkey mentality to one that portrays the professionalism and career potential represented by air carrier and other aviation pursuits. It is important for technical education institutions to instill in their students the pride and ethical roots of this emerging new professionalism, and for minority parents to encourage their children to consider aviation maintenance as a viable career option.

Deborah Douglas (1991) studied several women who encountered difficulties in learning certain specialties because of their lack of experience with tools and machinery prior to attending military technical schools. She notes that the military had not anticipated this situation and did not make any curriculum adjustments to compensate. The military's attitude, as viewed by Douglas, was that women would have to adjust to the existing standards the same way men did, whatever the consequences. She argued that this did affect the participation of women as some dropped out of the specialty, and others were discouraged from even trying.

Without exception, all maintenance schools have subliminal and covert barriers to learning for women. Baty (1999) notes that traditional maintenance training institutions and the industrial workplace have been wholly male; women in these environments have fewer (if any) designated facilities to support their involvement as maintenance professionals. Baty claims that rest room facilities for women were generally makeshift or were located at greater distances from the workplace or classroom, and often lacked adequate accommodations such as lockers, showers, and other important considerations.

The role of aviation organizations

Despite these unfortunate circumstances there are organizations attempting to recruit and retain minorities in aviation. These include: Women in Aviation International (WAI); the International Black Aerospace Council; the Organization of Black Aerospace Professionals (OBAP, formerly the Organization of Black Airline Pilots); and Negro Airmen International (NAI), which primarily recruits pilots. Louque (1994: 16) notes that the support group environment generated within minority organizations has resulted in a dramatic increase in the recruitment and retention success rate of minorities in traditionally white male-dominated fields. Schreiber (2001) affirms this concept, noting that African-Americans are highly responsive to organizations and efforts that give back to the community, viewing such efforts as a gateway to mutual respect rather than an attempt to exploit the market solely as a revenue opportunity.

Baty noted that WAI boasts of having more than 100 women from three aviation programs: aeronautical technology, flight, and administration. Louque supports the concept of critical mass (1994: 19), explaining that the more minorities enroll in a program, the more minorities the program can attract. The Association for Women in Aviation Maintenance (AWAM) is a non-profit organization formed with the purpose of championing women's professional growth and enrichment in the aviation maintenance fields. AWAM provides opportunities for sharing information and networking, education, fostering a sense of community, and increasing public awareness of women in the industry (Baty 1999).

Issues in hiring and retention

Adams (2000) notes that mechanics have long been the airline industry's backbone, invisible to passengers yet indispensable to their flight. Across the industry many mechanics are leaving their employers and sometimes the profession for better-paying jobs and better hours. Neither the military nor private schools that train mechanics are able to meet the demand for mechanics. Phillips (2000) notes that these schools have not seen the significant and necessary increase in enrollment which would be needed to attain the projected demand for mechanics.

During the mid-1970s, following the Vietnam War, many military mechanics filled jobs in the civilian work force. A factor, noted by Phillips, affecting both airplane and helicopter operators, was the impending retirement of Vietnam-era technicians, whose nearly 40 years of knowledge and experience could not easily be replaced. As a result, general aviation operators attempted to increase wages and benefits in order to retain these employees. Drucker (1999) indicates, however, that manual workers who have been working for 40 years are physically and mentally tired long before they reach traditional retirement age. Phillips predicts that many experienced professionals in their 50s and 60s, therefore, could be expected to retire soon. He anticipates that their retirement may not be as gradual or planned as the industry might like; but their departure from the work force is expected to create a need for another 4,500 aviation mechanics per year for 10 years. Although no one is forecasting a mass retirement, Phillips states that many aviation mechanics are eligible, and that the rate of retirement is expected to be higher than normal.

Marilyn Adams (2000) warns that if attrition does not slow and more new mechanics are not trained, the result will be flight delays, late packages, and aircraft sitting on the ground. Another area of interest noted by Adams is when other industries siphon off valued airline mechanics' skills. She noted that in 1998, aviation maintenance schools graduated 4,510 men and women with mechanics certificates, but only 3,338 took aviation jobs. One-fourth of the graduates were snapped up by other industries—often offering better starting pay and hours—such as power companies, electronics firms, the railroad and auto industries, and amusement parks. Adams stressed that these non-aviation industries recognize the technical skills an aviation mechanic or technician possesses and are willing to pay superior wages. Adams notes the reason they need these skilled aviation mechanics or technicians is to maintain sophisticated arcade rides, elevators, and electronic business systems.

Publicity and marketing involves public relations and communications skills, which are used to reach the largest possible audience of persons qualified for and interested in aviation maintenance as a career. Recruitment committees will need to approach publicity and marketing targeted at minorities with great care because Henry (1995) indicates that stereotypes, misperceptions, and myths are more prevalent in special marketing, such as ethnic or minority marketing, than almost any other field. He also warns that the images perpetuated by public relations have not fared very favorably, and that it is important to realize the effects such depictions have had on minorities' self-respect, self-esteem, self-concept, and self-identity.

Aviation maintenance training institutions and the industrial workplace traditionally have been made up mostly of white males, while minorities in these environments have fewer, if any, designated organizations to support their involvement as maintenance professionals. Fowler (1992) noted unequal expectations for minority students because his research found that, from kindergarten on, they are expected to be neither technical nor scientific. They are not encouraged to do science experiments in elementary school, and they are not encouraged to take physics in high school. Leach and Roberts (1988) suggested that there was a need to upgrade the basic skills of women and minority students and to provide the academic support needed for success in high-tech occupational training programs. They found three reasons women had problems articulating through technical programs: recruitment, academic readiness, and retention.

The 1993 FAA study showed that the lack of qualified women and minorities in the hiring pool may be primarily due to their lack of awareness of the diversity of lucrative aviation career opportunities open to them. The FAA 1993 Research Committee reported that additional barriers to pursuing nontraditional aviation careers are the lack of mentoring programs, management training, and opportunities for career development. Mentoring is a key step to managing cultural change within universities and creating alignment with their values and strategic direction. The mentoring process links existing and new programs in a manner that promotes overall professional growth throughout the aviation industry (Galbraith and Cohen 1995). The lack of mentors could be offset by the recruitment of more minority faculty members. Schreiber notes that mentoring programs have proven to be highly effective in assuring the long-term success and retention of a diverse work force. In such programs Schreiber notes that established and successful executives agree to mentor younger employees of their own ethnic background. He recommends assigning majority and minority mentors to advise and assist minority employees in working successfully with members of the traditional work force, and further recommends making successful mentoring a significant objective in performance appraisals.

Minorities in higher education

The hiring of minority faculty may provide some relief for the problem of mentoring up-and-coming minority students; however, Louque (1994) found that there would not be a large pool of minority professors until more bright, young minorities are attracted to the field. Opp and Smith (1995) found that, except for Native-American students, all minority groups are underrepresented in colleges when compared to their proportional representation in the overall US population. Nettles and Perna (1997) noted that African-Americans comprised only 4.9 percent of America's college and university teaching faculty, a considerable underrepresentation relative to their 12.5 percent share of the US population and nearly one-half of their 9.6 percent share of students enrolled in higher education. Dilworth (1984) reported that within the black community, people recognized and valued the significant contribution African-American teachers have made to progress, as these teachers have inspired their students to become physicians, lawyers, engineers, and other professionals. One could also argue that African-American teachers are needed to inspire African-American students in integrated schools.

McKenzie (1984) advocated that teachers' attitudes about themselves, their careers, their schools and assignments, and their students must be positive. He further stated that educators should feel good about being teachers and members of an honorable profession, not just believe that they are serving time with the students. Black teachers should be prepared to go beyond the call of duty in urban schools, or in any area where African-American students are in need. Not only is it essential to teach, but one also needs to serve as counselor, therapist of sorts (listener), surrogate parent, spiritual advisor, nurse, custodian, lay engineer, police officer, and role model. There lies a challenge for the black teacher: to be at all times in all places the kind of person students respect, appreciate, and admire. In short, educators teach even when they are not in the classroom.

Another barrier for minority students and maintenance professionals, as noted by Douglas (1991), has been the display of materials with offensive racial and sexual overtones, and outright hostility and harassment. The display or use of sexually or racially offensive material is a form of harassment causing people to feel threatened, humiliated, patronized, or deliberately excluded. Douglas also notes that in many cases, it creates an intimidating or threatening environment that can lead to intense stress. It is both demeaning and depressing, often leading to an uncomfortable atmosphere with adverse effects on health, and even to resignation. These attitudes, repeatedly expressed, eventually distort students' perceptions until stereotypes and myths about all minorities, and women, are accepted as reality. Words are not value neutral. They express concepts and ideas. Words reflect society's standards. If "colorphobia" is one of its most powerful standards, then emotionally leaden racist words easily reinforce and perpetuate stereotypes.

Recruitment process

Because of their historic exclusion from much of aviation, blacks and women have less of an aviation tradition than white men, so voluntary programs are less apt to attract them without special recruitment efforts. The FAA 1993 Research Committee reported that individuals from underrepresented groups needed to know that aviation offers career opportunities to which they can aspire.

Hansen and Oster (1997) note that enlarging the pool of people interested in and qualified for aviation careers can address two concerns simultaneously. It can increase the number of minorities and women available for employment. It can also forestall any future supply problems by ensuring that the nation's increasingly diverse work force is being fully utilized by the aviation industry. Henry (1995) states that in the recruiting process it is essential to determine the audience and be selective in how to communicate to them. In defining the audience, Henry notes that it may be necessary to segment it into submarkets or several audiences. He further emphasizes that it is imperative to identify your market and understand what you want to accomplish from publicity exposure in the media in order to reach that market. Schreiber affirms that building a solid relationship with African-American consumers requires a targeted media approach. He notes that marketers must make commitments to broadcast and print media that offer a special relevance to the African-American community, rather than hoping to succeed by just throwing media weight at them or incorporating African-American consumers in broad-based media plans. Schreiber (2001) also notes that marketing messages targeted to African-Americans required an elevated level of cultural awareness, otherwise the messages risk reinforcing ethnic stereotypes and alienating the very consumers they are intended to persuade.

One of the major concerns associated with the mission of an educational institution is the recruiting, selection, and admission of qualified students. The recruitment process, according to Higgins and Hollander (1987), is implemented through search committees, advertising, interviewing, and having full awareness of legal parameters created by various civil rights laws and regulations. Adhering to a recruitment process develops and strengthens public relations, communication skills, investigations, and writing. Schreiber summarizes that remarkable results can be achieved in multicultural campaigns when the right questions are asked and the right marketing plans and methodologies are put in place.

Recruitment involves public relations and communication skills. This is accomplished through advertising. Higgins and Hollander state that the purpose of advertising is to reach the largest possible audience of persons qualified for and interested in the major course of study. Bornheimer et al. (1973) note that one of the most common methods of recruitment for educational institutions is to have faculty members, department chairpersons, or deans attend career days at high schools. Rodriguez (1986) found that a majority of the respondents in his study developed an interest in aviation between the ages of 10 and 20 years, and nearly all of the respondents felt confident in their ability to succeed in the discipline they were pursuing.

Henry (1995) noted that when a special market must be reached in order to accomplish the objective of the marketing public relations program, it is important to involve a professional who is knowledgeable and experienced in the language, culture, and nuances of that audience. Visits from minority faculty members to prospective groups may be invaluable to the possible recruitment of minorities into aviation maintenance. Eiff (1991) found that males and females agreed that career selection was most influenced by visiting potential job sites, followed by career education programs, programs by industrial representatives, and individual counseling or discussion. She also found that females were significantly more influenced by females than males. She believed that female role modeling and career influences are important in encouraging women to pursue nontraditional career opportunities.

Nieto (2000) emphasizes that US society comprises many cultures and many subgroups, each of which attaches special meanings to certain words, phrases, gestures, and expressions. Nieto also notes that minority students and teachers from the same background are often on the same wavelength simply because they have an insider's understanding of cultural meanings and, therefore, they do not have to figure out the verbal and nonverbal messages they are sending. Henry (1995) claims that it is possible that a recruiting session could be destroyed because someone misunderstands a word or gesture, even though there was no intention of causing offense. In general, positive communication will be ensured if everyone maintains a professional posture and respects others as professionals. The communication styles, according to Nieto, are only the tip of the iceberg; but they help point out the sometimes subtle ways that culture, if not understood, can interfere with learning.

The underrepresentation of women and minorities in aviation maintenance careers has been a problem confronting the aviation industry throughout its history. A white male-dominated occupation, aviation maintenance has resisted encroachment despite national trends and pressures to include greater numbers of women and minorities in all occupational fields of aviation. Despite repeated efforts by the industry to institute minority recruitment programs, a great void remains between employment equity and the number of women and minorities employed in the field of aviation. See the textbox for a discussion of the importance of a diverse workforce.

What should *we* do?

We all live in the same world. The challenge facing us is to decide how to recruit women and minorities in a way that promotes respect, understanding, and professionalism. Of course there are no easy answers, but we can focus on some possible solutions that were mentioned in this chapter. These possible solutions are categorized as follows: primary/secondary schools; aviation maintenance schools;

the aviation industry; and professional associations. In order to encourage the participation of women and minorities in aviation maintenance careers, there is a need for early exposure to the opportunities in aviation maintenance while they are still in primary and/or secondary school.

IS A DIVERSE WORKFORCE IMPORTANT?

One may wonder if the goal of a diverse workforce is necessary in aviation. For some employers, diversity may simply involve an act of compliance, meaning observing state and federal employment laws to avoid any potential penalties. While US companies may feel compelled to be more diverse, especially since discrimination is illegal, there are positive aspects to diversity hiring, recruitment, and retention. According to the Aviation Institute of Maintenance (AIM 2014), companies have discovered that the industries that have embraced the cultivation of diversity "excel when it comes to ideas, creativity, and accomplishing clear advancements in pleasing the customer base." They also point out that the customer base itself is becoming increasingly diverse. Thus, not only may a diverse workforce create a more productive workplace, it may also attract more customers, providing a win-win scenario.

Primary/secondary schools

Schools should provide career information to women and minorities during the latter stages of primary education and throughout secondary education. There is a significant lack of awareness and promotion of aviation career fields, particularly the aviation maintenance occupations, in elementary and secondary schools. Finnegan (2000) noted that aviation must do a much better job of selling itself to the younger generation. He claims that if we do not attract kids by the time they are in high school, we will probably lose them completely. Fiorino (2000) supports this idea, noting that there used to be a mystique about working with airplanes; now the romance is in computer technology.

We should encourage counselors, teachers, recruiters, and administrators to promote the investigation of nontraditional careers, such as aircraft maintenance. Such an approach could increase the number of women and minorities seeking nontraditional careers. The increased numbers could help bring into balance the gender and ethnicity inequity of the aviation work force and assist in the reduction of stereotypical barriers that may still exist.

In order to shape a well-qualified, diverse pool of aviation maintenance professionals, all segments of the youth population must be exposed to opportunities in aviation maintenance. A number of career awareness activities can provide that early exposure young people need in order to make educated career choice decisions. The career awareness activities could range from short-term exposure events, such as field trips, to long-term activities, such as mentoring and internships.

Aviation maintenance schools

Aviation maintenance technician schools should develop programs to publicize career opportunities in aviation maintenance aimed specifically to attract more women and minorities to the field. One barrier faced by females, in primary and secondary schools, is the lack of encouragement to investigate or consider nontraditional careers. This lack of encouragement reinforces psychological and cultural barriers concerning nontraditional occupations and furthers the lack of information concerning these careers. The awareness program can begin with feature articles about women and minorities in the aviation maintenance workplace. Increasing the numbers in training eventually would result in a greater number of minorities and women qualified to pursue professional careers in aviation maintenance. This outcome would also help bring into balance the sex inequity of the aviation maintenance work force and assist in the gradual reduction of the stereotypical barriers that presently exist.

Aviation industry

The aviation industry could work aggressively to increase the pool of qualified applicants from underrepresented groups by establishing internships with historically black colleges and universities and other schools and colleges with large minority enrollments. Minority and female role modeling and career influences are important in encouraging women and minorities to pursue nontraditional career opportunities. Women employed in diverse areas of the aerospace industry have proven to be outstanding professionals. Nevertheless, female representation in various aviation career facets has remained disproportionately low. The reason for the perpetuation of this imbalance can be attributed to few females preparing for a nontraditional career in aviation.

Aviation maintenance organizations and professional associations should develop in-service training programs for counselors, teachers, recruiters, and administrators to encourage them to discuss aviation career options. Although certain aviation occupations, such as being a professional pilot, are relatively well known and respected, counselors typically make little aviation career information available to students. The aviation maintenance profession is not only less well known, it is also usually considered less attractive by persons with even a rudimentary knowledge of the air transportation industry. This lack of awareness and promotion of aviation maintenance may hinder student interest in an aviation maintenance career. Lack of such information serves as a barrier to entry into such careers. This is particularly disadvantageous for minorities and women who may not have aviation maintenance role models. If these students are not aware of the attractiveness and rewards of a career field, it is unlikely that they will gravitate toward it. Counselors, teachers, and recruiters need to encourage young women and minorities to consider nontraditional career options and help them in the investigation of these paths.

Professional aviation associations

Professional aviation associations should establish mentor groups within aviation maintenance organizations. Mentoring is a key step to managing cultural change within an organization and creating an alignment with its values and strategic direction. The mentoring process links existing and new programs in a manner that promotes overall professional growth throughout the aviation industry (Galbraith and Cohen 1995). The lack of mentors could be offset by the recruitment of more minority faculty members. Schreiber (2001) notes that mentoring programs have proven to be highly effective in assuring the long-term success and retention of a diverse work force. In such programs Schreiber notes that established and successful executives agree to mentor younger employees of their own ethnic background. He recommends assigning majority and minority mentors to advise and assist minority employees in functioning successfully with members of the traditional work force. Schreiber further recommends making successful mentoring a significant objective in performance appraisals.

Conclusion

The undertakings outlined here could be accomplished through good counseling techniques and exposure to representatives from aviation maintenance, and through programs using mentoring, job shadowing, and sponsorship. These programs would be effective because they unite potential aviation maintenance technicians with role models and introduce students to the career environment.

References

Adams, M. 2000. Airlines grapple with shortage of mechanics. *USA Today*, October 16.

Aviation Institute of Maintenance (AIM). 2014 (Mar 31). Available at: www.aviationmaintenance.edu/blog/women-in-aviation/women-in-aviation [accessed March 20, 2018].

Baty, P. 1999. Women in aviation: a decade of dreams. *Aircraft Maintenance Technology*, March: 86.

Bornheimer, D.G., Burns, G.P. Burns, and Dumke, G.S. 1973. *The Faculty in Higher Education*. Danville, IL: Interstate Printers.

Dilworth, M. 1984. *Teachers Totter: A Report on Teacher Certification Issues*. Washington, DC: Howard University.

Douglas, D. 1991. *United States Women in Aviation, 1940–1985*. Washington, DC: Smithsonian Institution.

Drucker, F. 1999. *Management, Challenge for the 21st Century*. New York: HarperCollins.

Eiff, M.A. 1991. A descriptive study of the differences in career choice dynamics among male and female aviation flight students. Unpublished research paper. Southern Illinois University-Carbondale.

Federal Aviation Administration (FAA). n.d. Standards for the certification of aviation maintenance technicians using a structured- experience program. Available at: www.faa.gov/. . ./maintenance. . .maintenance/kroes_-_apprentice_final_report.doc [accessed July 11, 2018].

Federal Aviation Administration (FAA). 2018. US airmen statistics. Available at: www.faa.gov/data_research/aviation_data_statistics/civil_airmen_statistics/ [accessed July 11, 2018].

Federal Aviation Administration (FAA). 1993. *Pilots and Aviation Maintenance Technicians for the Twenty-First Century: An Assessment of Availability and Quality*. (GPO: 1993-301-719/95802). Washington, DC: US Government Printing Office.

Finnegan, B. 2000. Out on the frontier of improving aircraft maintenance. *PAMA MX Magazine*, September.

Fiorino, F. 2000. Wanted: skilled airline mechanics. *Aviation Week and Space Technology*, April 17: 91–92.

Fowler, W. 1992. *An Analysis of Enrollment, Secondary Science Programs, and the Attitudes of Black and Hispanic Students towards these Programs*. Doctoral dissertation, Pepperdine University.

Galbraith, M.W. and Cohen, N.H. (eds) 1995. *Mentoring: New Strategies and Challenges*. San Francisco: Jossey-Bass.

Hansen, J.S. and Oster, C.V. 1997. *Taking Flight: Education and Training for Aviation Careers*. Washington, DC: National Academy Press.

Henry, R. 1995. *Marketing Public Relations: The Hows that Make it Work*. Ames: Iowa State University Press.

Higgins, J. and Hollander, P. 1987. *A Guide to Successful Searches for College Personnel: Policies, Procedures, Legal Issues*. New York: College Administration Publications.

Leach, J.D. and Roberts, S.L. 1988. A soft technology: recruiting and retaining women and minorities in high tech programs. *Community, Technical, and Junior College Journal*, 59(2): 34–37.

Louque, A. 1994. *The Participation of Minorities in Higher Education*. Doctoral dissertation, Pepperdine University.

McKenzie, F. 1984. Educational excuses. *Journal of Negro Education*, 53(2): 102–103.

Nettles, M. and Perna, L. 1997. *The African American Education Data Book. Volume I: Higher and Adult Education*. Executive summary. Fairfax, VA: Frederick D. Patterson Research Institute. (ERIC Document Reproduction Service No. ED 406 870).

Nieto, S. 2000. *Affirming Diversity: the Sociopolitical Context of Multicultural Education*. New York: Addison Wesley Longman.

Opp, R.D. and Smith, A.B. 1995. *Effective Strategies for Enhancing Minority Student Recruitment in Two-Year Colleges*. Lubbock: Texas Tech. University: College of Education. (ERIC Document Reproduction Service No. ED 383 396).

Phillips, E.H. 2000. Mechanic shortage raises growth, safety concerns. *Aviation Week and Space Technology*, April 17: 82–87.

Rodriguez, C.L. 1986. A descriptive study of females preparing for a nontraditional career in aviation. Unpublished research paper, Southern Illinois University–Carbondale.

Ruble, T.L., Cohen, R., and Ruble, D.N. 1984. Sex stereotypes. *American Behavioral Scientists*, 27: 339–356.

Schreiber, A. 2001. *Multicultural Marketing: Selling to the New America*. Lincolnwood, IL: NTC/Contemporary Publishing Group.

US Department of Labor (DOL). Federal Glass Ceiling Commission. 1995. A solid investment: making full use of the nation's human capital. Available as a pdf at: www.dol.gov/oasam/programs/history/reich/reports/ceiling2.pdf [accessed July 11, 2018].

17

SAFETY, ECONOMIC FAVORITISM, OR AGE DISCRIMINATION?

The story behind the FAA's Age 60 Rule

Michael Oksner

FAR 121.383, the Federal Aviation Administration (FAA) Age 60 Rule (the Rule), was one of the most debated and contested acts of federal governance ever, and for good reason. The Rule states:

> No certificate holder may use the services of any person as a pilot on an airplane engaged in operations under this part if that person has reached his 60th birthday. No person may serve as a pilot on an airplane engaged in operations under this part if that person has reached his 60th birthday. (FAA 1996)

There is a great deal of evidence that the Rule was never intended as a safety measure, never supported with any rational basis, and no attempt was ever made by the FAA to generate the data that would either support or refute this policy. Yet it persisted for almost five decades, surviving numerous challenges by individuals, organizations, and court appeals, and stymied Congressional efforts seeking change. What follows is a condensed overview of a questionable retirement policy once mandated by the US federal government.[1]

Historical background

Prior to 1958 pilot retirement plans were imposed by the airline and not through the working agreement. In 1958 three pilots from American, Western, and TWA Airlines respectively grieved their airlines' policy of forced retirement. In all three cases a "neutral" was used to break the deadlock in the grievance board. All three neutrals ruled in favor of the pilots and ordered the pilots returned to work with seniority. Western and TWA complied, but American Airlines refused which led, in part, to a strike over the 1958 Christmas holidays (Butler et al. 2002: 9).

Shortly after the conclusion of the American strike the CEO, Major General C.R. Smith, sent a personal letter (dated February 5, 1959) from his Park Avenue residence in New York to Lieutenant General Elwood "Pete" Quesada, the first Administrator of the newly formed FAA. The letter begins: "Dear Pete: During the course of our recent negotiations with the pilot's association we found it unwilling to agree to the company's policy concerning retirement of air line pilots at age 60." The final sentence concludes: "It may be necessary for the regulatory agency to fix some suitable agre [*sic*] for retirement." Quesada and Smith had worked closely together during World War II as officers with the Army Air Corps in Europe. This letter was typed at Smith's residence, not on American Airlines stationery, as a personal request. Prior to the Smith letter there was no interest in the issue of older pilots expressed by either the new FAA or its predecessor, the Civil Aeronautics Administration. A concurrent FAA memo on the proficiency of older pilots states in part: "Ability to perform adequately under realistic flight conditions would seem to be a reasonable test of adequacy of at least the psychomotor functions of airline pilots" (Butler et al. 2002: 9–10).

In follow-up correspondence Smith provided Quesada with training statistics for pilots going through transition training on the new Boeing 707. Smith reports that it takes longer for older pilots to make this transition than "especially selected" younger pilots. It would seem intuitive that Korean War veterans with jet aircraft experience would have an easier time adjusting to the higher speeds, slow engine response times, and wider turn radiuses of the new jet than the WWII veterans who never flew in jet aircraft. There is no record of any discussion about safety- or health-related concerns; rather, the issue was with the higher costs of training the older pilots.[2]

On June 3, 1959 Quesada convened a hand–picked panel to discuss capping the retirement age for pilots. Based on the Smith training data, the panel concluded that no pilot be allowed to transition to jets past age 55, and that all airline pilots be retired before age 60. In October FAA in–house lawyers advised Quesada that the training data would not support an age rule and that "scientific or factual justification" of the Rule was not possible.[3] Undaunted, a determined Quesada announced on December 5, 1959 that the Age 60 Rule would go into effect on March 15, 1960. Quesada resigned as Administrator the following January and assumed a position on the Board of Directors at American Airlines.

As discussed above, the FAA had no hard evidence to support a forced retirement rule. Rather, they justified the Rule on the basis that the incidence of disease among humans begins to accelerate somewhere between ages 55 and 65, and that 60 was a reasonable age to assume that commercial pilots become too risky for safe cockpit operations in large passenger aircraft. Court challenges began immediately, initially by the Air Line Pilots Association (ALPA). On April 28, 1960 the US Court of Appeals, 2nd Circuit, ruled that the Rule was not discriminatory and "The Administrator did not act unreasonably in placing greater limitations on the certificates of pilots flying planes carrying large numbers of passengers who have no opportunity to select a pilot of their own choice."[4]

Having established the legitimacy of the Rule, the hurdle facing pilots seeking waivers under the provisions of FAR 121.383, the Age 60 Rule, would prove impossible to surmount. Court challenges in 1978–79 by pilots seeking waivers raised questions about the FAA retirement policy that led to a Congressional inquiry with the intention of modifying or eliminating the Rule.[5] Concurrently, following the deregulation of the airline industry in 1978, ALPA reversed their stance on the Rule from seeking change to supporting it. With ALPA and the FAA now aligned against change, the effort was watered down to a Congress-mandated study to determine if there is a relationship between the incidence of disease and aging pilots.

The relationship between age and health

The way to determine if there is a direct correlation between aging and the incidence of medical catastrophic failure would be to collect and analyze relevant data, which in 1959 FAA lawyers advised did not exist at that time. Beginning in 1960 the FAA attempted to collect relevant data on two fronts: the internal Georgetown Clinical Research Institute (GCRI) (no relation to Georgetown University); and the National Institutes of Health (NIH)-sponsored Lovelace Foundation Study of Physiologic and Psychologic Aging in Pilots. The GCRI study, primarily focused on air traffic controllers (ATCs), was terminated in 1966 with no useable data, no statistical design for analysis for data, and no analysis despite excessive cost overruns.[6]

The Lovelace group recommended a program that would use the most exacting techniques available to produce a "Profile Aging Ratio" (PAR) that could replace the arbitrary age Rule with individual testing. By 1968 the Lovelace research found results favorable to the health and fitness of older pilots that troubled some individuals within the FAA. A memo from the FAA's Federal Air Surgeon to the Lovelace Foundation states in part:

> It is generally accepted that the impetus for initiating your study (and ours) was the adoption of the 60-year rule with the resulting pressure to develop a means of selecting pilots who might be able to fly beyond the age of 60 . . . Instead of a primary concern with the upper age bracket—selecting those who are above the average 60 year old with the idea of allowing them to fly past this age—we should [instead] concern ourselves with selecting out those in the lower age group who measure below average. (Butler et al. 2002: 30)

It becomes apparent from this letter that certain members of the FAA did not want to alter the retirement age; rather, they would refine the methods to better eliminate subpar younger pilots. This obfuscates the original mandate to replace an arbitrary age limit with a physiological profile that would permit determining qualified pilots who might remain in their cockpits as long as they satisfy medical standards.

There were numerous opportunities to study aging with older pilots, but perhaps the most comprehensive was assigned to the National Institutes of Health (NIH) by the Congress in 1981.[7] The NIH tasked the National Institute on Aging (NIA), which in turn contracted the Institute of Medicine (IOM) of the National Academy of Sciences, to collect the relevant data and conduct the study. The resulting IOM study was reviewed by an 18-member panel of experts assembled by the NIA which concluded: "there is no convincing medical evidence to support age 60, or any other specific age, for mandatory pilot retirement" (Butler et al. 2002: 20). The NIA panel expressed strong dissatisfaction over the fact that: 1) the FAA had never collected relevant data; 2) the alleged data was not available; and 3) some data that might be relevant (gathered for other purposes) was never analyzed for the purposes of the study (Butler et al. 2002: 20). The panel recommended that: 1) the Rule remain in place, but 2) waivers be issued to certain pilots to develop the missing relevant data; and 3) FAR Part 135 commuter operations be brought under the Part 121 umbrella.

The FAA initially agreed to all of these recommendations, but in 1983 withdrew the notice that it would comply by alluding to "retirement and insurance plans which conform to retirement at age 60," while asserting that safety was the reason (Butler et al. 2002: 23). Members of the NIA panel expressed strong disagreement and disappointment with the FAA's response and the lack of follow-through on their findings and recommendations (see Butler et al. 2002: 21–22). The NIA report mirrors the findings of the "neutral" in the Western Airlines grievance, the Lovelace study, and the FAA's lawyers, that there exists no evidence to support a medical rationale for forced retirement at any age.

The NIA panel cited data from a 1977 pilot accident study that showed a dramatic increase in accident rates beginning precisely at age 60 (Butler et al. 2002: 19). This is to be the first of many flawed and misrepresented studies used by the FAA to support retaining the Rule. The statistical methodology in question correlates accidents per hours of flight with the age of the pilots involved (that is, 4.5 accidents per 100,000 flight hours). When airline pilot statistics are included in any such study, this super-safe cohort is eliminated at age 60 by the Rule, causing the appearance of a dramatic increase in accident rates after age 60. If the airline pilot population is left out of the study data, the accident rate actually decreases with ages well past 60. The FAA withdrawal from the panel's recommendations in 1983 relied on a more recent study of aviation accident rates using the known misleading methodology to once again allege that based on the current data they would not tamper with the Rule (Golaszewski 1983). This study was rejected and publication refused by the FAA, yet they continued to reference it for many years thereafter whenever called upon to defend the Age 60 Rule (see Butler et al. 2002: 46–47).

Following the FAA's refusal to comply with the NIA recommendations, Dr Stanley Mohler wrote to the FAA Federal Air Surgeon, Dr Frank Austin:

> The Age 60 Rule is an operational rule, FAR 121. Today there is no medical basis for the Rule . . . If the operations people want to continue to fight for the Rule, let them make their own case, as there is no longer a medical basis for it.

Austin pencils the following onto Mohler's letter:

> I believe this and Adm. (Donald D.) Engen [the FAA Administrator] believes this. He wants to keep the Age 60 Rule now. I will support the admiral in his position. When it can be done—age 60 will be eliminated (I think!) It's an ECONOMIC Issue! (Butler et al. 2002: 42–43)

The Age Discrimination Act and the fight for waivers

In 1967 the Congress passed the Age Discrimination in Employment Act (ADEA) that protects individuals who are 40 years of age or older from employment discrimination based on age. Certain occupations might be deemed a Bona Fide Occupation Qualification (BFOQ) that allows employers to ignore the ADEA if age is determined to be reasonably necessary to the normal operation of the particular business. The Department of Labor (DOL), the original overseer of the ADEA, determined that the Rule satisfied BFOQ criteria and did not challenge the FAA's allegations that older pilots compromised airline safety. ADEA enforcement was later handed over to the Equal Employment Opportunity Commission (EEOC).

Representatives from the EEOC were included on the 1981 NIA panel and submitted comments in support of the ensuing FAA Notice of Proposed Rulemaking (NPRM) to accept the NIA panel recommendations. EEOC Chairman Clarence Thomas testified before the House Select Committee on Aging in 1985 that in the EEOC's opinion the FAA should adhere to the NIA panel recommendations, concurring with NIA Director T. Franklin Williams' testimony at the same hearings. Thomas followed up with a letter to Adm. Engen urging the FAA to grant a petition by 39 pilots for exemptions from the Age 60 Rule so they could participate in a controlled study envisioned by the NIA panel.

Since then the EEOC continued to petition the FAA and the Congress to either justify the Rule with empirical evidence or eliminate it. For example, in 1993 EEOC Chairman, Tony Gallegos, submitted comments: "The Age 60 Rule should be lifted by the FAA. Medical and proficiency tests on an individual basis are effective and nondiscriminatory ways to assure that commercial pilots maintain the highest standards of safety at all."[8] The letter reiterates the arguments of many other experts that medical technology exists to replace a blanket age rule with individual testing to allow qualified pilots the option to fly past 60. The EEOC had success against corporations with court findings that "the FAA's Age 60 Rule did not establish a BFOQ as a matter law."[9] The FAA, however, remained aloof by maintaining that, as the regulatory agency responsible for safe air carrier operations, they do not have to adhere to the ADEA when mandating safety rules, with the courts siding with the FAA.

In 1988 Captain Melvin Aman et al. were denied waivers to fly past 60 by the FAA and, per prescribed procedure, appealed to the US Court of Appeals. The petition cited the recommendations of a six member panel "with impressive qualifications in the fields of cardiology, aerospace medicine and neuropsychology" (Aman 1988). The panel devised a battery of tests for assessing the fitness of pilots over age 60 that coupled with "existing operational tests required by the FAA and the airlines (such as flight simulator testing), provided an adequate basis for exempting some older pilots from the Age Sixty Rule" (Aman 1988). Petitioners also assert "that an exemption must be granted because older pilots who satisfy the protocol and existing operational tests are safer than the average pilot because performance improves with experience" (Aman 1988). However, the *Aman* court finds that no amount or form of testing is sufficient to overturn the Rule or cause waivers to be issued. On the issue of experience the court finds:

> The FAA failed to set forth a sufficient factual or legal basis for its rejection of the petitioners' claim that older pilots' edge in experience offsets any undetected physical losses. We therefore vacate the denial of the exemptions and remand to the FAA for further proceedings to provide findings and explanations addressing the deficiencies we have noted and for other appropriate proceedings not inconsistent with this order. (Aman 1988)

There is no record of an FAA reply to the *Aman* remand regarding the benefit gained by the experience of older pilots.

Two years later the same court hears another appeal by pilots denied waivers by the FAA. This majority decision finds evidence from both sides "somewhat flawed" (Baker et al. 1990). The "flawed" evidence from the FAA is, once again, the debunked, unpublished Flight Time Study of 1983. The majority admonishes the FAA:

> The FAA should not take this as a signal that the age sixty rule is sacrosanct and untouchable. Obviously, there is a great body of opinion that the time has come to move on. The agency must give serious attention to this opinion. (Baker et al. 1990)

In a scathing dissent, Judge Herbert Will argues in part:

> Pilots with tens of thousands of hours of flight time . . . suddenly are grounded on their sixtieth birthdays, even though the day before they were flying, without restrictions, and were acknowledged to be qualified and, ironically, are still deemed qualified to pilot planes with thirty passengers or less . . . The pilots have plausibly alleged that the FAA's distinctions and exemption practices are inconsistent.[10]

Further studies

Probably because of the *Baker* decision finding, that "Obviously, there is a great body of opinion that the time has come to move on," the FAA commissioned Hilton Systems, Inc. to perform studies in consideration of whether to initiate rulemaking on the Age 60 Rule (Butler et al. 2002: 23). Hilton Consolidated Database research finds that:

> The data for all the various groups of pilots were remarkably consistent in showing a modest decrease in accident rate with age . . . Our analysis provided no support for the hypothesis that the pilots of scheduled air carriers had increased accident rates as they neared age 60. (Butler et al. 2002: 40)

Since the findings hardly support a 60-and-out policy, the FAA then asked the Hilton team to conduct further analysis to determine an age where accident rates might affect safe cockpit operations. In response the Hilton team reports:

> Taken together, these [new] analyses give a hint, and a hint only, of an increase in accident rate for Class III pilots older than 63 years of age. This suggests that one could cautiously increase the retirement age to 63. (Butler et al. 2002: 26)

The Hilton Consolidated Database Study carefully avoided mixing airline pilot data with Class III general aviation data to prevent the contrived increase in accident rates as a consequence of the Age 60 Rule. Additional Hilton Systems tests showed that advanced simulator testing was a valid, accurate predictor of pilot performance in the cockpit, and could be used to determine the pilot's ability to operate aircraft regardless of age.[11]

The early 1990s was a very active, turbulent time in the history of the Rule. A series of aircraft accidents in this period involving commuter aircraft prompted the NTSB to recommend that the demarcation for Part 121 and Part 135 commercial passenger operations be modified from the current 29 passenger threshold to 9 passenger aircraft. In 1993 the FAA issued a NPRM:

> that would require certain commuter operators that now conduct operations under Part 135 to conduct those operations under Part 121 (the "Commuter Rule"). Public hearings and written comments were considered which argued pro and con regarding the necessity of an age rule with the new Part 121 inductees. (Butler et al. 2002: 48)

Although the majority of comments were in favor of changing, or even eliminating, the Rule, the FAA dismissed support for individual testing. "While science does not dictate the Age of 60, that age is within the age range during which sharp increases in disease mortality and morbidity occur" (FAA 1996).

A second Flight Time Study was published in 1991 using the same methodology as the 1983 version (never officially accepted by the FAA and debunked in both the *Aman* and the *Baker* cases). Yet the FAA relied on these findings to justify not changing the retirement age as recommended in the Hilton Consolidated Database Study (Butler et al. 2002). Discussion about the relative importance of experience versus age-related medical decline, improved medical technology, the multi-crew concept, and "Suggested Protocol for Gathering Additional Data" were all put aside for the time being because "valid selection tests did not exist" (FAA 1996). There was also a brief comment regarding the economic impact on carriers and pilots if the Rule were to change; but since the Rule was not going to be amended at this time, the FAA had not evaluated "the economic impact of a proposed change" (FAA 1996). The FAA acknowledged a recent proposed change to the maximum age for pilots in Europe that would allow pilots to operate up to 65 as long as one crewmember was under 60, to be effective July 1999 (FAA 1996).[12]

ICAO and the growing debate on the Rule

Recognition of the recommended European pilot age limits led to a discussion of exceptions allowed by the agency to the Age 60 Rule. The agency allowed foreign first officers older than 60 to fly into US airspace per ICAO standards which establish a maximum age for the pilot in command of 60; however it merely "recommends" a limit of 60 for the first officer. The FAA allowed foreign first officers over age 60 to operate within US airspace per the ICAO standards, but would not allow domestic first officers the same opportunity.[13] In fact the FAA did issue waivers in the early 1990s to captains of certain international cargo carriers to allow named captains past age 60 to operate aircraft into US airspace.[14]

In 1995 the FAA passed the "Commuter Rule" "that would require certain commuter operators that now conduct operations under Part 135 to conduct those operations under Part 121 (Butler et al. 2002: 48). A great deal of discussion regarding the Rule arose during the rulemaking process for the proposed commuter policy changes. In their final analysis the FAA issued waivers to commuter pilots over age 60 to fly until the end of 1999 to moderate the economic hardship imposed on both the carrier and their older pilots. This announced policy decision ironically fulfilled all of the recommendations put forward by the NIA panel in 1981 but later rejected: 1) to retain the age rule in Part 121 operations; 2) to allow certain pilots to fly past age 60 to develop the missing data; and 3) Part 121 operations be extended to Part 135 carriers. So the FAA authorized foreign first officers (and some captains) to operate in US airspace past age 60, and accorded the same privilege to pilots of domestic commuter operations while still refusing to issue a single waiver to pilots with similar credentials and health while piloting Part 121 aircraft with 30 or more seats.

On March 10, 1995 the ICAO Air Navigation Commission requested member states "to complete a questionnaire on the age limits in their current licensing regulations, on envisaged changes, on the operational and medical requirements applicable when an age in excess of 60 years is allowed" (ICAO 1996). In summary:

> 65.2 percent of respondents operate with a de facto age limit of 64–65 years. Also in states with no upper age limits the operational age limit is kept lower by established airline policies and may be as low as 55. No state has experienced any particular operational problems caused by older pilots. There is no question that a majority of states prefer to have an upper age limit. It may seem reasonable for the ICAO to consider increasing the upper age limit.[15]

Brazil, China, France, Lebanon, Uganda, and the US were among the countries that refused overflights by pilots older than 60. However the US did relax this policy for first officers, applying the 60 limit to the pilot in command and no age limits to the first officer.

In 2005 ICAO published the findings of an upper age limit survey which found that 83 percent of member states supported upward movement in the retirement age. ICAO then asked the states to comment on changing the age standard. Responses resulted in the adoption of the European standard—of one pilot between ages 60 and 65 in multi-piloted aircraft as long as the other pilot is under age 60—as the international standard effective November 23, 2006 (ICAO 2005).

In 2001 Senator Frank Murkowski led an effort to alleviate a severe pilot shortage by relaxing the age 60 ceiling. Some relief appeared to be on the horizon until Senator Jim Jeffords chose to leave the Republican Party and become an Independent, leaving the Democrats to assume leadership. The Democratic leadership did not support the change, and thus the effort dissolved. In 2003 Senator James M. Inhofe authored a bill to up the age limit to 65. This time the bill for change reached the Senate floor for a vote. However, in the post-9/11 operating environment airlines had suffered severe economic trauma. Almost every airline had furloughed some of their junior pilots, and there was no relief in sight. Tremendous lobbying by certain carriers and unions persuaded US Senators to not consider changing the Rule at this time, and the proposed bill was defeated.[16]

In 2001 representatives from the FAA appeared before the Senate Committee on Commerce, Science, and Transportation to testify regarding the Age 60 Rule and to report on a four-part study produced by the FAA's Civil Aeromedical Institute (CAMI), as requested by the Senate Appropriations Committee in 1999.[17] Report one was a bibliography of relevant literature. Report number two was a rehash of a study conducted by Northwestern University in 1999 and reported in the *Chicago Tribune* (Schmeltzer 1999; Butler et al. 2002: 27–28; see also Broach 1999). Both the CAMI and the Northwestern studies refer to the same data which delineates the number of accidents and incidents that occurred in individual ages from 23 to 73 for airline pilots operating under Part 121 from January 1, 1990 until June 11, 1999.[18] The data shows that there were nine incidents and/or accidents that occurred with pilots aged 60–73, yet the CAMI discussion ignores the data for this cohort in their second report. Of the 450 accidents and incidents reported during the study period, no Part 121 pilot older than 59 was identified with an accident.

Report number four was designed and directed by the Senate Appropriations Committee as the defining study to determine if the Rule might be changed to 63 per the Hilton Systems recommendations. For unknown reasons the instructions from the Appropriations Committee inexplicably replicated the flawed methodology of prior Booze and Flight Time Studies. These instructions require the CAMI to comingle data for airline pilots with non-commercial pilots (as well as "Class I or Class II medical certificates") that will knowingly alter, upward, the accident rates at ages 60 and above. The CAMI, well aware of this flawed, dishonest methodology, sent this report, without protest or comment, to the Senate.

FAA officers reported to the Senate that the combined studies did not support changing the Rule. The following January Federal Air Surgeon Jordan published a clarification for report number four that did not appear in the CAMI testimony of March 2001. In the January 2002 Aviation Medical Report Jordan states:

> The fourth report in the CAMI series examined accident rates under 14 CFR, Part 121 and 14 CFR, Part 135 (air taxi regulations) for professional pilots holding air transport or commercial pilot and Class I or II medical certificates for the period 1988–1997. An overall "U"-shaped trend was found, with pilots aged 60–63 having a statistically higher accident rate than pilots aged 55–59. However, all of the accidents involving pilots over 60 occurred in Part 135 operations. Pilots flying under Part 135-regulated operations have historically had a higher accident rate and this difference could have influenced the overall distribution when the data are combined. Therefore, no definitive conclusions about the relationship of age to accident rates for pilots engaged in commercial operations can be drawn solely on the basis of the study. (Jordan 2002)

Was this omission intentional? Report four's methodology created quite a stir among the medical community, resulting in the CAMI producing a "Final Report" which cleaned up the data search and found that:

> The trend was best described by a linear (straight-line) rather than a quadratic ("U"-shaped) function reported in previous studies on pilot age and accident rates . . . On the other hand, the null hypothesis of no difference in accident rates for the 55–59 and 60–63 age groups could not be rejected. That is, the accident rate for these two age groups appeared to be statistically the same.[19]

The combined findings from reports two and four reveal zero accidents for airline pilots aged 60–71, and probably through age 73, as the FAA does not refer to any accidents in this oldest cohort in their reports. Had this knowledge been revealed at the Senate hearings, the Murkowski age bill may have resulted in a different outcome.

The new rule

On January 30, 2007 the FAA Administrator, Marion Blakey, addressed the National Press Association, announcing that "it's time to close the book on Age 60. The retirement age for airline pilots needs to be raised. So, the FAA will propose a new rule to allow pilots to fly until they are 65." Through this announcement Blakey finally answered the *Aman* court Remand to explain why experience does not outweigh the alleged effects of aging:

> the fact of the matter is that there's a heckuva lot of experience behind those captain stripes, and we shouldn't have to lose it as early as we do. I want our older captains to be around longer to help the younger pilots rising up through the ranks. (Blakey 2007)

She praised the few hundred pilots allowed to fly past 60 with an impeccable safety record when the Commuter Rule was implemented. As she stated:

> Under our current rules, we will have captains older than 60 carrying Americans on foreign carriers originating overseas, from countries such as Canada, Australia, Israel, Japan, about three dozen countries overall. They'll be coming here, picking up Americans, and then flying them elsewhere. So you have to ask: It's safe to fly with foreign pilots on our shores, but it's not safe with our own? [In closing, Blakey concludes]: We're moving forward because it's a change whose time has come. The objections of the past don't cut it anymore. This is the right thing to do. Experience counts, it's an added margin of safety, and at the end of the day, that is what counts. Isn't it? (Blakey 2007)

Per the Administration Procedures Act, Blakey would issue a NPRM so the public, the industry, and individual pilots would have the opportunity to comment before making a final decision, albeit a foregone conclusion. With this speech Blakey argued for every pilot who petitioned for a waiver, for the medical professionals who had advocated individual testing in lieu of firing every experienced pilot whose sole reason for leaving was the celebration of their 60th birthday.

Blakey clearly states that we were expunging the most experienced pilots, and very likely reducing the level of safety in doing so. Although the first to make such an admission while still acting as Administrator, she was not the first to admit that the Rule was an economic favor and not a medical necessity. Former Administrators Adm. Donald D. Engen, T. Allan McArtor, and David Hinson have publicly admitted that the Rule was an economic issue, though only after they left office.[20] Yet Blakey never issued a waiver following her speech even though she, or any of her predecessors, could have done so if it was "in the public interest."

We may never know what covert power perpetuated the Age 60 Rule. It certainly was not medically warranted. There is no accident data that indicts older pilots are the problem; in fact quite the opposite is true. Throughout the entire Age 60 Rule era the FAA refused to collect meaningful data even when mandated by the Congress or the courts. Rather, they went out of their way to obfuscate the facts with contrived studies that were acknowledged as flawed by FAA officials. The Congress and the courts failed in their oversight to require the agency to remain within the law when enforcing this regulation. When the Congress finally did away with the Rule, the ensuing law contained no medical provisos. Nor did the FAA, or the Federal Air Surgeon, suggest, much less insist on, changes to existing medical exams for the newly created older pilot population.[21] Had the Congress not intervened, the FAA would have squandered another year's worth of pilot experience, retired on their 60th birthdays, before wallowing through the bureaucratic NPRM process to a foregone conclusion. There is no medical evidence that age 65 warrants forced retirement. Eventually we will do away with any age restrictions unless some medical or operational evidence dictates caution. That evidence does not exist today. The conduct of certain FAA officials enforcing the Age 60 Rule as a medical necessity, with a hidden agenda, lacks moral or ethical standards expected of persons in a position of power.

AGE DISCRIMINATION: THE DEBATE CONTINUES

When the US Congress raised the pilot retirement age to 65 the new rule applied to pilots flying under Part 121 operations, not Part 91 (small, non-commercial aircraft) or 135 (commuter, on-demand, and some cargo operations). In 2006 the EEOC sued ExxonMobil for age discrimination. The case lasted eight years until 2014, when the Fifth Circuit Court ruled in favor of dismissing the case. In its ruling the court concluded that age was a BFOQ of being a corporate pilot at ExxonMobil. The court found that "ExxonMobil had established, through the testimony of several medical experts, that the risk of sudden incapacitation in flight significantly increased with age, and that there are no adequate means of individually testing each pilot to determine whether a pilot was at risk to suffer such an incapacitation" (Ripple 2014). However, Ripple states that Part 91 and Part 135 operators should not assume a mandatory retirement age will always be considered legal and it will require a case-by-case analysis of the company's operations.

Notes

1 Most of the information contained herein was the product of a Petition for Exemption (Butler et al. 2002) litigated by Attorney Anthony P.X. Bothwell representing petitioners seeking waivers from the Rule. This petition is the most thorough documentation of what transpired through the life of the Rule, without which much of the information herein would never have seen daylight. The Butler Court of Appeals ruled in favor of the FAA, and the Supreme Court denied the plaintiff's request for review. The wealth of information contained in the Butler filing was very influential in moving Congress to change the age to 65. Bothwell was retained for follow-on suits—*Oksner v Blakey 2009* and *Grant Adams v the US Government* 2012—claiming violations of the Federal Tort Claims Act. Much of the information contained in the publication was the product of research done by Captain Samuel Woolsey, who spent his post-airline life gathering, dissecting, and analyzing the evidence from both sides of the debate on the Rule.

2 Letter from C.R. Smith to "Pete," dated April 30, 1959, with a note in the upper-right corner—"Mail to home address"—and also stating "containing results of the 707 training program" (see Butler et al. 2002: 10–11).

3 Memorandum from Dr H.L. Reighardt, Federal Air Surgeon, October 9, 1959. The memo further stated: "most comments had to do with reasons why the subject material itself was inappropriate and indefensible" (in Butler et al. 2002: 12).

4 *Air Line Pilots Association, Intl v Elwood R. Quesada*, argued April 8, 1960, decided April 21, 1960 (see Butler et al. 2002: 16–17).

5 Hearings before the House Select Committee on Aging 1979.

6 US House Committee on government Operations, Better Management Needed of Medical Research on Aging, House Report No. 2080, September 26, 1966.

7 Public Law 96-171 enacted December 29, 1979, directed the NIH to conduct a study concerning mandatory age retirement for pilots.

8 Comments by Tony E. Gallegos submitted "in response to notices published in the Federal Register soliciting comments about whether the [FAA] should initiate rulemaking about its [Age 60 Rule]" October 14, 1993 (see Butler et al. 2002: fn 272, 68–69, also Exhibit II).

9 Other cases include *EEOC v Boeing*; also Rockwell Intl., Grumman Corp, Lockheed, McDonnell Douglas, and Exxon–Mobil.

10 Baker et al. (1990). Dissent by Judge Will.

11 Age 60 Project: Experimental Evaluation of Pilot Performance, Hilton Systems Technical Report, January 1993.

12 In addition, the Joint Aviation Authorities (JAA) in Europe proposed harmonizing the European rule to allow pilots who have not reached the age of 65 to operate in multi-pilot operations, provided no more than one pilot in the cockpit is over the age of 60.

13 "If foreign airlines operate in the U.S., the FAA requires that the carrier adhere to the ICAO standard" (FAA 1996).

14 One example is an "extension for compliance with FAA Bulletin Number HBAT 92-06" issued to 13 B747 captains for CorseAir until October 31, 1993 (dated January 6, 1993), which permits the named captains to operate in US airspace past their 60th birthday (also Cargolux and Icelandic Air).

15 ICAO (1996), excerpts from Section 5. Discussion.

16 Senator Inhofe proposed age 65 legislation again in 2006 and 2007 with no success.

17 Statement of L. Nicholas Lacey, Director of Flight Standards Service, FAA, and Federal Air Surgeon Jon Jordan, before the Senate Committee on Commerce, Science, and Transportation, on the Age 60 Rule, March 13, 2001.

18 Data for airline pilots past age 60 is provided by those issued waivers per the Commuter Rule of 1995.

19 Methodological Issues in the Study of Airplane Accident Rates by Pilot Age: effects of Accident and Pilot Inclusion Criteria and Analytic Strategy. Dr Dana Broach, CAMI, Final Report, May 2004.

20 Investigative Report of the Center for Public Integrity, August 27, 2008.
21 First-class medical criteria did not change; however, first officers over age 60 would be required to pass a first-class medical exam every six months as do captains.

References

Amam, M.A. 1988. *Melvin M. Aman et al. v FAA*. US Court of Appeals, 7th Circuit. Argued April 15, 1988, decided September 12, 1988.

Baker, J.H. et al. 1990. *John H. Baker et al. v. FAA*. US Court of Appeals, 7th Circuit. Argued May 7, 1990, decided October 31, 1990.

Blakey, M. 2007. Comments before the National Press Association, January 30.

Broach, D. 1999. *Pilot Age and Accident Rates: A re-analysis of the 1999 Chicago Tribune Report and Discussion of Technical Considerations for Future Analyses*. FAA Civil Aeromedical Institute, Report Number 2.

Broach, D. 2004 (May). Methodological issues in the study of airplane accident rates by pilot age: effects of accident and pilot inclusion criteria and analytic strategy. Final Report. CAMI.

Butler, Dallas E. et al. 2002. *Petition for Exemption*. Department of Transportation, June 10.

Federal Aviation Administration (FAA). 1996. *The Age 60 Rule*. Disposition of comments and notice of agency decisions. 14 CFR Part 121. [Docket No. 27264]. RIN 2120-AF96.

Golaszewski, R. 1983. *The Influence of Total Flight Time, Recent Flight Time and Age on Pilot Accident Rates*. FAA Order No. DTRS57-83-P-80750, June 30. Bethesda, MD: Acumenics Research and Technology.

International Civil Aviation Organization (ICAO). 1996. Air Navigation Commission. Working Paper on upper age limits for flight crew members.

International Civil Aviation Organization. 2005. Air Navigation Commission. Working Paper on upper age limits for flight crew members.

Jordan, J. 2002 (Jan). Age and accident studies. Aviation medical report.

Ripple, G.P. 2014 (May 2). Federal Court: mandatory retirement age for pilots is not age discrimination. National Business Aviation Association. Available at: www.nbaa.org/admin/personnel/age-65/20140502-federal-court-mandatory-retirement-age-for-pilots-is-not-age-discrimination.php [accessed February 22, 2018].

Schmeltzer, J. 1999. FAA data find older hands are steadier pilots near retirement have fewer accidents. *Chicago Tribune*, July 11.

PART V

Health and the environment

This part examines a variety of issues concerning health, especially regarding pilots, followed by investigations into the impact of aircraft on the environment. The first topic regarding pilot health is discussed in Chapter 18 by Warren Jensen. Through his analysis we find that health-related issues are more complex than may first appear. He first points out that when it comes to safety we can only minimize risk; we cannot eliminate it. One of the key challenges for aviation employees is to determine which conditions fall within acceptable risk and which do not. Much responsibility belongs to the pilots themselves to assess their medical states and consider to what extent the treatments may impact their ability to operate aircraft. Jensen also shows that different sectors of the aviation industry treat medical readiness differently. Unlike civil aviation, a commanding authority controls the military setting, thereby removing some of the pilot's discretion. Ultimately, as Jensen maintains, in many areas of aviation minimizing risk is a combination of guidance—in the form of information, rules, and procedures—coupled with good decision-making.

Chapter 19 addresses health and environmental concerns in general aviation. Bob Breidenthal focuses on two main concerns: aviation gasoline, also known as avgas, and noise pollution. Concerning avgas, one of the main issues is that it is comprised of tetra-ethyl lead, which is a well-known toxin that causes neurological damage in children. Lead has been eliminated from automobiles (except for some racing fuels), so why does it continue to be utilized in some general aviation aircraft? The chapter also focuses on aircraft noise, primarily from poorly muffled engine exhausts and propellers with sonic tips. As Breidenthal claims, neither of these pollutants is essential for flight. For Breidenthal the main issue behind the lack of change in policy is economic. In a society in which the polluter does not have to pay for the damage, it is cheaper for them to pollute. In aviation an additional complication is due to liability laws since a death in a Cessna pays out much more than a death in a Chevrolet automobile. Any change in our pollution regulations depends on the EPA, and Breidenthal points out that this agency is subject to the whim of both political appointees and temptation from lobbyists.

The final two chapters in this section are devoted to the impact of commercial aviation on the environment. While Chapter 20 addresses issues that arise in flight, Chapter 21 concerns the impact of aircraft on the ground, including pollution, invasive species, and emergent diseases. In the former chapter Steven Kolmes focuses on two key environmental concerns: greenhouse gas emissions and persistent contrails. While discussing the ways in which people can offset their carbon emissions, Kolmes demonstrates that this solution is insufficient. If someone tries to help the environment by buying energy-efficient cars, light bulbs, etc. that person will still create a negative impact on the environment simply by flying as a passenger. Because of this problem, Kolmes investigates alternatives such as traveling by rail for short trips, increasing aircraft fuel efficiency, and making ATC more efficient as well. He then shows us how persistent contrails impact the environment. Persistent contrails are long cirrus clouds, and post-9/11 studies reveal that they dissipate after ten hours, much longer than previously thought. While it remains uncertain to what extent they may contribute to climate change, these contrails may be more difficult to eliminate than greenhouse gases. Kolmes provides alternatives, such as flying at lower levels and changing daily flight schedules so that the possibility of forming persistent contrails is diminished.

In the follow-up chapter Kolmes begins by examining problems of ground-level pollution arising at airports, such as water pollution involved in aircraft de-icing. He states that significant reductions in water pollution due to de-icing can be achieved with appropriate investment in modernized treatment facilities and techniques. In addition to water quality, airports are also among the most significant sources of local air pollution. Kolmes claims that several solutions could help lessen the amount of air pollution, such as single-engine taxiing, using newer aircraft, policy options, etc. By revealing how aircraft contribute to problems with invasive species and emergent diseases, Kolmes further reveals the complexity of issues that arise on the ground. As Kolmes states, often the most devastating effect of invasive species is when they attack agricultural activities. With the number of passengers increasing, the number of inspectors lags behind, which in turn allows more invasive species to enter the US. The final issue concerns ill passengers who may carry potentially harmful diseases onboard. Kolmes focuses on SARS, avian influenza, and tuberculosis in order to examine possible ways to prevent the spread of such diseases via air travel. Because of these various factors at play, Kolmes claims that management of air travel needs to control them as best it can.

18

ETHICAL ISSUES IN AVIATION MEDICINE

Warren Jensen

Proper risk management in the aviation industry requires personnel in safety-sensitive positions to be mentally and physically capable of performing complex operations. It obligates individuals to self-evaluate their physical and mental state prior to duty. When considering other factors, including the potential for significant consequences, these decisions can be difficult or made improperly. Aviation professionals are frequently required to assess their ability to perform when faced with illness, medication and alcohol use, fatigue, stress, and other potentially distracting influences in their lives.

Professional aviators are medically screened every 6 to 12 months, depending upon their age and responsibilities. While these examinations look at their current state of health and tests for conditions that could impact their ability to perform in the future, many conditions that do not require medical consultation can alter their ability to perform between these required medical screenings. The process of evaluating their current medical state and making the appropriate decisions can be easy, difficult, or maddeningly unclear. A discussion of the ethical basis of these decisions is an appropriate aid for deciding on the correct course of action in determining their medical eligibility to perform aviation duties. These issues are also shared by other individuals performing safety-sensitive duties in the aviation industry.

Acceptable risk

Choices in life represent situations that have inherent risks. Water that comes from the tap is not sterilized, distilled water, but has impurities that (hopefully) meet standards to minimize the risk of disease. In the world of safety culture, the elimination of risk is not possible, so we must strive to identify and minimize risk. The medical process to evaluate pilots is not a thorough or exhaustive medical screening as much as a conscientious review and evaluation of the conditions that would lead to an incapacitation or impairment of the individual.

The concept of acceptable risk considers that some risk is inherent to medical conditions that may exist despite due diligence on the part of pilots and their medical examiners. For example, pilots at a certain age may experience heart disease leading to a heart attack without symptoms or medical evidence prior to the event. While this is a low percentage event, it can have an unforeseeable and unintended negative effect. It is a risk to the safety of operations; but such risk is acceptable in that medical testing to identify the risk does not appear to be appropriate for all pilots with this responsibility.

One approach to identifying individuals with increased risk of heart disease is to consider pilot age. All pilots applying for a medical certificate to act as pilot in command of an aircraft with more than 19 passengers (for example, a US first-class medical certificate) are required to submit an electrocardiogram (ECG, electrical tracing of heart rhythms) to evaluate risk for heart disease. Since the incidence of heart disease is higher in this group, one needs additional monitoring to assess the risk of heart disease in these individuals.

Other conditions, such as a head injury with loss of consciousness, have the potential for incapacitation following recovery from the injury. With proper and timely medical evaluations, the potential for risk can be identified, but not completely eliminated. If pilots meet current medical standards, they return to flight duty, even though their risk of incapacitation may be greater than individuals who have not had a head injury. The increased risk due to the prior injury has been determined to be a very low percentage event, similar to the inherent acceptable risk in individuals without a significant medical history, described above. With the knowledge that humans inherently have the potential for infrequent, but foreseeable, unintended medical events, they may be medically certified.

Federal Aviation Regulations (FARs) were (and continue to be) created to maintain standards for the purpose of managing identifiable risks. Aviation medicine standards provide rules and guidelines for identifying, characterizing, and managing risks due to human physiologic performance. This process leads to considerable discussion regarding the criteria that would constitute an acceptable level of risk of impairment for individuals in safety-sensitive positions. Specifically, the challenge faced by pilots, controllers, and their managers is often: "Does this condition, and its associated (and potential) performance problems, fall within an acceptable level of risk?"

Based on research, the FAA has developed programs to grant special issuance medical certificates to individuals with a diagnosis of depression who are currently on acceptable anti-depressant medications and meet well-established monitoring guidelines. These guidelines are designed to carefully monitor capability and stability of pilots participating in the program. Similar programs are used to assist pilots with substance abuse issues who wish to be reinstated to flight activities. Through the development of these programs, individuals with previously disqualifying conditions can report and manage these conditions within guidelines. It appears these cases are within the bounds of acceptable risk due to these programs.

Medical conditions and risk assessment

In the civilian aviation industry, individuals are guided by FAR 61.53, titled "Prohibition on operations during medical deficiency," that states:

> an individual cannot act as pilot in command . . . while that person knows or has reason to know of any medical condition . . . or is taking any medication that would make that person unable to meet the requirements for the medical certificate necessary for the pilot operation. (FAR 61.53(a) 1 and 2)

The responsibility rests with the pilot to determine if their condition meets aviation medical standards, namely the ability to hold an appropriate FAA medical certificate. The quality of this decision is based on the pilot's ability to assess their medical state and how they will be impacted by their problems and treatments, as well as knowledge of the applicable regulations. All of this information must be weighed to determine the pilot's readiness to perform safety-sensitive duties.

Basic guidance regarding medical conditions that affect human performance and flight safety is provided during initial flight training and through FAA publications. Rule-based decision guidance regarding some medical conditions, medications, and alcohol is found in the FARs and the Aeronautical Information Manual. An example of the rule-based decision process would be the drug and alcohol regulation, published in FAR 91.17. Part 91.17(a) reads that no individual may act or attempt to act as a crewmember of a civil aircraft:

- Within 8 hours after the consumption of any alcoholic beverage
- While under the influence of alcohol
- While using any drug that affects a person's faculties in any way contrary to safety; or
- While having .04 percent by weight or more alcohol in the blood.

In 91.17(a) (1) the rule is stated for an eight-hour limit, but it is a criterion in addition to statements (2), (3), and (4). While statements (2), (3), and (4) appear to be rules, unless pilots can correctly assess their performance or blood alcohol level, these rules appear to be intended more for enforcement than ethical decision-making.

Not all medications or medical conditions have clear guidance or rule-based decision aids. Medical evaluations and information regarding "non-rule based" conditions may not be readily available, or are limited by cost, convenience, or availability. While aviation medical examiners have excellent guidance from the aviation regulatory agencies and support organizations (FAA 2008b), there is a gap in the information chain for most pilots and air traffic controllers trying to assess the risks based on their current medical condition. Can we quantify the risk to the safety of air traffic if a controller is working with seasonal allergies? What is the risk to safety if the controller has taken over-the-counter (OTC) medications? For how many days should a pilot be removed from flight duty following abdominal surgery? How are these questions answered?

Suicide threats, ideation, and attempts are serious concerns regarding individual reliability and aviation safety. The medical certification community feels any individual who expresses threats or ideas concerning suicide should be evaluated. In response, individuals may feel this evaluation is a career-limiting step and, as a result, may underreport these events that relate to themselves or others. In this case, individuals who deem the actions of the certifying organization as being too harsh, or that their situation will be misunderstood, may elect to omit their medical history during evaluation. Other pilots may be reluctant to report these events so as to not "ruin the career" of a colleague. This presents a challenge when screening individuals for behavior risks.

A PILOT AVOIDS A MEDICAL EVALUATION

Ron is a 45-year-old captain who was preparing for bed one evening when he had the onset of chest pressure, nausea, and shortness of breath. His symptoms were mild and unassociated with activity, emotion, or eating. He noted he was sweating and feeling anxious about the pain when he informed his wife, Nada, of his symptoms. She insisted he be evaluated in the emergency room, and they left immediately for the hospital.

En route to the hospital, Ron noted his chest pressure and other symptoms began to resolve. When they arrived at the hospital parking lot, he felt completely normal. At this point, he refused to enter the hospital for an evaluation. He told Nada that his symptoms were completely gone, and therefore were unlikely to be caused by heart disease. He reasoned that he had no risk factors for heart disease, such as smoking, family history of heart disease, high blood pressure, obesity, diabetes, or elevated cholesterol. He attributed his previous symptoms to overeating, and wanted to go home.

Nada was unhappy with Ron's decision, and reasoned that he needed to be evaluated to be medically qualified to fly. He felt his medical certification was not in question because he had not been diagnosed with a disqualifying condition and he felt fine. He said he had no intention of seeking medical evaluation unless his symptoms returned. The ride home was very quiet.

—Warren Jensen

Despite efforts to provide guidance to pilots, many questions still exist regarding the risk evaluation of foreseeable, but unintended, problems due to medical conditions. The textbox example reveals some of the challenges facing commercial pilots. Other organizations (such as the military) take a different approach to address these questions, using different regulations and support structures.

Military applications

In military settings, the issue of medical readiness-to-perform is controlled by a commanding authority, removing some of the pilot's discretion. The duty of a military flight surgeon is to evaluate pilots who have consulted physicians and/or experienced a medical condition prior to their return to flight duty. This requirement is clearly stated and offers little room for interpretation. The relationship of the flight surgeon and their unit is a fine line of promoting the health, performance, and safety of the pilots while identifying conditions that represent unacceptable risk in this demanding aspect of aviation.

Even with this level of control, decisional errors still occur, and tend to be related to the interpretation of acceptable risk. The following example demonstrates differing assessments of an injury, the power of conflicting priorities, and how they influence the determination of acceptable risk.

> As a new flight surgeon in the unit, I was asked by the wing commander to see the lieutenant colonel in flight ops to "evaluate his hand." The lieutenant colonel was wearing a brace for what he insisted was a "minor crack in a bone" on his dominant hand. The obvious swelling and deformity of his hand led to me think otherwise, and I was told the lieutenant colonel had been flying with this injury for several days.
>
> When questioned about the injury, the lieutenant colonel insisted that the injury did not need flight surgeon clearance, as his assessment was that the injury was minor and did not have the potential to affect his performance. Upon evaluation, the hand was exquisitely tender to pressure, finger motion was limited, and grip strength decreased. In my opinion, the lieutenant colonel had failed to report an obviously disqualifying condition and violated regulations to report for evaluation after the injury and prior to the resumption of flight duty.
>
> When confronted with this information, the lieutenant colonel agreed to report to the clinic for evaluation, but on the condition he could fly that afternoon prior to the clinic visit. The reasoning for his request was that he was involved in a training exercise that afternoon. (Interview with flight surgeon, 1998)

In the military model, using flight medicine consultations has clear advantages, but it remains dependent upon the quality of decision-making of the personnel involved. In many areas of aviation minimizing risk is a combination of guidance—in the form of information, rules, and procedures—coupled with good decision-making.

Remotely piloted vehicle (RPV) or unmanned aircraft system (UAS) operations

With the 2016 implementation of Federal Aviation Regulations Part 107–Small Unmanned Aircraft Systems (sUAS), new questions are raised regarding the medical certification process for remote pilots and visual observers (VOs). The sUAS, more commonly referred to as drones, weigh less than 55 pounds. According to Part 107, the sUAS need to be kept in visual line of sight and fly at or below 400 ft. They should operate during daylight hours (or civil twilight) and travel at or less than 100 mph.

Though neither the person manipulating the controls nor the VO is required to obtain an airman medical certificate, they may not participate in the operation of a drone if they know they have a physical or mental condition that could interfere with the safe operation of the sUAS (FAA 2016: 5-3). The FAA then provides some examples of medical impairments such as a temporary loss of dexterity, blurred vision, any debilitating medical condition that would make the person unable to operate the controls, etc.

As for the operation of larger vehicles, primarily used for surveillance and intelligence gathering, the medical certification standards are identical to those involving professional pilots. Current training programs in the US Air Force and aviation universities for large RPV operations also require pilot certification. Current standards for commercial pilot medical certificates are used for these operations. Further reviews and research will likely influence any change to the medical certification process for RPV operators.

The move to create alternate medical certification processes

Recently legislation has been authored to alter the medical certification process for private pilots, adding some limitations to their activities (aircraft size, flight weather conditions, etc.) while decreasing the monitoring requirements of a third-class medical certificate (Tennyson 2015). The goal of this effort is to adopt a process currently used for pilots of hot air balloons, gliders, and light sport aircraft. It has been described as a screening similar to medical standards to obtain a state-issued driver's license. Generally, this involves specific questions regarding incapacitation-related medical problems (for example, heart disease, diabetes, seizures).

The debate regards the expense and restrictions of the current system to be balanced with the assurance of medical review of conditions beyond that of a self-reporting system. Will increasing ease and availability of a third-class medical certificate compromise the safety of US national airspace? Will this reporting system be adequate to identify individuals at risk?

The nature of fatigue and stress

Some conditions that have a significant impact on human performance are not generally considered an injury or illness. Fatigue and psychological stress can affect decision-making performance when operators begin to change their risk assessments and/or choose to deviate from standard operating procedures (Campbell and Bagshaw 2002). Stress and fatigue have also been known to affect the ability of the human operator to monitor the quality of one's performance. For these reasons, fatigue and stress not only impact the quality of performance, but also the ability to assess it.

Most individuals find it is difficult to assess the performance effects of stress and fatigue. A review of Aviation Safety Reporting system data indicated that 20 percent of safety reports cited crew fatigue to be a factor (Masters and Kohn 1996). Stress and fatigue are common occurrences in the aviation professional and, while their effects are often appreciated retrospectively, ethical decision-making requires that we understand these risks in real-time situations.

> We were preparing for our last leg after a long day with many delays. We knew we were tired, but I had done this before, so we prepared for another flight. After I lost my place on the checklist for the second time, I realized my experience taught me the wrong lesson. Even though I had done this before, I wasn't [as] capable as I needed to be. I cancelled the flight, got some sleep, and we arrived the next morning. The hardest part of the decision wasn't recognizing fatigue symptoms; it was realizing its effect on me. (Interview with commercial pilot, 2006)

In professional aviation, rule-based decision-making is aided by duty time regulations, limiting the amount of flight time before a required rest period. Due to the nature of fatigue, individuals can be at risk for fatigue-related errors even though they are in compliance with crew duty time regulations.

The nature of fatigue and stress is such that they are difficult factors in self-evaluation and in predicting performance degradation. The onset can be very subtle but has a significant impact on performance. Often humans approach fatigue and stress with the attitude that perseverance can overcome their effects. This paradigm has been successful in the past but can lead to poor assumptions and significant errors in the future.

The challenge of self-medicating

Pilots confronted with medical problems often seek answers or solutions to continue their original plan to complete their missions, when possible. Not only do they have to consider the nature of the medical problem (headache, sinus congestion, etc.) but also the risks associated with treatments that are readily available without aeromedical advice. While some OTC treatments provide guidance—"do not drive or operate machinery when using this medication"—others do not. Guidance for individuals as to when they can return to duty is also not clearly stated.

Even when prescribed medications are used, advice regarding the medication or time restrictions may not be readily available. The FAA is tasked with providing guidance in this area, but efforts to develop a database to this point have been unsuccessful. Other organizations—for example, the Aircraft Owners and Pilots Association (AOPA)—have provided unofficial resources to their members.

Ethical issues in aviation medicine decision-making

Most discussions of ethical issues in aviation medicine are quick to point to cases of clear violations or those in which obvious, correct choices did not occur. In this author's experience, however, many aviation medical questions can be resolved with available guidance. Without access to sufficient information/resources, decision-making becomes much more difficult.

As mentioned above, civilian pilots hold the primary responsibility to determine their capability to fly. When trying to determine the potential adverse risks due to their medical condition, pilots must have the ability to determine the foreseeable adverse risks to safety. Symptoms of medical problems are often viewed as raw data that can be difficult to interpret. It is difficult to evaluate objectively a medical condition when it is perceived by the individual to be mild, the symptoms are vague, or when it is unclear as to whether or not a more thorough evaluation is needed.

When an individual possesses an incomplete understanding of a medical condition, or their prior experience leads them to believe that their condition does not or will not impact their ability to perform, realistic appraisal of acceptable risk is often incorrect. The issue may very well indicate the need for additional education (for instance, a current popular acne medication is prohibited due to color vision alterations). It is also observed that individuals may misapply prior experience (namely, vicarious learning) to determine their ability to perform.

Even when seeking medical advice, pilots may find that physicians are unfamiliar with aviation medicine and applicable federal regulations, and therefore may be unable to correctly advise the pilots. Adding to the confusion, patients can experience differing opinions regarding the diagnosis or the changing recommendations of medical research: "Less than 48 hours after FAA learned the anti-smoking medicine Chantix might lead to safety problems, it ordered pilots and air traffic controllers to stop taking it immediately" (Federal Aviation Administration 2008a).

The concept of respect in ethical decision-making refers to the principle of treating others befitting the dignity and worth of those persons (Pantakar et al.

2005). When pilots and air traffic controllers fail to report a disqualifying condition, violate crew rest or alcohol policies, or use of a prohibited medication, it shows a lack of respect for themselves, their colleagues, passengers, and others. Often, the decision-maker does not perceive one's own circumstance as an unacceptable risk. Even when pilots are aware of the rule that would prohibit them from flight duty, they may reason that they are still following the intent of the rule to avoid unacceptable risk.

Some individuals fail to report medical conditions that they feel should not ground them from flight duties, with the intention to bypass what they view as the onerous nature of the regulatory agencies and the review process. The lack of respect individuals have for the process of medical review can lead them to violate ethical principles by not disclosing potentially disqualifying medical conditions. This concern can even lead to individuals failing to list conditions that are acceptable, due to their perceptions that they would be disqualified without recourse.

Pilots' interpretation of their symptoms can be biased by their desire to meet their personal and professional goals. The case of the lieutenant colonel with the hand fracture demonstrated egoism on the part of the pilot, with his intention to continue flying despite his duty to report a disqualifying injury. He was able to justify his decision because his assessment of the injury led him to believe that he was fulfilling the intention of reporting regulation (since he felt the injury did not represent an increased risk to himself, others, or the assets of the Air Force). While the decision to fly may have been self-serving, the pilot justified the decision as he was also training others, which he considered his duty. In essence, this pilot was engaged in the principle of double effect (Pantakar et al. 2005) in which he was debating his ethical duty to provide training as opposed to the potential negative effect of the hand injury.

While "calling in sick" may be a common issue in the workplace today, the opposite pressures are often true for pilots. Balancing family, work schedules, and outside interests, as well as maintaining the reputation as a reliable professional, can place pilots in the double effect principle. The issue may become even more difficult if a disqualifying condition now leads to job/income instability.

Conclusion

Ethical decision-making is critical in the safety-sensitive professions of aviation. These decisions are attempts by individuals to balance duty to their careers, employers, and families while showing respect for themselves and others. The impressive safety record of the aviation industry is evidence that many difficult but correct decisions are made every day. Proper training of safety-sensitive workers in ethical decision-making is an important step in their training.

Patankar stated it well when he wrote: "it is the moral obligation of the decision-makers to exercise due diligence in foreseeing the harmful side effects of a morally permissible action" (Patankar et al. 2005: 5). Pilots, air traffic controllers, and operators of remotely piloted vehicles must remain alert and prepared to make proper decisions regarding their health and performance.

Examples for discussion

The following three case studies are meant to aid in the discussion of the ethical dimensions of pilot health.

Example 1

The student pilot failed to report his use of Adderall (a stimulant) and the diagnosis of Attention Deficit Hyperactivity Disorder (ADHD) on his application for a medical certificate. He states the condition and medication do not interfere with his ability to perform piloting duties. He reasons that he has no other medical problems, his grades in school were adequate, he had an unrestricted driver's license, and his doctor stated he had no restrictions on his activity. While the student pilot is intent on listing reasons why his medical certificate should be granted, it does not negate the problem of failing to report the medical condition on the application. The application states: "Do you now have, or have you ever had any of the following conditions?" and "Do you currently use any medication?" It is clear that the competing goals of treating his ADHD and pursuing an aviation career have led to this ethical challenge.

Example 2

During a three-day trip, a commercial pilot began to get cold and sinus congestion symptoms, prompting him to self-medicate to relieve his symptoms prior to flight. The medication he had used in the past (which he knew was approved for flight) was unavailable, so he selected another OTC medication to take just prior to flight. He was uncertain if this substitute medication is approved for flight duties.

Medication use in pilots is carefully monitored. While some medications are approved for use, others may be restricted for reasons that are not obvious. Even when medications are approved for use in flight, the conditions they are treating need to be evaluated to determine if they would affect performance or have the potential to incapacitate the pilot. In this case, the alternate medication may be approved, but without that knowledge, use of the medication would violate the ethical duty of the pilot to be properly prepared to command the flight. The condition he was treating may be considered grounding as well.

Example 3

A commercial pilot suffered an accidental, traumatic amputation of his index finger. Following surgical repair of his hand, six months of physical therapy, and a consultation with his aviation medical examiner, he was issued a new medical certificate. His medical examiner forwarded the application to the FAA, with the only notes regarding "a successful surgical repair" and "no abnormalities" on his physical exam. No mention was made of the finger amputation and other injuries to the hand.

In this case the pilot appropriately sought renewal of his medical certificate following his injury, but the certificate was issued inappropriately by the medical examiner. While the pilot's injury was not necessarily disqualifying, the inadequate documentation forwarded to the governing body was misleading, not allowing them to make an adequate evaluation of his condition. Normally, a pilot would undergo a flight test to see if an injury resulted in an impairment that would compromise safety. It appears this pilot upheld his ethical duty to report the condition and present himself for evaluation, but the medical examiner improperly reported the injury in an attempt to bypass the standard evaluation procedure.

In recent guidance, the FAA addressed the process to evaluate obstructive sleep apnea (OSA) as a risk to the restorative sleep in pilots. An incident with an airline in 2008 occurred when pilots fell asleep during a flight, with a later finding prompting review of sleep apnea screening.

References

Campbell, R. and Bagshaw, M. 2002. *Human Performance and Limitations in Aviation*. Ames: Iowa State University Press.

Federal Aviation Administration (FAA). 2016 (Jun 21). Small unmanned aircraft systems. advisory circular 107-2. Available as a pdf at: www.faa.gov/uas/media/AC_107-2_AFS-1_Signed.pdf [accessed February 17, 2018].

Federal Aviation Administration (FAA). 2008a. Anti-smoking medicine Chantix banned. Available at: www.faa.gov/news/updates/?newsId=56363 [accessed March 4, 2015].

Federal Aviation Administration (FAA). 2008b. *Guide to Aviation Medical Examiners*. Available at: www.faa.gov/about/office_org/headquarters_offices/avs/offices/aam/ame/guide/ [accessed March 4, 2015].

Masters, R. and Kohn, G. 1996. Aeromedical support of airline and civilian professional aircrew, in *Fundamentals of Aerospace Medicine*, edited by R. DeHart. Baltimore: Williams and Wilkins, 776–802.

Pantakar, M.S., Brown, J.P., and Treadwell, M.D. 2005. *Safety Ethics: Cases from Aviation, Healthcare and Occupational and Environmental Health*. Aldershot, UK: Ashgate.

Tennyson, E.A. 2015. *Measure mirrors House General Aviation Pilot Protection Act*. Aircraft Owners and Pilots' Association. Available at: www.aopa.org/News-and-Video/All-News/2014/March/11/Senate-third-class-medical-bill [accessed March 4, 2015].

19

ENVIRONMENTAL CONCERNS IN GENERAL AVIATION

Avgas and noise pollution

Robert Breidenthal

Compared to the automobile, light aircraft have a relatively small environmental impact. An automobile requires a continuous road for the entire length of a journey, while a small airplane only needs a short runway at the beginning and at the end. Roads are bad news for the environment, breaking up habitat into smaller pieces and thereby reducing the diversity of species per unit of habitat. While general aviation runways are less damaging than roads, the aircraft emit two sources of pollution that dwarf those of the automobile. Over 167,000 piston-engine aircraft operate in the United States (FAA 2016) and rely on aviation gasoline, or avgas, which contains tetra-ethyl lead (TEL). TEL is a toxin that causes brain damage in children, and has been removed from automotive fuel, except for some racing fuel. The second pollutant is excessive aircraft noise, primarily from a poorly muffled engine exhausts and propellers with sonic tips. Neither of these pollutants is essential for flight. The technological solutions for these pollutants are straightforward, but expensive. In a society in which the polluter does not have to pay for the damage, it is cheaper for them to pollute. The legal system also discourages improvements to aircraft because of liability laws.

Automobiles vs. aircraft

On a long trip most people now travel by commercial airline. For short distances the automobile is usually preferred. With the development of relatively clean automobile engines, one may be tempted to think that the most damaging environmental impact of the automobile is disappearing. However, one major deleterious influence of the automobile is not from its emissions, but rather from its roads since they tend to exist for a long time. For example, one can still find Roman roads in Europe.

A new road has a major effect on the environment. As a network of roadways penetrates an area, the natural habitat is fragmented. It turns out that the number of viable species per unit area of habitat depends on the size of the contiguous habitat (MacArthur and Wilson 1967). As the pieces are fragmented, fewer critters per unit area can survive in the remaining habitat. The effect of fragmentation is surprisingly strong. It may preferentially affect the large predators, which might require more area. An elegant experiment by ecologist Robert T. Paine showed that the removal of a predator dramatically reduces the number of species (Mills et al. 1993). For a quarter of a century he repeatedly removed starfish from an inter-tidal slope of rock off the coast of Washington State to see what would happen to the other species in the region. When this "keystone species" starfish predator was eliminated, the rock was overrun with mussels, to the essential exclusion of all other species. Paine demonstrated that the predator was necessary to achieve a rich diversity of life. Because a road network fragments the habitat, the number of species is reduced.

In contrast, an aircraft only requires one runway at the origin and another one at the destination point. The combined lengths of the runways might only be a mile, even though the trip is two orders of magnitude longer. Also, the runways do not form a continuous swath which would cut off migration routes. From this viewpoint, an aircraft clearly has a much smaller impact on the environment than a car. It is no coincidence that photographers and hunters in search of large native animals frequently fly to their destinations. While general aviation may have less environmental impact per vehicle than the automobile, there is plenty of room for improvement. The two major sources of pollution are lead from avgas and noise.

Lead

Tetra-ethyl lead (TEL) was added to gasoline in the 1930s to raise the octane rating of the fuel, corresponding to a reduction in the tendency of the engine to knock. This allowed the compression ratio of the engine to be increased, thereby improving the fuel economy and efficiency of the engine. A measure of the resistance of a fuel to knock is its octane rating. By adding side branches to the originally linear hydrocarbon molecules, the octane rating of the fuel can be increased. However, this comes at the cost of additional refinement, which is more expensive than simply adding TEL.

It turns out that lead is a toxin that damages the neurological system of children. Lead dust from car exhausts contaminated the environment, especially the big cities. Brain function and the IQs of children exposed to lead are reduced, so much so that geochemist Claire Patterson asserted in the 1970s that urban centers were unfit for children (McGrayne 2002). Chemist Derek Bryce-Smith also issued an early warning. However, even at moderate lead burdens, the brain damage is pain-free and difficult to detect. How can one determine if a child is not as intelligent as he or she would have otherwise been without the lead? Only a statistical study of the correlation between blood lead level and IQ reveals the damage. Patterson showed that there was no safe threshold. Child IQ began to fall with even minuscule concentrations of lead in the blood.

It is fascinating to see the response of the US Environmental Protection Agency (EPA) to this tragedy. The official government position was that the ambient lead concentration was just below that which caused brain damage. In other words, the EPA found no problem. As discussed below, the EPA is in the business of defending polluters. The public can tolerate subtle but real brain damage for a long time. Fortunately, the introduction of the automotive catalytic converter to lower NO_x and hydrocarbon emissions forced the elimination of TEL from auto fuel. Otherwise lead would have ruined the converter.

Even though lead is no longer generally in auto fuel, it still lingers in aviation gasoline, known as 100 octane, low-lead (100LL) avgas. Most aircraft piston engines were originally designed with a relatively low compression ratio to run on fuel with a lower octane rating. A leaded fuel with a high-octane rating is overkill for them. However, some higher compression engines cannot operate on lower octane fuels. Because it is impractical to supply several different types of fuel at each airport, the industry has standardized a single fuel that works for essentially all gasoline piston engines, 100LL avgas. Once something is established in certified aircraft, it is difficult to remove. Safety and liability issues discourage change within the FAA and throughout the aviation industry. Without effective feedback from those who suffer the damage, the pace of change is glacial. Thus, lead remains in standard avgas until at least 2018.

There are a few current alternatives. It is possible to burn automobile fuel in some airplane engines. However, this requires a special "Supplemental Type Certificate" permit, and such fuel is not universally available. The quality of automobile fuel is not as high as avgas, so impurities such as water can be problematic. There can be issues with vapor pressure as well. Finally, some auto gas contains ethanol, which is not compatible with some rubber tubing and gaskets on aircraft. Aircraft diesel engines have been developed in Germany that can operate on jet fuel. They offer improved fuel economy, although the engine is relatively heavy, and the conversion to a new engine is expensive. New combustion technology and engine management systems would permit operation of high-compression gasoline engines with a lower grade of fuel. This would allow a lower grade of avgas to be run in higher compression engines. Turbine engines use jet fuel, essentially kerosene, which, like diesel fuel, is free of lead. However, small turbine engines are not as efficient as piston engines of the same power class. Thus, jet engines are not a viable replacement for piston engines in small aircraft.

Given what we now know about the dangers of lead pollution, one might think that leaded fuel would have quickly disappeared, even in the absence of government edict. Victims of lead pollution would sue to remove the lead from the fuel. It would become so expensive to use leaded fuel that aircraft owners would immediately switch to a lead-free fuel, modifying their equipment as necessary. Alas, the system does not work that way. The EPA is in the business of issuing pollution permits, both explicit and implicit. As long as a polluter's emissions are below the officially approved levels, the polluter has a government license to pollute. The injured party would find little recourse in court. Under current laws, the tort system does not really apply to pollution. Even if you had access to a courtroom, you would need to convince a jury that there was injury. We have seen that the neurological damage from lead exposure is subtle, not like a broken arm. To whom do you compare the victim? This type of scenario is difficult to prove in court. For example, Friends of the Earth put pressure on the EPA and the FAA by asserting in a 2012 lawsuit that a 2015 deadline to release a finding of accelerated endangerment due to general aviation emissions constitutes "an unreasonable delay by the agency in performing its statutory duty" under the Clean Air Act. However, a US District Court ruled that the EPA should not be forced to rush the issuance of its report on the public health effects of lead emissions from general aviation aircraft (NBAA 2013).

Fortunately, change is on the horizon with the elimination of avgas scheduled to be complete in 2018. In June of 2013 the FAA, in partnership with the EPA, asked for proposals for fuel options to 100LL in order to transition to a new unleaded fuel. Proposals were submitted in July of 2014. The request was in response to a July 2012 Unleaded Avgas Transition Aviation Rulemaking Committee (UAT ARC) final report that noted an unleaded replacement and corresponding technology was unavailable at the time. This problem led to the government initiative called the Piston Aviation Fuels Initiative (PAFI), which has been tasked with "facilitating the development and deployment of a new unleaded avgas with the least impact on the existing piston-engine aircraft fleet" (FAA 2013). With Phase 1 of PAFI completed in early January 2016, the FAA selected Shell and Swift Fuels to participate in Phase 2 (FAA 2016). One finalist was to be selected to replace 100LL by 2018.

While we should celebrate this change, the question remains as to why it took over three decades for the EPA and the FAA to finally do the right thing. It seems unlikely that concern for neurological damage was the driver. After all, damning medical data have been in the open literature for decades. Instead, the driving force for the change may have been the fact that the western world is down to one vendor selling TEL: Innospec (formerly known as Octel Corporation), a specialty chemicals company of Englewood, Colorado. Since lead was removed from automobile gasoline in most countries, the worldwide market for TEL has shrunk. At some point in the future, Innospec might decide that it is no longer profitable to manufacture the relatively small quantities of TEL. If Innospec were to quit manufacturing the additive, much of the general aviation fleet would quickly be grounded (see textbox for the bribery controversy involving Octel).

BRIBERY CASE AT OCTEL

According to a 2010 *Guardian* investigation, Octel was accused of bribing government officials in Iraq and Indonesia to continue the use of TEL in auto gas there (Leigh et al. 2010). Bribes were paid in 2007 in Iraq to block field trials of MMT, a non-lead alternative additive. In Indonesia money was paid for a "defense of lead" campaign. As a result, the phase-out of TEL was successfully blocked for five years. The company admitted that in its drive to increase profits, it bribed the officials with millions of dollars, despite the health risks associated with TEL. In the wake of these charges, members of the Octel management were fired. Both the CEO, Paul Jennings, and his predecessor, Dennis Kerrison, apparently obtained multi-million dollar payoffs from their company. The new management says that the incident was deeply regrettable and that nothing like it will ever happen again.

Aircraft noise

There are strict regulations on noise from large commercial aircraft. Boeing and Airbus must meet the acoustic requirements of the FAA in order to certify their aircraft. Some airports have even tighter limits, such as requiring quiet night operations. As a consequence, great effort has been made to meet these requirements. The noise from large jet engines has been reduced to such an extent that during the approach to landing, some aircraft make more noise from their landing gear and flaps hanging out in the breeze than from their engines.

In contrast to the large commercial jets, there are essentially no such requirements for general aviation. Interior noise is typically so great that headsets must be worn to prevent hearing damage from long-term exposure. Many aviators even use noise-canceling headsets to further attenuate the sound. Exterior noise levels are also high. Typically, a light airplane must be quite high for the noise to be insignificant on the ground. The main culprit is usually an inadequate engine muffler. In some cases, the speed of the propeller tips is excessive. If the Mach number of the propeller tip approaches one, shock waves strongly radiate noise in the plane of the propeller disk.

Of course, it is possible to use a better engine muffler. There have been dramatic improvements in muffler design in the last half-century. Modern motorcycle mufflers are remarkably small and light. One only needs to be a near a loud, customized motorcycle to appreciate modern commercial mufflers. Propellers can be designed and operated to minimize the tip Mach number and its associated noise. However, it costs money to develop and to certify improved equipment. There can be a small engine performance penalty from additional exhaust backpressure. If a muffler is retrofitted on an existing aircraft, there is the additional installation expense.

A subtler cost is the risk that such a retrofit will expose all involved parties in a lawsuit when an aircraft of that type suffers an accident, whether that particular aircraft was retrofitted or not. Attorneys have successfully argued in court that a vendor must have been making an inferior product if it needed to be upgraded in some other aircraft. For all of these reasons there is a perverse disincentive to improve a certified aircraft. There is no age limit for an aircraft to pass its annual inspection. Even when an aluminum structure reaches the end of its fatigue life, in principle it can be replaced with new. While rust is an issue, steel tubing has practically an infinite fatigue life. Because the cost of a new airplane is so high, old airplanes are kept in service for a long time.

Aviation liability

New airplanes have become very expensive, in part because of liability risks. The average settlement for an aviation fatality is approaching 2 million dollars per person. In contrast, the average settlement for an automobile fatality is much less. In other words, in the US a death in a Cessna is different than a death in a Chevrolet. If the same court decisions applied to automobiles, the US could hardly afford an automobile industry. This is a peculiar weakness in our usually admirable tort system.

Perhaps 80 percent of all general aviation accidents are attributable to pilot error. Maintained by third parties, the machine may have left the factory 40 years earlier. Having nothing to do with the accident, the manufacturer is nonetheless subjected to a lawsuit because of its deep pockets. Manufacturers frequently settled out of court to mitigate the risks of an unfavorable verdict. It reached the point that in 1986 Cessna completely shut down its piston production lines for a decade, until a new law limited the maximum period for which the manufacturer could be held accountable. Ironically, Cessna's aircraft enjoy the best safety record in the industry.

Faced with this situation, rational strategies for an aviation manufacturer are to either raise the price of its product to cover future anticipated lawsuits or reduce its assets so that it is no longer an attractive financial target for a lawsuit. These perverse liability costs make it more difficult to address the pollution issues discussed above. Old aircraft are not replaced because new ones are so expensive. Old aircraft are not upgraded because the upgrades would be cost prohibitive. New aircraft may not benefit from the latest improvements because the enhancements imply that the existing fleet needed improvement and was therefore defective. Thus, investment in new technology is discouraged because of the high parasitic costs.

Solutions

By spending money on research and development, these pollution problems almost certainly have technical solutions. We know how to avoid knock and how to design a good muffler. There will be additional costs associated with the transition to a lead-free fuel and to new mufflers. The efficiency of the engine may be slightly reduced, resulting in lower performance and higher operating costs. An improved muffler might weigh more. Thus, it will cost aviators more money, at least initially.

Someone always pays for the cost of pollution. The question is who. In the present system, the general public pays, as it is subjected to lead and noise exposure. However, in an ethical world, it would be the polluter who is responsible. Not only is that ethical and fair, but it also provides an economic incentive to reduce the pollution to an optimally low value. The optimum is automatically found by the self-interest of the polluter, at the balance point between the incremental reduction in the cost of damage and the incremental increase in the cost of the aircraft. Note that the economically optimal level of pollution is not in general zero. For example, some uses of lead may be sufficiently beneficial to justify the damage. It is difficult to imagine a practical piston engine with exactly zero noise emissions.

The issues are more legal than technical. Under current law the polluter has an implicit or explicit license from the EPA to pollute. Thus, any change in our pollution regulations depends on changes within the EPA. In other words, they are subject to the whim of political appointees and the temptations from lobbyists. Contrast that with scientific evidence presented in court, where a jury decides who was damaged, by whom, and by how much. While our tort system is the natural mechanism to decide issues of pollution damages and compensation, it seems unlikely that the EPA and major polluters would allow the EPA to disappear. The EPA is the friend of the status quo polluter.

Even if pollution issues were resolved in court, there is room for improvement. For example, many judges are technically illiterate. They have been to law school but may have never studied thermodynamics or vorticity. Consequently, judges may allow outrageous technical evidence from unscrupulous expert witnesses. This degrades the court system, imposing a high cost on society. Instead of deserving plaintiffs receiving fair compensation so that polluters and shoddy manufacturers are punished, our current system does not penalize the polluters, which in turn delays improvements in aviation. With better judges our courts would throw out junk lawsuits that raise the cost of aircraft while simultaneously lowering the airplanes' quality. Since neither the judge nor the jury is technically competent, the jury cannot fairly evaluate the testimony of experts. Without an effective feedback system, these disreputable expert witnesses are free to repeat their performance with impunity in subsequent trials. It has developed into a lucrative business for them.

Justice would be better served if there were some type of feedback system for the performance of expert witnesses. One such system has been developed in the medical community. A neurosurgeon, Jeffrey Segal, founded a membership-based organization, Medical Justice, which attempts to defeat frivolous lawsuits by filing countersuits against them (Segal and Sacopulos 2007). This organization uses the collective opinion of specialist medical societies to evaluate expert witness testimony in lawsuits. In a 2001 case the Seventh Circuit Court of Appeals decided that such societies had the power to police their own membership. Some of these societies now have panels to review the quality of expert testimony. If testimony contradicts what a panel would regard as reasonable, that testimony may be considered an ethical violation, resulting in possible expulsion of that witness from the medical society. This would presumably inhibit the employment prospects of that individual as an expert witness in future trials. Another approach used by Medical Justice is a patient–physician contract, requiring both sides in a legitimate dispute to restrict their technical witnesses to those experts who successfully maintain their membership in such societies.

In the aeronautics field, there are also specialty societies, such as the American Institute of Aeronautics and Astronautics. Perhaps the aviation industry should adopt the medical model. While most aeronautical trials involve aircraft accidents, other issues such as patent and pollution matters may also be addressed in courts filled with expert witnesses.

General aviation aircraft cause a relatively small impact on the environment. However, there is room for improvement. It seems the ultimate solution to pollution is a revision of our legal system.

References

Federal Aviation Administration (FAA). 2016. About aviation gasoline. Available at: www.faa.gov/about/initiatives/avgas [accessed June 17, 2016].

Federal Aviation Administration (FAA). 2013. FAA requests proposals for options to help general aviation transition to unleaded fuels. Available at: www.faa.gov/news/press_releases/news_story.cfm?newsId=14714 [accessed June 17, 2016].

Leigh, D., Evans, R., and Mahmood, M. 2010 (Jun 30). UK firm Octel bribed Iraqis to keep buying toxic fuel additive. *The Guardian*. Available at: www.theguardian.com/business/2010/jun/30/octel-petrol-iraq-lead [accessed June 16, 2016].

MacArthur, R.H. and Wilson, F.O. 1967. *The Theory of Island Biogeography*. Princeton: Princeton University Press.

McGrayne, S.B. 2002. Lead-free gasoline and Clair C. Patterson, in *Prometheans in the Lab Chemistry and the Making of the Modern World*. New York: McGraw-Hill, 168–197.

Mills, L.S., Soule M.E., and Doak, D.F. 1993. The keystone-species concept in ecology and conservation. *BioScience*, 43(4), 219–214.

National Business Aviation Association (NBAA). 2013. Court: EPA cannot be forced to issue report on 100LL effects. Available at: www.nbaa.org/ops/environment/avgas/20130403-court-says-epa-cannot-be-forced-to-issue-report-on-100ll-effects.php [accessed December 10, 2014].

Segal, J. and Sacopulos, M. 2007 (Jul 12). Editorial: do-it-yourself tort reform. *Wall Street Journal*, A15.

20

GREENHOUSE GAS EMISSIONS, PERSISTENT CONTRAILS, AND COMMERCIAL AVIATION

Steven A. Kolmes

Commercial aviation is an extraordinarily complex human venture that has environmental ramifications that are infrequently examined. This chapter brings together several ways in which commercial aviation has an environmental impact on the atmosphere. It will discuss the procedures and technologies that are candidates for reducing the impacts of commercial aviation on the atmosphere, including greenhouse gas (GHG) emissions and persistent contrails.

Carbon footprints and ways to offset them

We live in a world where increasing atmospheric levels of greenhouse gases, primarily carbon dioxide, are forcing global climate change and posing a threat to humanity's social and economic stability. It is vital that people begin to evaluate the greenhouse gas footprint (often shortened to "carbon footprint") of their activities, in order to modify them and diminish our greenhouse gas emissions. Our growing scientific understanding, combined with some easily accessible web-based resources, allow people to consider how their lifestyle could be altered to come more into line with a goal of climatic stability. Some things are intuitive, like driving smaller vehicles, using mass transit more, living in (and hence heating and cooling) smaller houses. Other things may not be immediately obvious, like eating less meat (especially feedlot beef which consumes many calories of fossil fuel and an enormous amount of water for each calorie of food produced). But even if a person lives in a modest-sized home, buys appliances rated highly for energy efficiency and compact fluorescent light bulbs, and drives a small biodiesel, hybrid gas-electric or electric car, if they fly around the world much at all (even giving lectures on environmental sustainability), unless they take mitigating action, they will have a huge deleterious carbon footprint because of both fossil fuel use and other factors involved in commercial air travel.

What can be done to derive the benefits of commercial aviation without making an excessive contribution to greenhouse gas emissions? Numerous individuals and businesses have decided to mitigate their global climate change impact associated with air travel by purchasing carbon offsets. Carbon offsets work by having individuals or businesses 1) calculate the greenhouse gas emissions from their use of air travel (or other activities), and 2) purchase sufficient carbon offsets to "zero out" the greenhouse gas contribution. This is done by providing funds for activities that reduce atmospheric greenhouse gas levels, such as reforestation projects, wind power, methane emission abatement, or energy efficiency projects. A considerable number of organizations allow travelers to compute and purchase carbon offsets.

Several comparisons of processes and programs for some of the carbon offset providers have been compiled. The Tufts Climate Initiative (2007) has analyzed offset quality and percentage of cost devoted to offsets (rather than administration or profits) in a report entitled "Voluntary Offsets for Air Travel Carbon Emissions."[1] While this is useful for context and for some information that remains usable, the carbon offset sector has had a number of companies disappear and others emerge over the last few years. The different companies involved focus on a variety of mechanisms for reducing or mitigating greenhouse gas emissions, and carbon offsets have proved to be a dynamic marketplace with new ideas emerging all the time.[2]

For example, *Native*Energy, majority-owned by Native Americans, provides this statement:

> *Native*Energy website purchases are helping build the Ghana Clean Water Project. By offsetting your carbon footprint, you will provide essential upfront funding for the installation of Hydraid(r) water filters—reducing greenhouse gas pollution and bringing clean water and a healthier more productive future to families across the Greater Accra Region of Ghana. In addition to the Ghana Clean Water Project, we have other Help Build(tm) projects available for businesses. By purchasing *Native*Energy's Help Build(tm) carbon offsets, you will provide upfront funding for a new carbon avoidance project. Once built, your project will cut greenhouse gas pollution on your behalf for years to come. (*Native*Energy 2014)

It may be that in the future most online booking of airline tickets will automatically ask the purchaser if they want to pay an additional fee to offset the carbon emissions of their travel.[3] More than one airline now claims to have been the first to offer carbon offsets. Indisputably among the pioneers in this sort of venture, Japan Airlines (JAL) makes offsets available via the CarbonNeutral Company and Recycle One; Delta Air Lines works with The Nature Conservancy; British Airways manages its own carbon offset program (the Customer Carbon Fund, which uses their charity partner Pure to manage the resources they collect); Scandinavian Airlines (SAS) also works with the CarbonNeutral Company.

Although a growing list of airlines offers travelers voluntary programs for carbon offsets, they still represent a minority of companies in the industry.

The International Air Transport Association (IATA) runs a carbon offset program that offers a standardized service and lets any member airline utilize an externally certified carbon offset with Quality Assurance Standard (QAS) at their external verifier. Thirty IATA members (over 260 airlines) use either the IATA carbon offset program or their own carbon offset program to offer this element of air travel to their customers (IATA 2016).[4] Airlines not already mentioned that offer carbon offsets either on their own websites or via a link to another provider include Virgin Atlantic, Lufthansa, and Air Canada, among others. United Airlines now offers an additional option by allowing passengers to redeem their airline miles for carbon offsets, although it should be noted that pressure has been brought to bear on the airline by a group of over 500 elite frequent flyers to stop acting in ways that delay or block actions against climate change.[5]

The GHG footprint of a traveler also depends on the type of seating they purchase. Business class and first-class passengers occupy much greater square footages on an airplane than economy travelers, and different classes of seats have different load factors. According to a World Bank analysis, on a wide body aircraft a business traveler would have roughly three times the GHG footprint of an economy-class traveler, and a first-class passenger would have roughly nine times the GHG footprint of an economy-class traveler (Bofinger and Strand 2013). A first-class passenger purchasing the GHG offset calculated for an economy-class passenger would not offset even 10 percent of their flight-associated GHG emissions. The potential for offsets eventually being purchased by the airlines themselves and treated as increased operating costs or purchased by a central entity on the behalf of airlines might in the future mitigate the sorts of complexities associated with determining just what GHG footprint each traveler accounts for (Cames et al. 2014).

Beyond carbon offsets, some tools exist that can help customers choose to fly on airlines that operate with lower GHG emissions. The International Council on Clean Transportation (ICCT) published a guide that ranks US domestic airlines on their fuel efficiencies (Zeinali et al. 2013). Zeinali et al. (2013: ii) say of their analysis:

> First, it uses airline-reported fuel consumption data, rather than modeled estimates, to account fully for all the ways in which airlines can reduce fuel burn (e.g., aircraft technology or operational practices). Second, it develops an efficiency metric that recognizes that airlines burn fuel to provide both mobility (measured in terms of passenger miles traveled) and access (frequency of service and number of airports served), allowing an equitable comparison between airlines. Third, the efficiency metric distinguishes productive from nonproductive miles flown by identifying those airlines that operate particularly circuitous routes. Finally, the study attributes the transport service provided by and fuel consumption of affiliate carriers to mainline

carriers in order to enable comprehensive comparisons across carriers' full business operations.

The ICCT analysis reveals a significant difference (overall 26 percent fuel efficiency range) among US carriers. Some airlines (for example, United Airlines) operate at about an average efficiency level for the US industry; others (for instance, American Airlines) are considerably less fuel efficient, while Alaska Airlines tops the efficiency chart (Zeinali et al. 2013). While GHG offsets are based on the number of miles flown, travelers can further lessen their personal environmental impact by both offsetting and additionally trying to fly on the most fuel-efficient airline possible. However, choosing the right airline is still only approximate with the currently available information, as Zeinali et al. (2013: iv) note: "The most efficient airline for a given itinerary varies by city pair, with the top performer on one route not necessarily being efficient on another. Thus, route-specific data are required for consumers to make more environmentally responsible decisions."

Greenhouse gases and commercial aviation

What amount of greenhouse gases does a passenger flying in a commercial aircraft actually produce, and are there really better alternative forms of transportation in terms of emissions? To give an idea of the scale of the carbon footprint of commercial air travel, an example TerraPass emissions report is given below:

- Seattle to Newark nonstop round trip
- Personal emissions: 1,867 lbs CO_2
- Miles flown: 4,787
- Personal fuel use: 95 gallons

Traveling the same distance in a typical bus would produce roughly 1/4 as much CO_2, while a train trip would produce considerably less. The value for CO_2 production due to this one round trip by air travel is not far off the amount that a person would generate driving a small car a typical distance over the course of a year. For comparison, you can save roughly 2,000 lb of personal CO_2 emissions annually by driving a fuel-efficient car weighing 2,000 lb instead of a medium-sized, 4,000-lb SUV. However, one long round-trip flight to another city across a continent or an ocean, and you have fully lost the carbon emission benefit gained by an entire year of environmentally responsible personal vehicle selection. Two round-trip flights of that duration cost the entire benefit gained from forsaking an SUV and commuting by bicycle for a year.

Duration of flight matters to the CO_2 emitted per km because the large amounts of CO_2 emitted during takeoff and landing make short flights proportionally worse overall. For example, an Airbus A320-200 uses 0.425 kilowatt hours per second-kilometer (kwh/s-km) on a 300-km route, but only 0.248 kwh/s-km over a 1,200-km route. In contrast, a high-speed rail line uses between

0.106 and 0.141 kwh/s-km for travel (over distances of 200 km and 1,200 km respectively (Janic 2003). Aggravating the lesser efficiency for shorter flights is the fact that air traffic control inefficiencies (because of using ground-based navigation points instead of direct routes) increase the distance traveled by a factor of 1.15 for short flights and only 1.05 for long flights (Peeters and Williams 2009). At any distance, traveling by high-speed train has a smaller energetic cost than a commercial flight; but this is especially pronounced over the shorter distances where the potential of substituting rail travel for air travel might be greatest. The Amtrak/Acela express train service in the Northeastern US works with some carriers developing an air/rail codeshare, and efforts to smooth transfer from rail to air are already operating in the Northeast US. In Germany, Deutsche Bahn (the national railroad) allows international travelers from more than 80 airlines to travel on a connecting railway line (NESCAUM and CCAP 2003). There is also the intriguing possibility of using more high-speed rail links directly between European airports to replace shorter and therefore more carbon-intensive flights, helping to mitigate some of the negative climatic impacts of interurban travel (NESCAUM and CCAP 2003).

The National Research Council (NRC 2010a) estimates that domestic air transportation in the US accounts for 7 percent of domestic energy use for transportation (needless to say the percentage is much higher for international travel), which is in line with the 6 percent of GHG emissions estimated for commercial aviation in the UK (Airports Commission 2013). However, the NRC notes that few large cities in the US are within 500 miles of one another compared to the spacing of cities in the EU and Japan, so high-speed light rail will not be as viable an alternative to reduce the energy expended on commercial aviation within the US, and we must look to reduced demand or increased efficiency rather than switching to alternative modes of transportation. As the NRC notes in a companion volume on limiting future climate change:

> Domestic air transportation currently accounts for about 12 percent of US passenger miles and ~3 percent of total US GHG emissions. Technological advances coupled with a highly competitive airline industry have prompted air carriers to steadily improve (by ~1–2 percent per year) the energy efficiency of their fleets and their operations during the past 40 years . . . New aircraft entering the fleet today, such as the Boeing 787, use about 20 percent less fuel per seat-mile than the aircraft they are replacing. Airline fleet turnover should thus over time lead to continued gains in aircraft energy efficiency. At the same time however, domestic airline passenger traffic is forecast to grow by between 2 to 3 percent per year over the next two decades, which is likely to offset the technological efficiency gains. Additional efficiency improvements can be found in navigation and control technology and changes in air traffic management practices. For instance, more direct routing and reduced taxiing and idling at airports represent potentially important areas of opportunity for future energy savings. (NRC 2010b: 62–63)

Changes in aircraft pose interesting environmental questions as well. The double-decker A380 is the biggest passenger airliner in the world and, according to calculations published in *The Economist* (2006), the versions of the A380 with four Trent 900 engines, despite the engines being individually as fuel efficient as any currently available, produce as much CO_2 emissions as a 14-km line of cars on a highway far below the airliner.[6] According to *The Economist*, IATA states that both an automobile and an airliner use 3.5 liters per 100 passenger-km of travel. However, those calculations are based on a fully loaded airliner and one person occupying an automobile: they ignore the potential CO_2 per passenger-km savings from having more than one person in a car; and they ignore the fact that trips by airliner in general are a great deal longer than trips that are feasible by automobile.

Air travel frees us to move as never before in human history, but at a real greenhouse gas cost we must acknowledge and incorporate into planning a future of responsible commercial air operations. Huge aircraft might operate at either higher or lower levels of greenhouse gas emissions compared to their smaller alternatives, depending on a variety of design and operational factors discussed in more detail below. However, from an environmental perspective it is not clear that "bigger is better" in terms of aircraft. A focus on changes that increase fuel efficiency per passenger would be a more environmentally responsible perspective for the commercial airline industry than a perspective based on designing the largest or fastest commercial airliners possible. With inevitable future increases in the cost of aviation fuel, the impetus towards efficiency ought to be easier to accept by corporate officers principally concerned with profitability, as well as those with environmental concerns. The Boeing 787 Dreamliner, built largely of composite materials and with new engine designs, should supposedly use approximately 20 percent less fuel than a comparably sized conventional passenger aircraft.

GHG emissions from commercial aviation are still only a small fraction of total anthropogenic emissions, but the share of greenhouse gases contributed by aviation is growing rapidly. The 1999 Intergovernmental Panel on Climate Change (IPCC) report requested by ICAO estimated that about 3.5 percent of all anthropogenic climate change was accounted for by commercial aviation. By 2050 the IPCC has projected that commercial air travel will account for between 3 percent and 7 percent of anthropogenic climate change. A report from the Tyndall Centre for Climate Change Research at Manchester projected that for the UK, increases in aviation and maritime operations could overcome all the efforts made by other industrial sectors to reduce GHG emissions intended to bring the UK into accord with its Kyoto Protocol targets (Bows et al. 2006). Aviation CO_2 emissions have been growing faster in the UK than any other source of greenhouse gases. The UK's CO_2 emissions in 1990 were estimated at 171–175 million tons, with that number falling in the 1990s as coal was used less and heavy industry reduced, but by 2004 the numbers had risen again to 172–177 million tons. With aviation and maritime operations excluded, the emissions over the same period would have fallen by 4 percent to 161–155 million tons. The Tyndall Centre estimates that at a 6.4 percent annual growth rate in aviation (its average for the 1990s), over the next half century this growth would offset by itself all reductions in carbon emissions planned by the British government to reach Kyoto Protocol goals (Bows et al. 2006). The UK aviation industry's emission of carbon in 2004 was estimated at 9.8 million tons, and it is projected to rise to 16–21 million tons by 2030. Between 1990 and 2004, the number of people using airports in the UK rose by 120 percent.

The European Emissions Trading Scheme (EETS)

In 2012 the European Union (EU) was planning to bring the aviation sector into the already operating European Emissions Trading Scheme, which would have effectively served as an indirect carbon emissions tax on airlines, and various sources projected ticket price increases would occur in the range of 3–7 percent. Several different scenarios for the commercial aviation sector were analyzed that would have allowed the EU to meet its carbon emission reduction goals. While those appeared to be possible under the auspices of the EETS, it was nonetheless a challenging goal to achieve (Bows et al. 2009). This approach favors more carbon-efficient airlines over less careful competitors or ones with less-efficient equipment. In terms of commercial air travel, in some ways the EETS would be correcting an historical peculiarity. Consumer choices about how they will travel have been tipped towards aviation by an invisible subsidy as a result of the Chicago Convention of 1944, which banned taxes on commercial aviation fuel used for international air travel. Recent attempts by the EU to reverse this ban have failed. Especially in the case of low-cost carriers (LCCs), which have competed with ground transportation for customers taking short- or moderate-distance trips, the lack of fuel taxes has moved more travelers into the air and away from trains and other modes of transportation (Nilsson 2009).

In any event the implementation of the EETS did not go as planned. Significant opposition to the EETS by countries as diverse as the US, Russia, China, and India materialized. Some argued that by including non-European airlines in the EETS, the process infringed on national sovereignty elsewhere, and that by including emissions outside EU airspace, the EETS was in addition extraterritorial. Twenty-six countries signed a statement opposing the EETS. In the US a law entitled "The European Union Emissions Trading Scheme Prohibition Act of 2011" was passed, since US airlines participating in the EETS would effectively be a tax which, according to the trade group Airlines for America, would be removing $3.1 billion from US airlines and transferring them to the UK government between 2012 and 2020 (Rosenthal 2013). In November of 2012 the EU decided to "stop the clock" on enforcing the EETS when flights between the EU and airports elsewhere were involved (Airports Commission 2013). Between 2013 and 2016 only flights within the European Economic Area (EEA) were included in the EETS system.

This decision followed the ICAO Assembly in 2013[7] at which it decided to develop a global market-based approach to aviation emissions (as part of a "basket of measures") that is to be applied by 2020.[8] Further progress along these lines from ICAO was anticipated at their 2016 Assembly. However, this step has not been without its critics; the ICAO "Decision to agree in three years to do something in seven years, if it's not too hard, while shutting down attempts to do something in the meantime" was heavily criticized by the Flying Clean Alliance.[9] The airline industry, in contrast, views the decision to move towards an international agreement and the "basket of measures" as an indication of clear commitment and leadership in the arena of climate change mitigation.[10]

International treaties

Aviation emissions, in common with shipping, mainly take place outside of national borders. This gives them an unusual and challenging status in relationship to international treaties. As a United Nations Environment Programme (UNEP) Synthesis Report (2011: 11) notes:

> Emissions from the aviation and shipping sectors are a special case compared with other sectors because a large fraction of global civil aviation and shipping emissions are "international" and not fully attributable to a particular country. International emissions have not been included in the Kyoto Protocol targets for Annex I countries and they do not fall under country pledges . . . As of 2006, 62% of the emissions from aviation were international . . . The 2005 emissions from global civil aviation were about 0.6 $GtCO_2$ per year . . . Business-as-usual projections for 2020 are about 0.6 to 1.2 $GtCO_2$ per year from aviation . . . Options for reducing emissions . . . include improving fuel efficiency and using low-carbon fuels.

An analysis of the legal status of the contesting parties as related to the ongoing ICAO deliberations, jurisdictional issues associated with the EETS with reference to the European Court of Justice (CJEU) decision in the suit brought by the Air Transport Association of America (ATAA), and potential for future legal review by entities such as the World Trade Organization (WTO) can be found in the International Centre for Trade and Sustainable Development (ICTSD) issue paper written by Gehring and Robb (2013). The paper identifies possible political alternatives that could be followed:

- Refraining from further regulatory action to reduce GHG emissions from aviation;
- Accepting a weak multilateral outcome as preferable to regional/unilateral action;
- Pressing for agreement on a strong multilateral outcome, including the developing countries; and
- Pursuing unilateral action while working to convince aviation partners that they need to address climate change and aviation either within or outside of ICAO. (Gehring and Robb 2013: 15)

They also raise the precautionary principle and question of intergenerational equity, and indicate that the goal of future negotiations should be sustainable development that requires a balance between environmental, economic, and social dimensions. A discussion of alternative proposals too lengthy to include in this chapter is included in the issue paper. They discuss the risks associated with current situation as follows:

> At the moment, we are at a cusp in the international negotiations where negotiations (and some countries and politicians) seem to favour the business-as-usual scenario for aviation and climate change. There are several risks involved in this first option: the biggest risk involves the exponential growth of GHG emissions with the related climate risks involved. The second, perhaps less obvious, risk is that, just like big tobacco, aviation might lose or endanger its "social licence" to exist once climate concern and demand for action grows. (Gehring and Robb 2013: 15)

A report from the Manchester Metropolitan University Centre for Aviation Transport and the Environment evaluated 23 mitigation scenarios for aviation-induced GHGs and found that the EETS option had the greatest incremental reduction in GHG emissions by 2050, followed by a scenario based on technological and operational improvements. In part the EETS option had the greatest positive impact because it could potentially be implemented quickly, and the report stressed that fast action is crucial in effective CO_2 mitigation because of the cumulative effects that will occur between now and 2050 (Lee et al. 2013).

Havel and Sanchez make the case that aviation is especially appropriate for a separate sector-specific greenhouse gas reduction treaty. They find that commercial aviation is one area in which a unique alignment in the interest of the industry and the interest of the government exists. From the perspective of the industry, there is already a comfort level with "exceptionalism" in legal terms, and has been since the Chicago Convention (which was the origin of ICAO) was signed in 1944, and a sense that such a treaty could promote the economic stability of the sector. For governments, a treaty dealing with greenhouse gas emissions from what Havel and Sanchez call "arguably the world's most visible services industry" would have political value. In terms of the symbolism of such a treaty they note: "Through whichever lens it is scrutinized by public opinion, aviation has huge symbolic purchase. Emissions reductions pursued by one of the great enablers of globalization would have powerful demonstration effects for other industries, as well as for states" (Havel and Sanchez 2012: 355).

However, any such treaty will require strong and widespread member state governmental support because ICAO itself has no authority in its charter to deal with greenhouse gas emissions. As Havel and Sanchez note (2012: 358–359):

> ICAO's direct authority to regulate aviation emissions can be characterized as weak or, more realistically, nonexistent. Annex 16 to the Chicago Convention . . . lays out limited standards for aircraft engines with respect

> to discharges of hydrocarbon, carbon monoxide, and nitrogen oxides. It does not cover carbon dioxide, which is the primary man-made contributor to global warming . . . Notwithstanding these limitations to its authority, ICAO has not been locked out of efforts to reduce aviation emissions at the international level . . . At its triennial General Assembly meetings in 2007 and 2010, 36 ICAO member states passed Resolutions reaffirming the Organization's legitimacy as the lead international body charged with developing a global response to aviation's role in climate change.

The Chicago Convention has a "cooperative ethos" that seems at odds with unilateral actions to reduce greenhouse gas emission (Havel and Sanchez 2012), but because climate change was not an issue of concern when the Convention was signed, its relationship to contemporary issues of emission control was never addressed in the original language and remains ambiguous. Havel and Sanchez develop ideas about how an international greenhouse gas reduction agreement for commercial aviation might be developed that would require too much space in this setting, but which are worth consulting.

Aviation industry actions

In terms of action from the industry itself, IATA, representing roughly 240 airlines globally, has adopted targets that include a cap on aviation emissions from 2020 onwards, and before that a 1.5 percent improvement in performance in terms of fuel efficiency every year from 2009 to 2020. IATA targets also include a 50 percent reduction below 2005 emission levels by 2050 (Airports Commission 2013). The question is not whether these goals and plans are commendable but whether the time frame has rendered them largely irrelevant. Various airports, airlines, and manufacturers are also taking steps to reduce aviation's impacts on GHG levels.[11]

It is worth noting that the impact of commercial aviation on climate change is not a one-way affair. Extreme weather events increasing in frequency—which promise things like snow disruptions, heat waves that may damage runways and other airport facilities, increased flooding events due to rainfall extremes, and changing wind patterns associated with a changing climate—are all likely to impact aviation to a greater or lesser extent (Airports Commission 2013). Signs that the jet stream is becoming less stable could presage initiatives that would necessitate changes in transoceanic routing. As surely as commercial aviation has the potential to impact climate change, climate change has the potential to impact commercial aviation (Williams and Joshi 2013).

How to reduce aviation's impact

What other possible avenues for reducing the carbon emissions of commercial air travel exist? The IPCC and IATA both point to improved air traffic control as a means to significantly reduce the CO_2 emissions associated with commercial air operations. International agreement to centralize air traffic control into a more efficient set of fewer entities, to straighten routes of air travel, and to avoid delays on takeoffs and landings would significantly reduce the fuel consumed per km of travel. *The Economist* (2006) reports that IATA estimates a 12 percent reduction in carbon emissions from commercial air operations by changes such as consolidating the 35 separate European national bodies responsible for air traffic control into one international organization.

The government of the UK supports not just the ICAO and the EETS scheme, but also the "Single European Sky" (SES) initiative, which would move from airspace defined by national boundaries to airspace organized by "functional airspace blocks" (FABs), intended to reduce CO_2 emissions by making air traffic control more efficient. The government's 2013 Aviation Policy Framework says of this:

> Within the UK, our commitment to SES has been demonstrated by the establishment of the first FAB in the EU, with Ireland in 2008. This is delivering real benefits including CO_2 reduction through greater flight and fuel efficiency. It is estimated that the UK–Ireland FAB has provided approximately £37.5 million of savings from 2008 to 2011, including around 48,000 tonnes of fuel (around 152,000 tonnes of CO_2) . . . Based on the current work programme, it is estimated that by 2020 annual savings could reach . . . 35,000 tonnes of fuel and 111,000 tonnes of CO_2 . . . The UK–Ireland FAB is working actively to enhance its links with air navigation service providers in other European countries with a view to further improving efficiency in the future, potentially leading to a FAB covering a wider area. The UK is also working closely with the adjoining FAB Europe Central (FABEC) States. (UK Department of Transport 2013: 45)

The IPCC (1999) estimates that improved air traffic control could reduce airline carbon emissions by 6–12 percent, and that operational improvements (optimizing speed, reducing additional weight, improving load factors, reducing nonessential fuel onboard, limiting use of auxiliary power units, and reducing taxiing) would allow an additional 2–6 percent in carbon emission savings. For example, in 2014 Air France began to evaluate TaxiBot systems that can be operated by a pilot and tow aircraft to their gate without the main engines operating (Green Air 2014a). Dramatic reductions in GHG emissions, in the order of 25 percent, might be achieved by using sophisticated modeling tools to make flight trajectory changes to optimize flight path according to both GHG emissions and economic impacts, and yet increase operating costs by less than 0.5 percent (Grewe et al. 2014). There are probably many other simple opportunities for greenhouse gas emission savings that have not yet occurred to anyone in the airline industry, but which undoubtedly will.

Other aircraft emissions, while not as significant as CO_2, are nonetheless also greenhouse gases. Oxides of nitrogen, usually referred to as NO_x (initially principally

NO, but this is converted to NO_2 in the atmosphere), undergo a complex series of chemical reactions in the atmosphere that can increase ozone (O_3) levels (worsening global warming) and decrease methane (CH_4) levels (reducing global warming) (Lee and Raper 2003). The overall effect of these simultaneous changes in O_3 and CH_4 levels remains to be elucidated, but it would be naïve to assume that they will simply cancel one another out or neatly compensate for CO_2 emissions. Water in vapor form is also emitted in aircraft exhaust gases, although calculations indicate that the direct effects of this water vapor on climate change are small. Other effects of water vapor are discussed in a separate section below. There are also soot and sulfate particles in aircraft exhaust which may also have implications for climate change, although these are likely to be less significant than the influence of CO_2 and NO_x in the exhaust streams (Lee and Raper 2003; IPCC 1999).

The carbon offsets discussed previously do not attempt to incorporate information about the climate effects of the less significant exhaust gases. Precisely computing the overall climatic impact of the mixture of gases in commercial aviation emissions is a challenging problem to which people are paying much attention (Peeters and Williams 2009). The General Accounting Office (GAO 2009) estimates that CO_2 accounts for 49 percent of the global climate change impacts from the emissions of commercial aviation, nitrogen oxides account for 22 percent, contrails 20 percent (as will be discussed at length later in this chapter), water vapor 4 percent, and soot 5 percent. Given the complexity of the situation and the large contribution of CO_2, it is understandable and probably appropriate that we often speak in terms of carbon emissions and carbon offsets.

Technological advances have reduced the fuel consumption rates of commercial aircraft by well over one half since the 1960s, and further fuel efficiency should be possible by increasing core thermal efficiency, increasing aerodynamic efficiency of the airframes and engine nacelle inlet section, and by reducing aircraft and engine weight by increasingly using new materials (Lee 2003). Use of extended fly-by-wire and eventual fly-by-light (fiber optic) technology could improve aircraft fuel efficiency by 1–3 percent, and active center of gravity control could provide another 1–2 percent in savings (Lee 2003). Winglets, wing extensions that reduce drag, are beginning to have widespread use and can often be retrofitted. More radical possibilities such as blended wing-body aircraft in which the wings and body are both part of the same airframe might produce very significant fuel savings (General Accounting Office 2009). There is a powerful environmental impetus to aircraft manufacturers investing in research to make these improvements, and this impetus is consistent with the economic advantages of reduced fuel use in a world with decreasing fossil fuel supplies. More speculative environmental benefits from technological advances might be made by producing sulfur-free kerosene for fuel from biomass using the Fischer–Tropsch (FT) synthesis process, or by developing hydrogen-powered aircraft, although the latter could also exacerbate the environmental impacts of contrails discussed later (Lee 2003).

Alternative fuels

The possibility of using biofuels to power commercial aircraft in order to mitigate their greenhouse gas effects is real but fraught with complexities. Two biofuels potentially suitable to the task are biodiesel and FT kerosene, and deciding whether or not the use of farmland to make fuel would have a positive impact compared to the use of fossil fuel is far from straightforward (Upham et al. 2009). In theory biofuels—which combust to release CO_2 that was recently in the atmosphere and which can again be removed from the atmosphere by the next season of crop growth—are quite different from fossil fuels, which release long-sequestered carbon into the atmosphere. However, farming for biofuels can replace farming for food, and various aspects of biofuel production have their own environmental impacts. The critical question about the usefulness of biofuels may be twofold: what sort of land is being used to produce them (Fargione et al. 2008), and what sort of dietary decisions are people willing to make to free croplands for use in fuel production without diminishing human food supplies (Kolmes 2008a). Biofuels might also have impacts on climate change forcing agents other than CO_2. A Canadian test of a commercial flight with 100 percent biofuel derived from canola (a vegetable oil derived from rapeseed) found a 50 percent reduction in aerosol emissions compared to conventional fuel. Tests performed on a static engine showed a reduction in particles of up to 255 percent and in black carbon emissions of up to 495 percent. These tests showed typical engine performance with an improvement of 1.5 percent in fuel efficiency in the steady state operations (NRC Canada 2013).

The overall proportional benefit of biofuels may prove to be lessened when not only lifecycle CO_2 emissions are considered, but non-CO_2 emissions and effects as well (Stratton et al. 2011). Many uncertainties remain to be clarified in terms of the impacts of potential biofuels. In the face of this complexity some airlines are moving ahead to try new approaches, as with Southwest Airlines which, beginning in 2016, agreed to purchase 3 million gallons per year of renewable biofuel derived from waste material from the forest products industry to blend and use at its San Francisco operation facility (Green Air 2014b).

A new and different approach to sustainable jet fuel has emerged in Virgin Atlantic working with LanzaTech to utilize waste carbon monoxide gases from steel production facilities by capturing it and fermenting it into ethanol, and then chemically converting it for use as jet fuel. This uses gas normally flared and released into the atmosphere as carbon dioxide, therefore recycling the carbon (Sustainable Aviation 2012). It remains to be seen whether or not this innovative approach will be both economically practical and scalable, but it is a novel effort of the sort that ought to be encouraged.

British Airways is moving ahead with planning to convert many tons of household garbage annually into jet fuel in a facility that will use a plasma to produce synthetic gas that can be converted into fuel for commercial aviation using the Fisher–Tropsch reaction. The pilot project is scheduled to have an initial engineering and design study done, and a proposed site selected, by 2019. Ultimately this type of facility could allow hundreds of thousands of tons of post-recycled waste to be converted into jet fuel, with an anticipated reduction in greenhouse gas emissions of 70 percent. There would also be an anticipated reduction of 90 percent in particulate emissions (Green Air 2018).

Complicating the process of reducing the GHG impact of aviation is that commercial aviation is likely to decarbonize more slowly than other forms of transportation. Other forms of energy switching, in different segments of the transportation sector, are more obvious than a clear choice for a low GHG fuel for commercial aviation, and commercial aircraft have a long service life compared to many other sorts of vehicle (Airports Commission 2013) so that, even if a new fuel was developed, the existing fleet of commercial aircraft would persist for a long period of time.

Other means of GHG reduction

In 2012 Sustainable Aviation—a joint effort in the UK of commercial airlines, airports, aerospace manufacturers, and air navigation service providers—published a "roadmap" of various things that they believed could contribute to an overall 50 percent reduction in the UK's commercial aviation GHG footprint by 2050, and the variety of opportunities for improvement they identify is worth considering. Their potential contributing steps to GHG reductions are:

> Increasing the energy-per-unit-mass and/or energy-per-unit-volume of the fuel used; Increasing the efficiency with which fuel energy is converted into thrust by the engine(s); Increasing the aircraft's aerodynamic efficiency (lift to drag ratio); Reducing the structural weight of the aircraft for a given payload-range requirement; Reducing engine weight; Reducing the weight of aircraft systems and cabin infrastructure; Cruising at reduced speed; Adapting aircraft design to maximise the advantages of reduced cruise speed; Using an aircraft whose range capability does not greatly exceed the mission's requirement; Making full use of the aircraft's mass-carrying capability at the desired range; Optimising stage-length; Adopting the most fuel efficient route, taking weather and prevailing winds into account; Avoiding queuing or holding, whether on the ground or in the air; Replacing ground-based APU [auxiliary power unit] usage with lower-carbon power from airport infrastructure; Taxiing using fewer engines, or by towing the aircraft; Ensuring the aircraft and all its systems are operating at their most efficient state of maintenance; Using a lower carbon intensity sustainable fuel derived from biomass or waste; Capturing carbon-dioxide from the air and using it to artificially manufacture aviation fuel; Capturing carbon dioxide from the air and sequestering it; Reducing net emissions through funding more cost-effective emissions reduction in other economic sectors. (Sustainable Aviation 2012)

Sustainable Aviation estimates what each component of its vision of a more efficient and lower impact version of commercial aviation might contribute to the industry in the UK, but how these estimates would differ from the situation in other countries is doubtless highly variable. Nonetheless, while on one level this could be read as a list of current inefficiencies in the industry, on another level it demonstrates the many constituencies already making efforts to make aviation more sustainable. Sustainable Aviation calls for government support for research and development efforts but is opposed to UK unilateral regulatory targets and measures which it believes would be counterproductive as well as costing the UK aviation industry money lost to international competition. This perspective, which underlies the formation of the organization, makes the inevitability of growth in commercial aviation and the technological optimism expressed seem potentially biased; but there is no question that the organization is correct that international rather than national efforts to mitigate climate change in a variety of sectors will be needed.

Research priorities

At the behest of the US National Aeronautics and Space Administration (NASA), the National Academies of Sciences, Engineering, and Medicine (NASEM) constituted a committee to propose priorities for a national research agenda to reduce CO_2 emissions from commercial aviation, and specifically on propulsion and energy technologies. The report identified four especially promising technological avenues to be followed as the US invests in future research efforts over the next 10 to 30 years. These "high-priority approaches" identified were: (1) advances in aircraft-propulsion integration, (2) improvements in gas turbine engines, (3) development of turboelectric propulsion systems, and (4) advances in sustainable alternative jet fuels (NASEM 2016).

Other potential research fields were considered but rejected as high-priority research areas. Hybrid-electric and all-electric systems were not recommended because batteries with the power capacity and specific power needed are not likely to be developed to a point that would satisfy FAA certification requirements within 30 years. The same reasoning was applied to technologies including superconducting motors and generators, fuel cells, and cryogenic fuels (NASEM 2016).

Ethical considerations

Finally, this topic calls for a brief comment about ethical reservations. In concert with a thorough attempt to reduce carbon emissions in every other way possible, purchasing carbon offsets allows a concerned person to mitigate the carbon emissions unavoidable in their domestic or business life. The carbon offsets ought to be a last resort, not a first thought; and consumer attention is called for to make certain that the offsetting activities being funded are authentically ones that would not have taken place had the carbon offset not been purchased (the principle of additionality). Lacking a prior effort to otherwise reduce carbon emissions as far as possible, the purchase of carbon offsets amounts to a form of environmental indulgence, permission to continue emitting excessive quantities of environmentally damaging greenhouse gases without guilt. History shows that the purchase of indulgences by the wealthy is not a strategy that tends to be widely acceptable in society. In the twentieth century Mahatma Gandhi taught that we must become the change we wish to see in the world, and I can think of no attribute of human life today where this is more true than in reducing greenhouse gas emissions.

This position is consistent with that of the United States Conference of Catholic Bishops (USCCB) taken after the 2001 IPCC "Third Assessment Report" was released.[12] They stated in a pastoral letter:

> According to reports of the IPCC, significant delays in addressing climate change may compound the problem and make future remedies more difficult, painful, and costly. On the other hand, the impact of prudent actions today can potentially improve the situation over time, avoiding more sweeping action in the future. (USCCB 2001: 6)
>
> Defined as the ability to exercise moral responsibility for the care of the environment . . . Stewardship requires a careful protection of the environment and calls us to use our intelligence to discover the earth's productive potential and the many different ways in which human needs can be satisfied . . . True stewardship requires changes in human actions—both in moral behavior and in technical advancement. (USCCB 2001: 8)
>
> The common good calls us to extend our concern to future generations. Climate change poses the question "what does our generation owe to generations yet unborn"? (USCCB 2001: 9)

Interestingly the Bishops' focus on the principle of prudence and on intergenerational equity is echoed in the issue paper from the International Centre for Trade and Sustainable Development that has already been discussed (Gehring and Robb 2013). A conviction that both prudence and the ethical demand of future generations are central to the question of aviation and GHG emissions can be reached from both theological and secular approaches. Gehring and Robb speak of "Precaution, inter-generational equity and the principle of common but differentiated responsibilities and respective capacities." The concept of common but differentiated responsibilities and respective capacities (CBDRRC) was expressed in the UN Framework Convention on Climate Change (Article 3) in May 1992.[13] It may be that employing this concept will be required for an ethically defensible approach to the GHG emissions from commercial aviation to be negotiated and broadly accepted. However even the meaning of CBDRRC is open to debate. As Scott and Rajamani (2012: 477) note:

> Although the CBDRRC principle has come to play a pivotal role in international environmental law, the core content of the CBDRRC principle, as well as the nature of the obligation it entails, is deeply contested. There are differing views on whether the basis for differentiation lies in differences in the level of economic development and capabilities, in contributions to GHGs in the atmosphere, or both. We would argue that it refers to both. If CBDRRC refers to differentiation based on capability alone the use of the term "respective capabilities" would be superfluous . . . There are also disagreements as to the nature of the obligation the CBDRRC principle entails. While some argue that it obliges states to act in particular ways, others contend that it is merely a consideration that should be taken into account in the decision-making process. The disagreements over this principle's content and the nature of the obligation it entails have spawned debates over its legal status.

It seems likely that the Bishops would insist that there is a real obligation entailed by the concept of CBDRRC, and that those nations capable of acting, and responsible for the greatest share of carbon dioxide in the atmosphere, have an ethical obligation to take a leadership role in mitigating climate change as they previously took a leadership role in promoting climate change.

The "Scientific Basis, Summary for Policy Makers" of the IPCC's "Fourth Assessment Report" (2007) was addressed by the USCCB's Department of Social Development and World Peace. They said in a message to congressional leaders:

> From an international perspective, climate change is in large part an issue of "sustainable development." The poor have a need and right to develop to overcome poverty and live in dignity. More affluent nations have a responsibility to encourage and help in this development . . . This priority for the poor in climate policy cannot be a marginal concern, but rather must be a central measure of future choices. If we do not address climate change and global poverty together, we will fail both morally and practically. (USCCB 2007: 5)

When the Fifth IPCC Assessment Report was released in 2013, the USCCB made its stand on the ethical implications of the concept of CBDRRC very clear in a letter to the Bicameral Task Force on Climate Change. The USCCB Committee on Domestic Justice and Human Development wrote:

> For the USCCB, a fundamental moral measure of any policy to address climate change is how it affects the poor, in our country and around the world. Well-designed policies can both reduce the severity of climate change and protect the most vulnerable. The USCCB supports strong leadership by the United States in enacting policies that protect poor and vulnerable people from bearing the impacts of climate change and from the human and

> economic costs of any proposed legislation to respond to climate change. The USCCB asks the US Congress and the federal government to consider the following principles as they shape policies and measures to address climate change:
>
> - Prudence requires us to act to protect the common good by addressing climate change at home and abroad.
> - The consequences of climate change will be borne by the world's most vulnerable people and inaction will worsen their suffering.
> - Policies addressing global climate change should enhance rather than diminish the economic situation of people in poverty.
> - Policies should create new resources to assist poor and adversely affected communities to adapt and respond to the effects of global climate change in the US and in vulnerable developing countries.
> - Policies to address climate change should include measures to protect poor and vulnerable communities from the health impacts of climate change, including increased exposure to climate-sensitive diseases, heat waves and diminished air quality.
> - Participation by local affected communities in shaping policy responses to address climate change and programs for adapting to climate change is essential.
> - Technology should be made available to people in the most vulnerable developing countries to help them adapt to the effects of climate change (adaptation) and reduce their greenhouse gas emissions (mitigation). (USCCB 2013)

These documents make the movement towards carbon offsets and international treaties something we must view from the perspective of both individual and company travel decisions, and from the developing thought of international law and agreements. We must do more than offset negative impacts of our actions; we must above all attempt to reduce and subsequently mitigate those negative impacts. We can neither afford inaction nor the privatization of the commons implied by personal carbon offsets unpaired with other corporate, national, and international actions.

Pope Francis, in his encyclical "*Laudato Si'*: On Care for Our Common Home" (2015)—which rapidly became viewed as a major moral and ethical reflection on the environment and social justice by a wide array of people and organizations (for example, see the eight articles and the editorial in the Nov./Dec. 2015 issue of the journal *Environment, Science and Policy for Sustainable Development*)—mentions aircraft only in one section:

> 103. Technoscience, when well directed, can produce important means of improving the quality of human life, from useful domestic appliances to great transportation systems, bridges, buildings and public spaces. It can also

> produce art and enable men and women immersed in the material world to "leap" into the world of beauty. Who can deny the beauty of an aircraft or a skyscraper?

The Pope is clear throughout the document that climate change is a great moral and ethical as well as a practical challenge, and he includes systems of transportation when he says:

> 26. Many of those who possess more resources and economic or political power seem mostly to be concerned with masking the problems or concealing their symptoms, simply making efforts to reduce some of the negative impacts of climate change. However, many of these symptoms indicate that such effects will continue to worsen if we continue with current models of production and consumption. There is an urgent need to develop policies so that, in the next few years, the emission of carbon dioxide and other highly polluting gases can be drastically reduced, for example, substituting for fossil fuels and developing sources of renewable energy. Worldwide there is minimal access to clean and renewable energy. There is still a need to develop adequate storage technologies. Some countries have made considerable progress, although it is far from constituting a significant proportion. Investments have also been made in means of production and transportation which consume less energy and require fewer raw materials, as well as in methods of construction and renovating buildings which improve their energy efficiency. But these good practices are still far from widespread.

We are late to come to acting effectively on climate change due to denial campaigns, and climate change impacts will strike most strongly at the poorest and most vulnerable persons (Kolmes 2011, 2016). Nonetheless, the ethical imperative to act comes along with the potential for climate change mitigation to produce new technologies that have powerful positive economic benefits in a variety of sectors (Kolmes 2008b), including commercial aviation, which has the potential to become more efficient, integrated, and profitable as a result of innovation that will simultaneously reduce greenhouse gas emissions to serve the common good.

Persistent contrails

This chapter would not be complete without a commentary on persistent contrails. Unlike other forms of commercial transportation, aviation poses a second threat to global climatic stability through the formation of persistent contrails. At 33,000 to 35,000 feet in altitude, the only clouds naturally present are cirrus clouds. Commercial jetliners fly at this altitude and release exhaust that includes water vapor. From 10 to 20 percent of the time, a commercial jetliner will be flying through conditions of atmospheric humidity and temperatures that produce ice crystal formation in the exhaust, and a visible contrail forms in the condensation nuclei (soot and sulfate particles) that are also present in the engine exhaust stream. It may even be that the condensation nuclei from aircraft exhaust could lead to the formation of cirrus clouds considerably later, when atmospheric conditions of temperature and humidity have changed (Lee and Raper 2003). Contrails are essentially linear cirrus clouds, and they are more persistent than was previously thought. While contrails are sometimes short-lived, under some atmospheric conditions (for example supersaturation with respect to ice) they can spread and last for long periods of time (Lee and Raper 2003).

Researchers used satellite pictures collected during the aircraft grounding after 9/11 as what they called a "tarnished golden opportunity" to collect critical data on contrail persistence (Travis et al. 2002). Contrails left by six military aircraft flying through an area west of Washington, DC, spread out over several hours to cover 20,456 square kilometers. The contrails dissipated after 10 hours. They flew through an area normally traversed by 70 to 80 jetliners an hour. On a typical day there would be an upper atmospheric cloud layer generated by aircraft that lasted all day. Unequivocal proof that aircraft generate cirrus clouds on a large scale has proven difficult to obtain, in part because of the complexity of the global climate system. But, despite the technical difficulties involved, evidence for a correlation between high air traffic density and high cloud amounts is slowly emerging, and an extensive recent review of this literature is available (Lee 2009).

Contrails that form cirrus clouds or that add to already existing cirrus clouds have a small daytime shading effect (blocking sunlight) and a larger heat-trapping effect (blocking heat loss from the earth by re-radiation) (Lee 2009). The shading effect only operates during daylight hours, while the heat-trapping effect operates all night as well. Contrails are therefore contributors to global warming. However, the precise amount that contrails contribute to global warming is uncertain. Different modeling approaches produce different estimates.[14] Contrails could produce a radiative forcing resulting in a range of 0.2 to 0.3 degrees C per decade, which would constitute a significant contribution to climate change (Minnis et al. 2004). Recent analyses indicate that contrail cirrus clouds are the greatest component of aviation contributing to climate change, and that they have a much greater radiative forcing effect than "young" linear contrails. As Burkhardt and Kärcher (2011: 56) write:

> The global net radiative forcing of contrail cirrus is roughly nine times that of young contrails, making it the single largest radiative forcing component connected with aviation. It is important to note that contrail cirrus have a much shorter lifetime than long-lived greenhouse gases. This difference in lifetime influences the relative importance of contrail cirrus and other forcing agents for climate change.

We will probably see more contrails in the sky as time goes on unless some fundamental changes take place in commercial aviation.

While persistent contrails are an issue that we know we need to pay attention to, they might be less resistant to being reduced than are greenhouse gas emissions. If jetliners flew at no more than 31,000 feet in the summer and no more than 24,000 feet in the winter (when contrails form more readily) they would burn slightly more fuel and emit slightly under 4 percent more CO_2, but they would produce many fewer persistent contrails (Williams et al. 2002). This contrail reduction also depends on latitude and other factors, so contrail avoidance plans would need to be very carefully developed to be effective (Gierens et al. 2008). However, this restriction would also reduce air space capacity in some areas, and require additional controller resources with flights restricted to a smaller airspace. It might also make it appreciably more difficult to provide turbulence-free flights. Such tradeoffs warrant serious consideration as we look for multiple approaches to reducing global climate change.

Night flights and winter flights disproportionately impact global climate because of different effects on cirrus cloud formation. Night flights account for only 25 percent of daily air traffic, but may account for 60 to 80 percent of the warming impact of commercial aviation because they lack the partial warming offset provided to contrails formed during the day by their blocking sunlight from reaching the earth's surface (Stuber et al. 2006). Winter flights account for only 22 percent of annual flights, but they may account for up to 50 percent of the warming impact of commercial aviation because of the greater tendency for contrails to form in winter (Stuber et al. 2006). Modifications of daily flight schedules could reduce the warming impact of commercial aviation, although issues of airport capacity and passenger travel schedules considerably complicate this possibility. If fewer flights were to take place in late afternoon or evening, and more early-morning flights took advantage of the partial contrail warming offset available due to shading effects, the global climate change impacts of contrails might be reduced to some degree (Duda 2008).

Conclusion

Commercial aircraft both provide a vital component of our transportation sector and have a negative impact on the atmosphere in several ways. The 2007 *Fourth Assessment Report* of the IPCC estimated that aviation accounts for between 2 and 8 percent of the radiative forcing driving global climate change, and numerous researchers are continuing to attempt to refine our understanding of exactly what effect commercial aviation has had (and will have) on the global climate (Lee 2009).

A variety of potential mitigation actions have been proposed to help control the negative aspects of air travel, ranging from aircraft technical improvements, to carbon offsets, to changes in flight scheduling, to broader societal responses such as emissions trading schemes (Staniland 2009) like the one adopted by the EU. Challenging trade-offs exist; for example, more-efficient engines will probably produce lower exhaust temperatures as they emit the same amount of water vapor, suggesting that they will form contrails at higher air temperatures and over a large altitude range (Fahey et al. 1999; Schumann 2000) and exacerbate the issue of persistent contrails while simultaneously reducing greenhouse gas emissions due to greater fuel efficiency. The development of high bypass-ratio turbofan engines to increase fuel efficiency has produced increased NO_x formation, so improving performance in terms of one pollutant coming at the cost of a second pollutant. Decisions may end up having to be made on the basis of the atmospheric residency time of different pollutants—much longer for CO_2 than for NO_x (Lee 2009). Concepts like combining wing and fuselage into a blended wing body design or developing ultra-high capacity aircraft to enjoy economies of scale, or increased use of composite materials, all have their own challenges in terms of economics and existing aviation infrastructure design (Peeters et al. 2009). Very careful consideration of economic and environmental trade-offs will have to take place in order to make efforts to mitigate the climate impact of air travel effective.

Notes

1 Another report from the Stockholm Environment Institute is available at: www.terrapass.com/images/blogposts/Air_Travel_Emissions_Paper.pdf.

2 Responsible companies that provide retail or wholesale offsets include: the German organization Atmosfair at www.atmosfair.de/en; the CarbonNeutral Company, managed by Natural Capital Partners (www.carbonneutral.com); terrapass (www.terrapass.com); Climate Care (www.climatecare.org); and Climate Trust (www.climatetrust.org) [all accessed May 1, 2016]. They invest in varied energy efficiency and renewable energy projects, including things like traffic signal optimization to reduce fuel use, carpool matching by internet, riparian reforestation, rainforest preservation, fuel switching, and many other activities.

3 However, there is also an argument made that aircraft operators, rather than consumers, should purchase offsets for their flights and pass these costs on as increased operating expenses. See www.greenaironline.com/news.php?viewStory=1992 [accessed October 20, 2014].

4 See also http://qascarbonneutral.com/airline-carbon-offset-programs/.

5 See the 2013 press release on this at the Flying Clean Alliance website: www.flyingclean.com/frequent_flyers_to_united_airlines_news.

6 If the 1,500 A380s envisioned by Airbus were ever in operation, *The Economist* reports that they would be producing as much CO_2 as 5 million automobiles.

7 See www.icao.int/Meetings/a38/Documents/Resolutions/a38_res_prov_en.pdf [accessed October 20, 2014]; www.un.org/climatechange/summit/wp-content/uploads/sites/2/2014/09/TRANSPORT-Aviation-Action-plan.pdf [accessed May 1, 2016].

8 See http://ec.europa.eu/clima/policies/transport/aviation/index_en.htm [accessed May 1, 2016].

9 See www.flyingclean.com/flights_delayed_icao_expected_to_put_off_to_tomorrow_what_it_should_have_done_16_years_ago; and www.flyingclean.com/blog [accessed May 1, 2016].

10 See for example Airports Council International's (ACI's) article: www.airport-business.com/2013/11/aviation-industry-leads-the-way-in-addressing-climate-change/ [accessed May 1, 2016].

11 At London's Heathrow Airport, a new take-off procedure for A380s in use since 2010, developed cooperatively by the airport and several airlines, reduced fuel use by 300 kg per flight and carbon dioxide emissions by roughly 1,000 kg per flight. See Airports Commission (2013, section 3.24) for more information.

12 IPCC reports are available at: www.ipcc.ch/publications_and_data/publications_and_data_reports.htm.
13 For more information see: http://unfccc.int/resource/docs/a/18p2a01.pdf [accessed October 20, 2014].
14 See Ponater et al. (2002); Minnis et al. (1999, 2004); and Lee and Raper (2003).

References

Airports Commission. 2013. Discussion paper 3: aviation and climate change. Available at: www.gov.uk/government/uploads/system/uploads/attachment_data/file/186683/aviation-and-climate-change-paper.pdf [accessed May 1, 2016].

Bofinger, H. and Strand, J. 2013. Calculating the carbon footprint from different classes of air travel. World Bank Policy Research Working Paper 6471. Available at: www-wds.worldbank.org/external/default/WDSContentServer/IW3P/IB/2013/05/31/000158349_20130531105457/Rendered/PDF/WPS6471.pdf [accessed May 1, 2016].

Bows, A., Anderson, K., and Footitt, A. 2009. Aviation in a low-carbon EU. In *Climate Change and Aviation*, edited by S. Gossling and P. Upham. London: Earthscan, 89–109.

Bows, A., Anderson, K., and Upham, P. 2006. Contraction and convergence: UK carbon emissions and the implications for UK air traffic. Tyndall Centre for Climate Change Research Technical Report.

Burkhardt, B. and Kärcher, B. 2011 (Apr). Global radiative forcing from contrail cirrus. *Nature Climate Change*. 1: 54-58. Available as a pdf at: www.pa.op.dlr.de/~pa1c/NatureCC1_54-58_2011.pdf [accessed July 13, 2018].

Cames, M, Gores, S., Graichen, V., Keimeyer, F., and Farber, J. 2014. An Aviation Carbon Offset Scheme (ACOS) Version 3.0: Update. Environmental Research of the Federal Ministry for the Environment, Nature Conservation, Building and Nuclear Safety. Project No. (FKZ) 3713 14 102. Available at www.umweltbundesamt.de/sites/default/files/medien/378/publikationen/sonstige_an_aviation_carbon_offset_scheme_acos_komplett.pdf [accessed May 1, 2016].

Duda, D. 2008. Contrails: a double sided sword? *The Controller*. 47: 12.

Economist. 2006. Aircraft emissions: the sky's the limit. Special report, June 8.

Fahey, D.W., Schumann U., Ackerman S., Artaxo P., Boucher O., Danilin M.Y., Kärcher B., Minnis P., Nakajima T., and Toon, O.B. 1999. Aviation-produced aerosols and cloudiness. In *Aviation and the Global Atmosphere*, edited by J.E. Penner et al. Cambridge: Cambridge University Press, 65–120.

Fargione, J., Hill J., Tilman D., Polasky S., and Hawthorne, P. 2008. Land clearing and the biofuel carbon debt. *Science*. 319: 1235–1238.

Gehring, M.W. and Robb, C.A.R. 2013. Addressing the aviation and climate change challenge: a review of options; ICTSD Programme on Climate Change and Energy; Issue Paper No. 7. International Centre for Trade and Sustainable Development, Geneva, Switzerland. Available at: www.ictsd.org/downloads/2013/08/addressing-the-aviation-and-climate-change-challenge.pdf [accessed May 1, 2016].

General Accounting Office (GAO). 2009. Aviation and climate change: aircraft emissions expected to grow, but technological and operational improvements and government policies can help control emissions. GAO-09-554. Available at: www.gao.gov/assets/300/290594.pdf [accessed May 1, 2016].

Gierens, K., Lim, L., and Eleftheratos, K. 2008. A review of various strategies for contrail avoidance. *Open Atmospheric Science Journal*. 2: 1–7.

Green Air. 2014a (Oct 10). Air France to evaluate operational and environmental benefits of semi-robotic TaxiBot system at Paris CDG. *GreenAir Online*. Available at: www.greenaironline.com/news.php?viewStory=1993 [accessed May 1, 2016].

Green Air. 2014b (Sep 30). Southwest Airlines agrees to purchase three million gallons per year of renewable jet fuel sourced from forest residues. *GreenAir Online*. Available at: www.greenaironline.com/news.php?viewStory=1988 [accessed May 1, 2016].

Green Air. 2018 (Jun 22). British Airways waste-to-fuels project with Velocys secures £4.9m funding including a UK government grant. *GreenAir Online*. Available at: www.greenaironline.com/news.php?viewStory=2497 [accessed July 16, 2018].

Grewe, V., Champougny, T., Matthes, S., Fromming, C., Brinkop, S., Søvde, Irvine, E.A., and Halscheidt, L. 2014. Reduction of the air traffic's contribution to climate change: a REACT4C case study. *Atmospheric Environment*. 94: 616–625. Available at: www.sciencedirect.com/science/article/pii/S1352231014004063 [accessed May 1, 2016].

Havel, B.F. and Sanchez, G.S. 2012. Towards an international aviation emissions agreement. *Harvard Environmental Law Review*. 36: 351–385. Available at: http://harvardelr.com/wp-content/uploads/2012/09/Havel-Sanchez.pdf [accessed May 1, 2016].

Intergovernmental Panel on Climate Change (IPCC). 1999. *Aviation and the Global Atmosphere*. Cambridge: Cambridge University Press.

Intergovernmental Panel on Climate Change (IPCC). 2001. Third Assessment Report. Available at: www.ipcc.ch/publications_and_data/publications_and_data_reports.htm [accessed May 1, 2016].

Intergovernmental Panel on Climate Change (IPCC). 2007. Fourth Assessment Report. Available at: www.ipcc.ch/publications_and_data/publications_ipcc_fourth_assessment_report_synthesis_report.htm [accessed May 1, 2016].

Intergovernmental Panel on Climate Change (IPCC). 2013. Fifth Assessment Report. Available at: www.ipcc.ch/report/ar5/ [accessed May 1, 2016].

International Air Transport Association (IATA). 2016. IATA Carbon Offset Program. Available at: www.iata.org/whatwedo/environment/pages/carbon-offset.aspx [accessed May 1, 2016].

Janic, M. 2003. The potential for modal substitution. In *Towards Sustainable Aviation*, edited by P. Upham, J. Maugham, D. Raper, and C. Thomas. London: Earthscan, 131–148.

Kolmes, S.A. 2008a. Food, land use changes, and biofuels (E-letter). *Science*. 319: 1238–1240. Available at: www.researchgate.net/publication/303720340_Food_Land_Use_Changes_and_Biofuels [accessed May 1, 2016].

Kolmes, S.A. 2008b. The social feedback loop. *Environment: Science and Policy for Sustainable Development*. 50, 2: 57–58.

Kolmes, S.A. 2011. Climate change: a disinformation campaign. *Environment: Science and Policy for Sustainable Development*. 53, 4: 33–37.

Kolmes, S.A. 2016. Environmental policy choices: the importance of the preferential option for the poor in *Laudato si'*. *Environment: Science and Policy for Sustainable Development*. 58, 3: 14-17.

Lee, D., Lim, L.L. and Owen, B. 2013. Mitigating future aviation CO_2 emissions: timing is everything. Report from the Manchester Metropolitan University Centre for Aviation Transport and the Environment. Available at: www.cate.mmu.ac.uk/docs/mitigating-future-aviation-co2-emissions.pdf [accessed May 1, 2016].

Lee, D. and Raper, D. 2003. The global atmospheric impacts of aviation. In *Towards Sustainable Aviation*, edited by P. Upham, J. Maugham, D. Raper, and C. Thomas. London: Earthscan, 77–96.

Lee, D.S. 2009. Aviation and climate change: the science. In *Climate Change and Aviation*, edited by S. Gossling and P. Upham. London: Earthscan, 27–67.

Lee, J. 2003. The potential offered by aircraft and engine technologies. In *Towards Sustainable Aviation*, edited by P. Upham, J. Maugham, D. Raper, and C. Thomas. London: Earthscan, 162–178.

Minnis, P., Ayers, J.K., Palikonda, R., and Phan, D. 2004. Contrails, cirrus trends, and climate. *Journal of Climate*. 17: 1671–1685.

Minnis, P., Schumann, U., Doelling, D.R., Gierens, K., and Fahey, D. 1999. Global distribution of contrail radiative forcing. *Geophysical Research Letters*. 26: 1853–1856.

National Academies of Sciences, Engineering, and Medicine (NASEM). 2016. *Commercial Aircraft Propulsion and Energy Systems Research: Reducing Global Carbon Emissions*. Washington, DC: The National Academies Press. doi:10.17226/23490.

National Research Council (NRC). 2010a. *Advancing the Science of Climate Change*. Washington, DC: National Academies Press.

National Research Council (NRC). 2010b. *Limiting the Magnitude of Future Climate Change*. Washington, DC: National Academies Press.

National Research Council (NRC) Canada. 2013 (Jan 7). Analysis of 100 percent biofuel tested in flight by NRC reveals reduction in admissions. Available at: www.nrc-cnrc.gc.ca/eng/news/releases/2013/biofuels.html [accessed May 1, 2016].

*Native*Energy. 2014. Ghana Clean Water Project. Available at: https://nativeenergy.com/project/ghana-clean-water-projecthb/ [accessed May 1, 2016].

Nilsson, J.H. 2009. Low-cost aviation. In *Climate Change and Aviation*, edited by S. Gossling and P. Upham. London: Earthscan, 113–129.

Northeast States for Coordinated Air Use Management (NESCAUM) and Center for Clean Air Policy (CCAP). 2003. Controlling airport-related air pollution. Available at: www.nescaum.org/documents/aviation_final_report.pdf/ [accessed May 1, 2016].

Peeters, P., and Williams, V. 2009. Calculating emissions and radiative forcing. In *Towards Sustainable Aviation*, edited by P. Upham, J. Maugham, D. Raper, and C. Thomas. London: Earthscan, 69–87.

Peeters, P., Williams, V., and de Haan, A. 2009. Technical and management reduction potentials. In *Climate Change and Aviation*, edited by S. Gossling and P. Upham. London: Earthscan, 293-307.

Ponater, M., Marquet, S., and Sausen, R. 2002. Contrails in comprehensive global climate model: parameterization and radiative forcing results. *Journal of Geophysical Research*. 107: 4164.

Pope Francis. 2015. *Laudato Si': On Care for Our Common Home*. Huntington, IN: Our Sunday Visitor. Available at http://w2.vatican.va/content/francesco/en/encyclicals/documents/papa-francesco_20150524_enciclica-laudato-si.html [accessed May 1, 2016].

Rosenthal, E. 2013 (Jan 26). Your biggest carbon sin may be air travel. *New York Times*. Available at: www.nytimes.com/2013/01/27/sunday-review/the-biggest-carbon-sin-air-travel.html?_r=0 [accessed May 1, 2016].

Scott, J. and Rajamani, L. 2012. EU climate change unilateralism. *European Journal of International Law*. 23, 2: 469–494.

Schumann, U. 2000. Influence of propulsion efficiency on contrail formation. *Aerospace Science and Technology*. 4: 391–401.

Staniland, M. 2009. Air transport and the EU's Emissions Trading Scheme: issues and arguments. European Union Center of Excellence European Studies Center, University of Pittsburgh, Policy Paper 13. Available at: http://aei.pitt.edu/11425/1/2009-Air_Transport_Emissions.pdf [accessed May 1, 2016].

Stecker, T. and Pyper, J. 2014. Garbage fuel will power British Airways planes. *Scientific American*. Available at www.scientificamerican.com/article/garbage-fuel-will-power-british-airways-planes/. [accessed May 1, 2016].

Stratton, R.W., Wolfe, P.J., and Hileman, J.I. 2011. Impact of aviation non-CO_2 combustion effects on the environmental feasibility of alternative jet fuels. *Environmental Science and Technology*. 45: 10736–10743.

Stuber, N., Forster, P., Rädel, G., and Shine, K. 2006. The importance of the diurnal and annual cycle of air traffic for contrail radiative forcing. *Nature*. 44: 864–867.

Sustainable Aviation. 2012. *Sustainable Aviation CO_2 Road Map*. Available at: www.sustainableaviation.co.uk/wp-content/uploads/2015/09/SA-Carbon-Roadmap-full-report.pdf [accessed May 1, 2016].

Travis, D.J., Carleton, A.M., and Lauritsen, R. 2002. Contrails reduce daily temperature range. *Nature*. 418: 601.

Tufts Climate Initiative. 2007. Voluntary offsets for air travel carbon emissions. Available at: http://sustainability.tufts.edu/wp-content/uploads/TCI_Carbon_Offsets_Paper_April-2-07.pdf [accessed May 1, 2016].

UK Department of Transport. 2013. *Aviation Policy Framework*. London: HMSO. Available at: www.gov.uk/government/uploads/system/uploads/attachment_data/file/153776/aviation-policy-framework.pdf [accessed May 1, 2016].

United Nations Environment Programme (UNEP). 2011. *Bridging the Emissions Gap, United Nations Environment Program (UNEP) Synthesis Report*. Available at https://wedocs.unep.org/handle/20.500.11822/7996 [accessed July 13, 2018].

United States Conference of Catholic Bishops (USCCB). 2013 (Feb 21). Letter to Representative Waxman and Senator Whitehouse. Available at: www.usccb.org/issues-and-action/human-life-and-dignity/environment/upload/Letter-to-Bicameral-Task-Force-on-Climate-Change-2013-02-21.pdf [accessed May 1, 2016].

United States Conference of Catholic Bishops (USCCB). 2007 (Jun 7). Religious and moral dimensions of global climate change. Available at: www.usccb.org/issues-and-action/human-life-and-dignity/environment/environmental-justice-program/upload/written-and-oral-testimony-before-senate-by-john-carr-on-climate-change-2007-06-07.pdf [accessed May 1, 2016].

United States Conference of Catholic Bishops (USCCB). 2001. Global climate change, a plea for dialogue, prudence, and the common good. Available at: www.usccb.org/issues-and-action/human-life-and-dignity/environment/global-climate-change-a-plea-for-dialogue-prudence-and-the-common-good.cfm [accessed May 1, 2016].

Upham, P., Tomei, J., and Boucher, P. 2009. Biofuels, aviation and sustainability: prospects and limits. In *Climate Change and Aviation*, edited by S. Gossling and P. Upham. London: Earthscan, 309–328.

Williams, P.D. and Joshi, M.M. 2013. Intensification of winter transatlantic aviation turbulence in response to climate change. *Nature Climate Change*. 3: 644–648.

Williams, V., Noland, R.B., and Toumi, R. 2002. Reducing the climate change impacts of aviation by restricting cruise altitudes. *Transportation Research D*. 7: 451–464. Available at: www.cts.cv.ic.ac.uk/documents/publications/iccts00186.pdf.

Zeinali, M., Rutherford, D., Kwan, I. and Kharina, A. 2013. US domestic airline fuel efficiency ranking 2010. International Council on Clean Transportation. Available at: www.theicct.org/sites/default/files/publications/U.S.%20Airlines%20Ranking%20Report%20final.pdf [accessed May 1, 2016].

21

GROUND-LEVEL POLLUTION, INVASIVE SPECIES, AND EMERGENT DISEASES

Steven A. Kolmes

This chapter follows up on the previous one by bringing together several ways that commercial aviation has an environmental impact on ground level, whether from air pollution near airports, water pollution due to de-icing, spread of invasive species, or emergent diseases. These considerations deal with the local areas around airports, and also at times with much broader geographic areas where commercial aviation can have a discernible impact.

De-icing aircraft, runway de-icing/anti-icing, and other water pollution issues

Regarding airports, one of the principal environmental concerns is the quality of runoff water, with the main challenge being pollution from aircraft and runway anti-icing and de-icing operations at northern facilities during the winter months. Runway anti-icing/de-icing is a more tractable problem than aircraft de-icing, both because there is a greater variety of chemicals available that are safe for use on runways than there are for aircraft and because the runway management is done by the airport operators themselves as opposed to the multiple airport tenants individually making decisions about aircraft de-icing. While airports can use relatively benign materials like potassium acetate or calcium magnesium acetate on runways, ethylene-glycol and propylene-glycol-based materials are widely used on aircraft (GAO 2000). Ethylene-glycol is both more effective as a de-icer and more toxic to humans, with 3 ounces constituting a lethal dose and smaller amounts causing kidney damage. Propylene-glycol is less toxic to humans, but both of these materials pose a threat to the quality of surface water and ground water (Holtzman 1997).

In addition to the direct toxicity of de-icing chemicals, biodegradation of these materials in freshwater diminishes the dissolved oxygen (DO) level in the water, greatly reducing water quality and potentially killing fish and other aquatic organisms. Techniques to reduce the impact of de-icing chemicals include vacuum sweeper trucks, storm water drainage systems, de-icing pads, and tanks or ponds to store glycol-rich waste so that it is released during periods of high water flow and its environmental impact is reduced by dilution (GAO 2000). Some airports, like Denver International, have facilities to collect and recycle glycols on the airport grounds.

The US Environmental Protection Agency (EPA 2000) estimates that prior to 1990 and the implementation of the Phase I Storm Water Discharge Permits, there were approximately 28 million gallons of 50 percent concentration aircraft de-icing fluid released annually to US surface waters. This has already been reduced to 21 million gallons annually, with an estimate of eventually reaching 17 million gallons annually when all of the airport storm water permits have been fully implemented. The EPA projects that if all US airports could reach 70 percent efficiency in collecting aircraft de-icing fluid, surface water discharges could be reduced to 4 million gallons annually. Significant reductions in water pollution due to aircraft de-icing can be achieved with appropriate investment in modernized treatment facilities and techniques.

After de-icing chemicals, a second concern for airport-generated water pollution is fuel spills. These can be due to leaks, improper fueling connections, or poorly monitored storage tanks (GAO 2000). Fuel spills can contaminate soil and/or groundwater. This is, in fact, exactly the same environmental issue that plagues gasoline stations for automobiles or boats.

A "Voluntary Pollution Reduction Program" (VPRP) has been undertaken by a group comprised of Airlines for America, Airports Council International–North America, the Regional Airlines Association, and the American Association of Airport Executives (AAAE). They have identified a defined set of airports responsible for 80 percent of the de-icing in the US. At these airports the VPRP is committed to documenting de-icing chemical use and the technologies used for pollution reduction, encouraging experimentation and adoption of new pollution reduction technologies, and developing quantitative goals for de-icing chemical pollution. They look to modified de-icing compounds with lower oxygen demand in freshwater, new techniques that use fewer de-icing compounds, and stormwater management techniques are areas of special attention. Their initial report was released in 2012. Reports in 2014 and 2017 then promised to establish quantitative pollution reduction goals, and compare the benefits of different pollution reduction technologies.

Meanwhile the EPA (2012) released airport de-icing effluent guidelines, for both new and existing airports with 1,000 departures or more annually, which banned the use of urea or, alternatively, required a numerical effluent limit for ammonia. No uniform set of rules for discharges at existing airports was established, and their limitations will be set in permits on a case-by-case basis. New airports in cold climates with 10,000 or more annual departures will need to collect 60 percent of their de-icing fluid. The AAAE (2017) considers the lack of a uniform standard a "major victory for the airport industry." It may well be that the combination of EPA action and voluntary industry action is able to significantly reduce the impact of aircraft de-icing over time.

Ground-level air pollution at and near airports

Airports are among the most significant sources of local air pollution. Airplanes on the ground emit literally hundreds of millions of pounds of toxic air pollutants annually. Idling and taxiing planes emit volatile organic compounds (VOCs) and oxides of nitrogen (NO_x), which are precursors of ozone, a major health-related air pollutant that interferes with lung function. John F. Kennedy International Airport is the second largest emitter of VOCs in New York City, and LaGuardia is among the major NO_x sources. Chicago's Midway Airport has been identified as a major source of the carcinogens benzene and formaldehyde (Holtzman 1997). Logan Airport in Boston produces two times as much benzene as the second worst emitter in Massachusetts (NESCAUM and CCAP 2003). Thirty of the nation's 50 largest airports are in ozone nonattainment areas, where air quality is considered substandard for human health. Other airport pollution includes unburnt hydrocarbons, particulate matter, and carbon monoxide. Particulate matter emitted in the UK, from a category that includes aircraft, trains, and ships, is estimated to account for 1,800 early deaths annually. To put this into context, the total annual early deaths from particulate matter contributed by all combustion sources (including cars, trucks, industry, and power plants) is 9,000 (Yim and Barrett 2011). Aircraft typically account for the majority (45–85 percent) of total airport air pollutant emissions.

However, not all airport pollution is from aircraft. Ground access vehicles and ground service vehicles also emit VOCs and NO_x, as do the airplanes to which they are providing access and services (Holtzman 1997). In one study of three airports in the Northeast (Logan International, Bradley International, Manchester Airport) 85 percent of airport NO_x came from aircraft, and auxiliary power units and ground service equipment emitted 15 percent of aviation-related NO_x (NESCAUM and CCAP 2003). Ground access vehicles such as taxis and buses have been reported in some instances to emit almost as much NO_x and more VOCs than the actual airport operations (Holtzman 1997).

Many airports have become more congested over time, and that directly impacts local air pollution. As taxi time increases, with airliners using their engines to taxi, a very significant impact on air quality and human health occurs. As Schlenker and Walker report, quoting Bureau of Transportation Statistics:

> Average airplane taxi time, measured by the amount of time that an airplane spends between the gate and runway, has increased by 23 percent from 1995 to 2007 . . . This increase in average congestion, combined with increased number of flights, translates to an aggregate increase of over 1 million airplane hours per year spent idling on runways over this time period . . .

> Our estimates suggest this increase also leads to significantly higher levels of ambient air pollution. We find that a one standard deviation increase in daily airplane taxi time at LAX increases pollution levels of carbon monoxide (CO) by 23 percent of a standard deviation in areas within 10km (6.2 miles) of the airport. The marginal effect of taxi time is largest in areas adjacent to an airport or directly downwind, and the effect fades with distance. (Schlenker and Walker 2011: 2; internal citations omitted)

They go on to describe how increased taxi time at major airports is associated with increased local pollution and therefore adverse health effects for people living within 10 km of airports. These effects include increased hospital admissions for respiratory problems and cardiovascular problems, with infants accounting for approximately one-quarter of the respiratory admissions and elderly people accounting for most of the heart-related health impacts (Schlenker and Walker 2011).

Toxic plume contamination from jet engines stretches 12 miles for approaches to airports, and 6 miles for takeoffs (Donaldson 1970). Many homes, schools, churches, daycare facilities, etc. are within those distances and beneath the plumes of our urban airports. Airports contribute highly variable proportions of the total air pollutants present in the counties housing them. In the preliminary estimates reported in one study of US airports and their surroundings (Ratliff 2007), the average proportion of NO_x in the surrounding counties due to airports was 1.73 percent; but for five airports the NO_x contribution ranged from 10 percent to over 30 percent of the NO_x produced in the surrounding county.

A very similar story exists for airport contributions to oxides of sulfur (SO_x) and production of dangerously small particulate matter that can lodge deep in human lungs (PM 2.5). At Santa Monica Airport (a small, single runway, general aviation facility) ground-level ultrafine particulates (UFPs) produced by aircraft were measured; elevations of UFPs were increased 10-fold at 100 meters downwind and 2.5-fold at 660 meters downwind. These elevated UFP concentrations pose a health threat to occupants of a large downwind residential neighborhood (Hu et al. 2009). Near-airport impacts of pollutants and noise are disproportionately high compared to the impacts to the general population, and policy makers often focus on an aggregate cost–benefit analysis that ignores impacts on local populations (Wolfe et al. 2014). Airports are not all equivalent: small ones as well as large ones can impact surrounding communities; and they are not the only contributors to air pollution in their vicinities, but they are real contributors nonetheless.

Airport personnel can be exposed to a variety of pollutants, and among these the genotoxic effects of polycyclic aromatic hydrocarbons (PAHs) have been studied. PAHs are greatest in concentration outside on the airport apron, but still occur at elevated levels in terminals and office spaces. Genetic damage due to PAH exposure among airport personnel has been documented (Cavallo et al. 2006). Airport personnel may face a greater risk from pollutants associated with air travel than do even frequent travelers. The textbox provides an example of how environmental problems at airports may have a significant impact on marginalized groups.

LONDON CITY AIRPORT: WHO BEARS THE CONSEQUENCES?

According to Heuwieser (2017): On 6 September 2016, a dozen activists of the Black Lives Matter group blockaded a runway at London City Airport. Their message: "Climate Crisis is a Racist Crisis". This act of civil disobedience was directed against the expansion of the business airport, located in a workers' district of London. People living in the flight paths of the airport—many of whom are Black British Africans—have incomes that are far lower than those of the passengers in the aircraft above. In Great Britain, Black British Africans are exposed to particulate levels in the air they breathe that are 28% higher than those to which white Britons are exposed, for white people are more likely to be able to afford housing in less polluted areas. Black Lives Matter also highlighted through its action that Great Britain contributes substantially to exacerbating the climate crisis yet is scarcely affected by its impacts. Africa, in contrast, is the continent most jeopardized by the climate crisis.

Nationally the implementation of stringent control measures is reducing air pollution from most economic sectors, but increasing air travel and lack of stringent controls on aircraft engines means that this is one economic sector whose contribution to air pollution is increasing (NESCAUM and CCAP 2003). Several measures could reduce airport pollution. Single-engine taxiing, possible with some aircraft, saves fuel and reduces emissions, but it could be hazardous under some conditions (for instance, wet runways) (NESCAUM and CCAP 2003). Newer airplane engines pollute significantly less than older ones, and replacing airplanes when economically feasible can reduce pollution. The EPA has adopted two tiers of more stringent rules for maximum NO_x output from jet engines (one for higher thrust engines and one for lower thrust engines), which were adopted in cooperation with a broader effort by the International Civil Aviation Organization (ICAO).[1] The EPA notes that ozone is formed when volatile organic compounds interact with NO_x in the presence of sunlight, and that both NO_x and ozone have significant human health impacts in terms of exacerbating asthma, causing pulmonary disease, and increasing cardio-pulmonary mortality (EPA 2007, 2006).

Replacing gasoline or diesel ground-service equipment with electric vehicles or ones that run on compressed natural gas or liquid propane could significantly reduce pollution and potentially save airports money in the long run (NESCAUM and CCAP 2003). A 2000 FAA pilot project called the Inherently Low-Emission Airport Vehicle Pilot Program (ILEAV) provided financial incentives to selected airports and was an exploration of possibilities as well as an example of providing financial incentives for this sort of transition (NESCAUM and CCAP 2003). Some airports were able to make much greater emissions reductions than others, with factors such as a larger proportion of airport-owned ground equipment facilitating the implementation of the grant (ILEAV 2005).

Other policy options with the potential to reduce airport pollution include: cap-and-trade programs or the purchase of emissions credits (as already discussed for the EU in the European Emissions Trading Scheme/EETS); variable landing fees such as those used in Switzerland and Sweden, where the landing fees are emissions based; and voluntary agreements at the local level to reduce emissions from aircraft taxi and gate operations (NESCAUM and CCAP 2003).

Technological options with the potential to reduce airport pollution include increases in jet engine efficiency and specifically designed systems by General Electric (GE), Pratt & Whitney, and others that reduce NO_x emissions. However, these new systems tend to produce more carbon monoxide and unburnt hydrocarbon pollution and use more fuel than earlier designs (NESCAUM and CCAP 2003). As already discussed, at this time there are tradeoffs in reducing NO_x and CO_2 emissions. Research is underway to further clean the emissions of jet engines, and doubtless the potential for a great deal of technological progress on this front remains. Improved aerodynamic design and new aircraft body materials also have potential to make contributions to pollution reduction. A study published in the UK in 2000 categorized potential improvements in aircraft technology in terms of both their contribution to pollution reduction and how far in the future such improvements might be; and strategies like lean pre-mixed, pre-vaporized combustion technologies to reduce NO_x, the use of composite materials in aircraft bodies, blended wing bodies, micro-electro mechanical systems, and others were highlighted as having strong potential (ADL 2000).

Commercial air travel and invasive species

Air passenger baggage is a pathway for non-indigenous invasive species to spread from one part of the world to another. An unintended consequence of increasing human mobility, invasive species are responsible for significant economic and biotic loss by replacing native species and spreading out of control numerically in habitats that lack their normal predators, pathogens, and competitors. Often the most devastating effect of invasive species is when they attack agricultural activities. Invasive species transported by commercial aviation tend to be small, hidden "hitchhikers": insects and arachnids are prominent among them. I will focus on an especially important and well-documented problem of those insect invaders that have appeared on commercial flights.

Analysis of records of invasive insect species intercepted as they entered the US show that the great majority arrive with air passengers (85 percent), with far less at border stations (14 percent) or by sea (under 1 percent) (McCullough et al. 2006). A review in an entomology journal indicated that between 1984 and 2000 there were 290,101 invasive insect interceptions recorded from air baggage at US airports (Liebhold et al. 2006). Many of these insects were found in fruit, most commonly mangoes but also guava and citrus fruits, among others. Less often they entered the US in leaf material, seeds, or other plant parts. There were 316 countries of origin, most commonly from countries with developing economies (including Nigeria, El Salvador, Haiti, Vietnam, the Philippines, the Dominican Republic, India, and others). Flies, beetles, moths, thrips, planthoppers, and other destructive invasive species have been recorded.

Insects that have caused significant agricultural disruption at times when they have become temporarily established in the US, such as the Mediterranean fruit fly, have been intercepted repeatedly and frequently over a period of years. Other less-heralded invasive species that have reached the US include the emerald ash borer, which is presently consuming forests in 12 Midwestern states and two Canadian provinces, a change from its Eastern Asian diet. Major points of interception for incoming invasive species include New York, Honolulu, Miami, Los Angeles, and several other airports. The number of invasive insects intercepted annually has held approximately steady over the years as domestic passenger enplanements have risen, suggesting that the lack of increase in numbers of inspectors has limited the rate of interceptions (Liebhold et al. 2006).

Climatically similar but geographically distant airports are connected by flights posing the greatest risk of transporting an invasive species capable of establishing itself at its destination (Tatem and Hay 2007). Analysis has shown that climatic similarities of distant airports in the Northern Hemisphere are generally greatest in the period from June to August, when commercial air travel peaks and also when potentially disruptive insect species are at their highest population levels and fruit and other plant material is ripe for transport. This timing would be offset by six months for movement of invasive species between Southern Hemisphere locations. However, specific instances of more complicated patterns of high-risk climatic similarity exist (like that of the Hawaiian Islands with high climatic similarity that switches over several months to sequentially match East Asia, Central America, and the Caribbean). It may be that the strict fumigation procedures already used in Australia and New Zealand will eventually have to be put in place at airports in the US as travel back and forth to regions with developing economies increases.

Countering the global spread of Ebola virus disease by commercial aviation

The great medieval epidemics of bubonic plague are associated in public memory with "plague ships," and in the days when wind was the power that moved people across water, that association was true. Sailing ships were confined environments; they moved slowly enough for plague to develop to high levels, and had hiding places where rats, the carriers of plague fleas, could hide. An early description of a plague ship of Marseilles captures both the initial infection and the beginning "lag period" of a typical exponential growth curve which has characterized not just the plague but the epidemic of Ebola in African nations with a slow growth and then an enormous bound upwards in infections:

> At this time [588 AD] it was reported that Marseilles was suffering from a severe epidemic of swelling in the groin and that this disease quickly spread to Saint-Symphorien-d'Ozon, a village near Lyon . . . I want to tell you exactly how this came about . . . a ship from Spain put into port with the usual kind of cargo, unfortunately also bringing with it the source of the infection. Quite a few of the townsfolk purchased objects from the cargo and in less than no time a house in which eight people lived was left completely deserted, all of the inhabitants having caught the disease. The infection did not spread through the residential quarter immediately. Some time passed and then, like a cornfield set alight, the entire town was suddenly ablaze with pestilence. (Gregory of Tours 1974: 509–11)

Today sailing ships are no longer viable commercial transportation, and bubonic plague can be treated quite successfully in its early stages with appropriate antibiotic therapy. But as new technologies emerge and globalization becomes the theme of the century, Ebola virus disease has emerged in Africa as a new version of an incurable plague; and the limited spread to date that has posed challenges to public health authorities around the world has been either by apparently asymptomatic persons who had recently contracted the disease using commercial aviation, or by transport of patients via air ambulances (Blake 2014). Commercial aviation has also allowed healthcare workers and supplies to be sent to Ebola-stricken regions quickly in order to work to eliminate the spreads of the infection at its source and prevent it from becoming a widespread global threat. It is worth a separate section to see what governments and the aviation industry are doing to prevent the spread of this emergent disease even as they provide a service necessary to the fight against it.

In 2014 IATA issued a statement stressing that Ebola virus disease transmission is unlikely on an airliner. It lays out the framework which the major global organizations involved have agreed upon for an Ebola virus disease response both in the air and at airports. Such response intends to limit the spread of the disease, reassure passengers, and permit commercial aviation to play a vital role in helping medical personnel move internationally to combat this emergent threat.

> On 8 August 2014, the World Health Organization (WHO) declared the Ebola virus disease [EVD] outbreak in West Africa a Public Health Emergency of International Concern (PHEIC) in accordance with the International Health Regulations (2005). In order to support the global efforts to contain the spread of the disease and provide a coordinated international response for the travel and tourism sector, the heads of the World Health Organization (WHO), the International Civil Aviation Organization (ICAO), the World Tourism Organization (UNWTO), Airports Council International (ACI), International Air Transport Association (IATA) and the World Travel and Tourism Council (WTTC) decided to activate a Travel and Transport Task Force which will monitor the situation and provide timely information to the travel and tourism sector as well as to travelers.

> The risk of transmission of Ebola virus disease during air travel is low. Unlike infections such as influenza or tuberculosis, Ebola is not spread by breathing air (and the airborne particles it contains) from an infected person. Transmission requires direct contact with blood, secretions, organs or other body fluids of infected living or dead persons or animals, all unlikely exposures for the average traveler . . . The risk of getting infected on an aircraft is also small as sick persons usually feel so unwell that they cannot travel and infection requires direct contact with the body fluids of the infected person . . . Affected countries are requested to conduct exit screening of all persons at international airports . . . for unexplained febrile illness consistent with potential Ebola infection. Any person with an illness consistent with EVD should not be allowed to travel unless the travel is part of an appropriate medical evacuation . . . Non-affected countries need to strengthen the capacity to detect and immediately contain new cases, while avoiding measures that will create unnecessary interference with international travel or trade. The World Health Organization (WHO) does not recommend any ban on international travel or trade. (IATA 2014)

The US Centers for Disease Control (CDC 2016) has released Ebola guidance for airlines that reiterates the fact that commercial airlines may deny boarding to passengers with "serious contagious diseases" (which Ebola virus disease certainly qualifies as). It calls on cabin crews to follow infection control procedures which are described, to practice hand hygiene, and treat all body fluids as infectious, and says that Universal Precaution Kits should be carried by all airliners travelling to Ebola virus disease-stricken counties. Since Ebola virus disease is not known to be transmitted in an airborne manner, separating an ill passenger from others is mentioned as a worthwhile effort. Pilots of international flights to the US are legally bound to report ill passengers with a list of certain symptoms (including those of Ebola virus disease) to the CDC; notification of airline cleaning personnel is also required, and the CDC provides guidance for their actions.

The WHO is explicit in stating that travel bans associated with the Ebola virus disease outbreaks could actually be counterproductive to slowing and eventually stopping the spread of the disease. It states that:

> WHO does not recommend general bans on travel or trade, or general quarantine of travelers arriving from Ebola-affected countries, as measures to contain the outbreak. Such measures can create a false impression of control and may have a detrimental impact on the number of health care workers volunteering to assist Ebola control or prevention efforts in the affected countries. Such measures may also adversely reduce essential trade, including supplies of food, fuel, and medical equipment to the affected countries, contributing to their humanitarian and economic hardship. (WHO 2014)

As of 2016 several experimental Ebola virus disease drugs and vaccinations were being tested in different countries. In the long term the control of this virus will depend on treatment and hopefully prevention because, even were an outbreak to be successfully controlled, clearly there would be no barrier to prevent other outbreaks from occurring in the future.

Global spread of other emergent diseases by commercial aviation

Commercial air travel can rapidly spread diseases by transporting infected people long distances so rapidly that, given the incubation period for diseases, their most serious symptoms have not yet become apparent during their travel period, and by retaining a large number of people in very close proximity to one another, sharing very limited restroom facilities and breathing recycled air. Several diseases have come to the fore as worthy of special concern in terms of air travel, such as: severe acute respiratory syndrome (SARS), extensively drug-resistant tuberculosis (XDR TB), and the possibility of a variant of the avian influenza (H5N1) emerging that allows easy human-to-human transmission. These diseases have all emerged as threats because of a global environment characterized by areas of overcrowding and poverty, very high-density farming of animals to supply a rapidly growing global hunger for meat, and the overuse or poor implementation of antibiotic therapies. We did not simply find these diseases; rather, to some extent our actions and lifestyles created them, and their threat to us is real.

Analysis of data for the typical spread of influenza in the US annually from 1996 to 2005 found that commercial air travel in November (mostly around Thanksgiving) was a significant predictor of influenza spread. The timing of influenza mortality was related significantly to international airline travel. The air travel restrictions after September 11, 2001 resulted in a delayed and prolonged influenza season (Brownstein et al. 2006). An analysis of several scenarios for travel restrictions as a tool for delaying the spread of an influenza pandemic indicate that airline restrictions would do little to slow the spread of the disease unless most air travel were to cease almost immediately after the beginning of epidemic spread was detected. Interventions to reduce local spread of the epidemic were found to be more effective, but nonetheless would do little to delay the spread compared to the time needed to develop stocks of a vaccine (Cooper et al. 2006). The CDC (2017) has developed guidelines and a fact sheet for how to deal with an outbreak of H5N1 avian influenza or H1N1 swine flu should one of them ever mutate to spread rapidly from person to person; but, given the preceding analyses, the lack of such a mutation is the only positive condition on which to hope.

SARS is a viral disease first reported in Asia in 2003. It was responsible for hundreds of deaths and thousands of infections in Asia, but did not spread around the world as was initially feared it might rapidly do. SARS transmission from person to person on international flights has been documented after some instances of an infected individual taking a commercial flight, but not in others (Olsen et al. 2003; Breugelmans et al. 2004). It is likely that the exact stage of the disease's incubation and course are crucial to the risk level involved for fellow passengers. When a highly infectious individual is on board an aircraft, proximity within a few rows of their seat considerably increases risk of transmission, but more distantly seated passengers are still at risk as well. Prevention of the spread of SARS might be greatly facilitated if contact information for passengers after their flights was more complete (researchers could not find all of the passengers to contact after the flights). Some technologies, like the infrared scanners employed in Singapore and elsewhere to detect passengers with a fever as they walk through, might also be of significant use.

More recently, the possibility of widespread transmission of XDR TB via commercial aviation arose in 2007 after an infected individual was not prevented due to weak public health measures in the US from boarding a flight to Europe and then returning. The infected person took long flights on Air France/Delta and Czech, and neither US domestic nor international procedures were sufficient to impede his travel and prevent widespread exposure of fellow travelers. Hundreds of individuals were potentially exposed to XDR TB, a disease with such widespread drug resistance that few treatment options exist, and for which the mortality risk to infected individuals is very high.[2] In the US 49 cases of XDR TB were reported between 1993 and 2006; but elsewhere in the world, where people live in especially crowded conditions with poor sanitation, XDR TB is much more common.

A 2006 WHO report on tuberculosis and air travel provides a detailed discussion of risk factors, with flights longer than eight hours considered the circumstance of greatest risk. Of a number of potential instances for transmission of non-XDR TB on commercial aircraft, the WHO could identify only two instances where it looked like TB transmission had occurred: once from a crew member to other crew members, and once from a passenger to nearby passengers. In both of these cases, close proximity and long exposure times were involved.[3]

Zika virus, and the potential for crippling or fatal microcephaly among babies born to mothers infected during pregnancy, is widespread as a new global concern. Even though Zika is not transmitted casually from person to person, there is a realistic possibility that infected individuals of either gender could extend the range of the virus by traveling with the virus in their blood to previously Zika-free regions, where Zika could spread if the *Anopheles* mosquito vector was already present. Zika is especially concerning because there is no procedure or treatment at present to protect unborn children from extensive brain damage. The WHO (2016) has promoted guidelines for preventing the spread of Zika virus which include the possibility of disinfection of aircraft (spraying for any live mosquitoes) as well as vector (mosquito) controls at airports. Doubtless the story of Zika virus and how various sectors of our society need to respond to it is far from over.

Other than preventing someone with an infectious disease from boarding a commercial aircraft, what can be done in an increasingly crowded and globalized world to prevent the spread of disease from one region to another? Increasing cabin ventilation in regions with disease outbreaks would reduce the risk of disease transmission. Older assumptions—that flights needed to be of long duration and passenger proximities needed to be within two rows of an infected individual for transmission of any disease—were disproven in 2003 when a single passenger with SARS on a 3-hour flight infected at least 22 fellow passengers as far as 7 rows of seats away. Although normal cabin ventilation is probably adequate for a typical flight, analysis shows that increasing ventilation would help protect passengers during an outbreak, although it would increase fuel demand (Mangili and Gendreau 2005). In addition, increased efforts to disinfect aircraft (including the parts of the ventilation system that are normally inaccessible), insisting that all commercial aircraft begin to have high-efficiency particulate air (HEPA) filtered ventilation systems, and various speculative but interesting techniques for using VOCs or smaller lipophilic compounds associated with early stages of disease infection to screen passengers provide techniques that might someday reduce the chance of disease transmission aboard commercial flights (Barker et al. 2006).

Conclusions

Ground-level environmental impacts of commercial aviation are varied and can affect airport personnel, neighborhoods, and broader geographic regions. Everything from emergent human diseases to invasive species to particulate pollutants needs to be considered, and management of air travel needs to control all of these factors as best it can. Commercial air travel ushered in unprecedented mobility for people, and along with that there are new challenges to consider and face.

Notes

1 See www.gpo.gov/fdsys/pkg/FR-2012-06-18/pdf/2012-13828.pdf; and http://nepis.epa.gov/Exe/ZyPDF.cgi/P100EIJ3.PDF?Dockey=P100EIJ3.PDF [both accessed May 1, 2016].

2 See: www.cdc.gov/tb/pubs/tbfactsheets/xdrtb.htm; www.cdc.gov/tb/xdrtb/travellerfactsheet1.htm; and WHO (2006).

3 This report also cites instances where other diseases, meningococcal disease and SARS, were apparently transmitted on commercial aircraft. The report also addresses the importance of ventilation during ground delays, and cites an instance of an influenza outbreak due to a three-hour ground delay during which the ventilation system did not operate and passengers received no supply of outside air. A similar report has been produced in the EU (ECDC 2009).

References

American Association of Airport Executives (AAAE). 2017. Industry deicing pollution voluntary reduction program. Available at: www.aaae.org/aaae/AAAEMemberResponsive/Advocacy/Regulatory_Affairs/Issues/Industry_Voluntary_Deicing_Pollution_Reduction_Program.aspx [accessed December 1, 2017].

ADL. 2000. Study into the potential impacts of changes in technology on the development of air transport in the UK. Arthur D. Little. Final Report to DETR. DETR Contract No. PPAD 9/91/14. Available at: http://webarchive.nationalarchives.gov.uk/+/www.dft.gov.uk/pgr/aviation/airports/ctat/changesintechnologyandairtra2835 [accessed May 1, 2016].

Barker, I., Brownlie, J., Peckham, C., Pickett, J., Stewart, W., Waage, J., Wilson, P., and Woolhouse, M. 2006. Foresight. Infectious diseases: preparing for the future. A vision of future detection, identification and monitoring systems. Office of Science and Innovation, London. Available at: https://assets.publishing.service.gov.uk/government/uploads/system/uploads/attachment_data/file/294794/06-757-infectious-diseases-systems.pdf [accessed 5 July, 2018].

Blake, M. 2014 (Oct 18). The private jet NOBODY wants to fly on. *Mail Online*. Available at: www.dailymail.co.uk/news/article-2796852/the-private-jet-wants-fly-inside-aircraft-used-transport-ebola-victims-america.html [accessed May 1, 2016].

Breugelmans, G., Zucs, P., Porten, K., Broll, S., Niedrig, M., Ammon, A., and Krause, G. 2004. SARS transmission and commercial aircraft. *Emerging Infectious Diseases*, 10(8), 1502–1503.

Brownstein, J.S., Wolfe, C.J., and Mandl, K.D. 2006. Empirical evidence for the effect of airline travel on inter-regional influenza spread in the United States. *PLoS Med*, 3(10), e401. DOI: 10.1371/journal.pmed.0030401.

Cavallo, D., Ursini, C.L., Carelli, G., Iavicoli, I., Ciervo, A., Perniconi, B., Rondinone, B., Gismondi, M., and Iavicoli, S. 2006. Occupational exposure in airport personnel: characterization and evaluation of genotoxic and oxidative effects. *Toxicology*, 223, 26–35.

Centers for Disease Control (CDC). 2017. Information on avian influenza. Available at: www.cdc.gov/flu/avian/ [accessed May 1, 2017].

Centers for Disease Control (CDC). 2016. Quarantine and isolation. Available at: www.cdc.gov/quarantine/air/managing-sick-travelers/ebola-guidance-airlines.HTML [accessed May 1, 2016].

Cooper, B.S., Pitman, R.J., Edmunds, W.J., and Gay, N.J. 2006. Delaying the international spread of pandemic influenza. *PLoS Medicine*, 3(6), e212. DOI: 10.1371/journal.pmed.0030212.

Donaldson, W.R. 1970. Air pollution by jet aircraft at Seattle–Tacoma Airport. US Department of Commerce, ESSA Technical Memorandum, WBTM WR 58.

Environmental Protection Agency (EPA). 2012. Airport deicing effluent guidelines. Available at: www.epa.gov/eg/airport-deicing-effluent-guidelines [accessed May 1, 2016].

Environmental Protection Agency (EPA). 2007. *Review of the National Ambient Air Quality Standards for Ozone, Policy Assessment of Scientific and Technical Information*. Washington, DC: US EPA.

Environmental Protection Agency (EPA). 2006. *Air Quality Criteria for Ozone and Related Photochemical Oxidants (Final)*. Washington, DC: US EPA.

Environmental Protection Agency (EPA). 2000. Preliminary data summary: airport deicing operations. Revised, EPA-821-R-00-0016.

European Centre for Disease Prevention and Control (ECDC) 2009. *Risk Assessment Guidelines for Diseases Transmitted on Aircraft. Part 2: Operational Guidelines for Assisting in the Evaluation of Risk for Transmission by Disease*. Stockholm: ECDC.

Federal Aviation Administration (FAA). 2005. Inherently Low Emission Airport Vehicle Pilot Program: final report. Available at www.faa.gov/airports/environmental/vale/media/ileav_report_final_2005.pdf [accessed May 1, 2016].

Government Accountability Office (GAO). 2000. Aviation and the environment: airport operations and future growth present environmental challenges. Report to the Ranking Democratic Member, Committee on Transportation and Infrastructure, House of Representatives, GAO/RCED-00-153.

Gregory of Tours. 1974. *The History of the Franks*. Translated by Lewis Thorpe. London: Penguin.

Heuwieser, M. 2017. *The Illusion of Green Flying*. Vienna: Finance & Trade Watch. Available at www.ftwatch.at/wp-content/uploads/2017/10/FT-Watch_Green-Flying_2017.pdf [accessed February 26, 2018].

Holtzman, D. 1997. Plane pollution. *Environmental Health Perspectives*, 105, 1300–1305.

Hu, S., Fruin, S., Kozawa, K., Mara, S., Winer, A.M., and Paulson, S.E. 2009. Aircraft emission impacts in a neighborhood adjacent to a general aviation airport in Southern California. *Environmental Science and Technology*, 43, 8039–45. Available at http://pubs.acs.org/doi/abs/10.1021/es900975f [accessed May 1, 2016].

International Air Transport Association (IATA). 2014 (Aug 18). Joint statement on travel and transport in relation to Ebola Virus Disease (EVD) outbreak. Available at: www.iata.org/pressroom/pr/Pages/2014-08-18-02.aspx [accessed May 1, 2016].

Liebhold, A.M., Work, T.T., McCullough, D.G., and Cavey, J. 2006. Airline baggage as a pathway for alien insect species invading the United States. *American Entomologist*, 52(1), 48–54.

McCullough, D., Work, T.T., Cavey, J.F., Liebhold, A.M., and Marshall, D. 2006. Interceptions of nonindigenous plant pests at US ports of entry and border crossings over a 17-year period. *Biological Invasions*, 8, 611–30.

Mangili, A., and Gendreau, M. 2005. Transmission of infectious diseases during commercial air travel. *Lancet*, 365, 989–96.

Northeast States for Coordinated Air Use Management (NESCAUM) and Center for Clean Air Policy (CCAP). 2003. Controlling airport-related air pollution. Available at: www.nescaum.org/documents/aviation_final_report.pdf/ [accessed May 1, 2016].

Olsen, S.J., Chang, H.L., Cheung, T.Y.Y., Tang, A.F.Y., Fisk, T.L., Ooi, S.P.L., Kuo, H.W., Jiang, D.D.S., Chen, K.T., Lando, J., Hsu, K.H., Chen, T.J., and Dowell, S. 2003. Transmission of the severe acute respiratory syndrome on aircraft. *New England Journal of Medicine*, 349, 2416–22.

Ratliff, G.L. 2007. *Preliminary Assessment of the Impact of Commercial Aircraft on Local Air Quality in the US*. Master's thesis, Massachusetts Institute of Technology. Available at http://dspace.mit.edu/handle/1721.1/40893 [accessed May 1, 2016].

Schlenker, W., and Walker, W.R. 2011. Airports, air pollution, and contemporaneous health. National Bureau of Economic Research Working Paper Series, No. 17684. Available online at www.nber.org/papers/w17684.pdf [accessed May 1, 2016].

Tatem, A.J., and Hay, S.I. 2007. Climatic similarity and biological exchange in the worldwide airline transportation network. *Proceedings of the Royal Society B: Biological Sciences*, 274(1617), 1489–96. DOI:10.1098/rspb.2007.0148.

Wolfe, P.J., Yim, S.H.L., Lee, G., Ashok, A., Barrett, S.R.H., and Waitz, I.A. 2014. Near-airport distribution of the environmental costs of aviation. *Transport Policy*, 34, 102–8.

World Health Organization (WHO). 2016 (Mar 8). WHO statement on the 2nd meeting of IHR Emergency Committee on the Zika virus. Available at: www.who.int/mediacentre/news/statements/2016/2nd-emergency-committee-zika/en/ [accessed May 1, 2016].

World Health Organization (WHO). 2014 (Nov 7). Statement from the travel and transport task force on Ebola virus disease outbreak in West Africa. Available at: www.who.int/mediacentre/news/statements/2014/ebola-travel/en/ [accessed May 1, 2016].

World Health Organization (WHO). 2006. *Tuberculosis and Air Travel, Guidelines for Prevention and Control*, second edition. Geneva: WHO. Available at www.who.int/tb/features_archive/tb_and_airtravel/en/index.html [accessed May 1, 2016].

Yim, H.L., and Barrett, S.R.H. 2011. Public health impacts of combustion emissions in the United Kingdom. *Environmental Science and Technology*, 46, 4291–6. Available at http://pubs.acs.org/doi/pdf/10.1021/es2040416 [accessed May 1, 2016].

PART VI

Other ground-level issues

While Part V ended with a discussion of environmental problems that occur at airports, this final section of the book addresses other key topics that arise on the ground: land use, small airport initiatives, aviation mapping, and air traffic control. The ever-changing world of aviation not only concerns ethical dilemmas that arise during flight operations but also areas that might not sound initially relevant to the industry, such as land use and mapping. As one can see in the following chapters, ground-level issues in aviation continually change and develop, especially due to new technologies as well as the impact of events such as 9/11. Along with these developments, the ethical dimensions of ground-level issues continue to be a source of debate.

The first two chapters examine pertinent topics on airport usage and small airport service development. Chapter 22 by Brian Ohm focuses on legal cases involving land use arguments between private property owners and the nearby airports. While communities enjoy the economic benefit of being located close to an airport, the landowners have rights to use their property without interference, and thus a legal issue may arise. Ohm examines some of the important court cases that show the complexity of determining "land" use for aircraft flying over someone's property. As Ohm demonstrates, these court cases present interesting insights into the balancing act engaged in by the courts as they evaluate the correct course of action for resolving conflicts that arise between airports and private property owners. Such cases recognize the importance to society of a properly functioning aviation system.

Chapter 23 addresses the ethics involved in small markets, especially through the Air Service Development (ASD) initiatives. Small markets have been a special focus of concern in the deregulated market in which the question of profitability continues to be a major factor. As author Russell W. Mills points out, the ASD incentives are designed to reduce the air carrier's risk of serving a small community with an unproven market by insuring the financial success of new service provided by an air carrier. While these types of program sound promising, one question is how effective they are. He also provides case studies on several ASD programs, including what became known as "the Chairman's flight," which resulted in the resignation of United Airlines CEO Jeff Smisek. Mills ends this chapter by discussing the ethics involved in the incentive programs. One key issue is that incentives involve power relations, and for these programs to be considered ethical, they should be voluntary and have an appropriate purpose, unlike the Chairman's flight.

Chapter 24 provides perhaps a surprising approach to aviation ethics, namely, the ways in which aviation mapping inform ethical practices. Peter Nekola first discusses the legacy of the Second World War, which was based on assertions that nations, not individuals, were the sources of moral agency. After the war a new optimism for peace focused on the underlying structure of the world as one of global standardization. Aviation maps allowed the visualization of this potential by providing new possibilities for seeing the world. Based on his analysis of Immanuel Kant, Nekola contends that human agency has more to do with human reason and its responsibilities than any sense of national, linguistic, or religious identity. Nekola ends with a discussion of the World Aeronautical Chart (WAC), developed after the Second World War. The new maps and charts explored the idea that the proper sphere of human ethical action could be understood in the networks that play across that surface, no longer restricted by national borders.

Chapters 25 and 26 examine the past and future of air traffic control (ATC) respectively. Chapter 25 provides the reader with an analysis of the 1981 air traffic controllers' (PATCO) strike in which all striking employees were fired. Michael S. Nolan's personal experience gives us insight into the complexity of the situation by examining the different roles of the key groups involved. In addition to President Ronald Reagan and the controllers, other significant players include the FAA as well as military personnel who took over some of the controller positions. Experts have argued that unions in the US have had less power since this strike, and thus this case remains an important topic for ethical discussions on labor issues.

Addressing the future of ATC, Chapter 26 provides an overview of some of the key issues for ATC in a post-9/11 world. Bill Parrot argues that with the industry's unquestioned commitment to safety, adequate oversight and focused supervision will continue to be essential. By highlighting the ever increasing importance of technology, Parrot also claims that human factors related to the ATC model will continue to occupy center stage, with daunting challenges for the future. Additional points, including morale problems within the controller workforce, will demand increased vigilance by the FAA in its efforts to promote safety. A key debate in the US concerns the privatization of ATC, and this chapter briefly addresses both sides. While general aviation tends to oppose privatization, most US air carriers support it.

In ending with a look toward the future and its uncertainties, this chapter and volume ends by demonstrating the continued need for ethical vigilance in all aspects of the aviation industry, from the level of personal responsibility to international treaties. For aviation to remain safe and efficient, all parties need to remain mindful of the ethical standpoints that impact their decision-making processes.

22

THERE GOES THE NEIGHBORHOOD

Conflicts associated with the location and operation of airports

Brian W. Ohm

Issues related to the use of land often focus on separating incompatible land uses. It should be fairly obvious that airports (and their associated aviation activities) present some unique land use compatibility challenges. Communities view airports as important for the economic vitality of the area. Landowners located near airports, however, have the fundamental right to use and enjoy their land without unreasonable interference. Aviation activities associated with airports can interfere with the use and enjoyment of neighboring lands, especially when those lands are used for residential purposes. Neighboring landowners can also do things on their land that can impact the aviation operations of airports.

In the US conflicts over land use compatibility often are left to the courts to resolve. Over the last century the conflict between airports and adjacent non-airport uses has resulted in hundreds of court cases. This chapter provides an overview of a few of the most significant ones. The cases present interesting insights into the balancing act engaged in by the courts as they evaluate what is the right (or wrong) course of action for resolving conflicts that arise between airports and non-aviation related land uses located near airports. These cases recognize the importance to society of a properly functioning aviation system. Within that context the courts have struggled to provide guidance for when aviation activities cross the line and constitute an unreasonable interference with a landowner's property rights.

Redefining American property law: what do landowners near an airport own?

American principles of property ownership are based on the judge-made, or common law, traditions of England. These traditions were developed centuries before the Wright Brothers made their historic first flight at Kitty Hawk, North

Carolina. Until the advent of modern aviation, property law in the US followed the principle that when someone owned land they were entitled to absolute and undisturbed possession of it, including everything below the surface and the airspace above the surface to the ends of the universe. The Latin phrase for this concept is *cujus est solum ejus est usque ad coelum* (whose is the soil, his it is up to the sky). Someone who interfered with those rights, like a person operating an airplane, could be liable for trespass. Aviation, therefore, posed a direct challenge to the historical notion of what it meant to "own" land in the US.

With the passage of the Air Commerce Act of 1926, the US Congress declared that the federal government, as opposed to the governments of the individual states, had exclusive control over US airspace. It was not clear, however, how this declaration fit with the ancient doctrine of land ownership. This conflict ultimately made it to the US Supreme Court for resolution. In the 1946 decision *United States v. Causby*, 328 U.S. 256, the Supreme Court first attempted to integrate aviation into American property law.

The facts of the case illustrate the challenges posed by aviation to the use and enjoyment of adjacent lands. In 1928 early aviators established an airfield about 8 miles from Greensboro, North Carolina. In 1934 the Causby family purchased 2.8 acres about one-third of a mile from the airport. Located on the property were a dwelling house and various outbuildings which were mainly used for raising chickens. The end of the airport's northwest–southeast runway was 2,220 feet from the Causby barn and 2,275 feet from their house. The glide path to this runway—100 feet wide and 1,200 feet long—passed directly over the Causby property. The 30 to 1 safe path of glide angle to the northwest–southeast runway of the airport passed directly over their property at 83 feet, which was 67 feet above the house, 63 feet above the barn, and 18 feet above the highest tree. Due to prevailing winds, this runway was used about 4 percent of the time in taking off and about 7 percent of the time in landing.

In April, 1942 the airport was taken over by the Greensboro-High Point Municipal Airport Authority for operation as a municipal airport. That same year the US government also obtained the right to use the airport. Military bombers, transports, and fighters, which started to fly over the land in May 1942, created a greater disturbance for the Causby family than previous activity at the airport because the planes were larger and made a louder noise, flights were more frequent, and the glare from the lights on the planes brightly lit up the Causby property at night. The noise and light from the military planes frightened their chickens, causing them to fly against buildings. About 150 chickens died as a result. Egg production also fell off. As a result the Causbys gave up their chicken business. This noise and glare also disturbed the Causby family's sleep, frightened them, and made them nervous. Although there were no airplane accidents on their property, there were several accidents near the airport and near the Causby property.

The Causbys sued the federal government, the operator of the military flights. At issue was whether the operation of the aircraft constituted a taking of property under the Fifth Amendment to the US Constitution. The Fifth Amendment prohibits the taking of private property without the payment of just compensation. For example, the government must pay property owners when it takes someone's land to build a highway. The question in this case was: did the government similarly need to compensate property owners for flights that passed over their properties?

In its opinion, the US Supreme Court recognized that the "airplane is part of the modern environment of life" and declared that the ancient doctrine extending ownership of the land to the periphery of the universe had no place in the modern world. The Court acknowledged that the air is a public highway. If the ancient doctrine were in place, the Court concluded that every flight would subject the operator to countless trespass suits. In a few sentences the US Supreme Court reduced private ownership of the airways by placing at least part of the airways under public ownership, thereby rewriting the law of property that had been embedded for centuries. One can only imagine the mess that would have ensued if the Court had decided otherwise, thereby encouraging hundreds, if not thousands, of trespass lawsuits against airplane operators.

Yet, while the Court announced this general rule of public ownership of the airways, it also recognized limits to the rule. According to the Court, a landowner must have exclusive control of the immediate reaches of the enveloping atmosphere. The landowner owns at least as much of the space above the ground as one can occupy or use in connection with the land. Otherwise buildings could not be erected, trees could not be planted, and even fences could not be run. If, by reason of the frequency and altitude of the flights, landowners cannot use their land for any purpose, they would suffer a complete loss similar to if the government had entered upon the surface of the land and taken exclusive possession of it. The Court opined that this type of interference with the enjoyment and use of the land would be unreasonable. It would be similar to the government taking an easement to the land. In light of the level of interference with the Causbys' enjoyment of their property, the Court agreed that an avigational easement (an easement allowing the use of air rights above someone's property) had been taken by the government requiring compensation. It is interesting to note that the Court did not view the interference strictly as a trespass. Rather, the Court viewed it as the physical acquisition of an easement which allowed the aviation activities to continue.

The Court in the *Causby* case, however, did not provide any bright line formula for how low and how frequent flights must be to significantly interfere with someone's property to constitute a taking. Note that the Court did not require that Congress compensate private landowners for taking their property rights when the Court declared that the airways were public highways. In the Court's opinion, this interference with property rights was not unreasonable. From a practical standpoint, it would have been difficult to figure out how to compensate landowners. It was also not clear in the rapidly developing field of aviation—where the federal government controlled the airways, private companies operated the airlines, and local governments began to own many of the airports—who should be responsible for paying for the acquisition of avigational easements.

Who pays compensation to the property owners?

With the beginning of the jet age in the 1950s, noise became an even more prominent issue creating conflicts between airports and their neighbors. This was a factor in the next significant aviation case decided by the US Supreme Court, *Griggs v. Allegheny County*, 369 U.S. 84, in 1962.

The case involved residential property, owned by the Griggs family, near the Greater Pittsburgh Airport, owned by Allegheny County in Pennsylvania. The slope gradient of the approach area to one of the runways left a clearance of 11.36 feet between the bottom of the glide angle and the Griggs' chimney. The planes taking off from the runway observed regular flight patterns ranging from 30 to 300 feet over the Griggs' residence. On landings, the planes were within 53 to 153 feet. On take-off, the noise of the planes was comparable "to the noise of a riveting machine or steam hammer." During these flights it was often impossible for people in the house to converse or to talk on the telephone. The Griggs were frequently unable to sleep even with ear plugs and sleeping pills. The windows of their home would frequently rattle, and at times plaster fell down from the walls and ceilings. Their house was so close to the runways or glide paths that the spokesperson for the members of the Air Line Pilots Association (ALPA) admitted: "If we had engine failure we would have no course but to plow into your house."

Following the rationale of the *Causby* case, the Court agreed that the interference with private property interests was so great as to constitute the taking of an avigational easement. However, a central issue in the *Griggs* case was who should pay for the taking of the easement. Paying for these easements could cost millions of dollars. The *Causby* case involved wartime flights where the US military was the operator. This case seemed to indicate that it should be the operators of the airplanes who are liable. Allegheny County argued that the operators of the flights—in this case private airline companies—should pay for the taking of the easement. The Court in *Griggs* disagreed with Allegheny County.

Allegheny County also argued that the US government should be liable for the taking of the easement. After all, it was the federal government that claimed the responsibility for creating the national aviation system and placed the airspace in the public realm. Again, the Court disagreed with Allegheny County. According to the Court, Allegheny County—the promoter, owner, and lessor of the airport—was the entity responsible for taking the avigational easement. The Court analogized the situation to a county that designed and constructed a bridge. The county would not have a usable bridge unless it had at least an easement over the land necessary for the approaches to the bridge. Likewise, a county that wants its airport to be usable should be in the same position of needing to pay for the avigational easements. Airports, like bridges, serve a public use for which the public (through the government) must pay when there is an unreasonable interference with the use and enjoyment of someone's land.

Who can regulate noise?

The *Griggs* case highlights the fact that the authority for attempting to deal with airports and their associated activities is split between the federal government and state and local governments. While the federal government, using its authority to regulate interstate commerce, gave itself broad authority to control the use of navigable airspace, the control of airports (and the liability associated with the operation of airports) was left to the states and local municipalities.

With the continued expansion of aviation activities in the 1960s, conflicts over noise within adjacent neighborhoods increased. The federal government began to take action on the issue with the passage of the Aircraft Noise Abatement Act of 1968 and the Noise Control Act of 1972. Some local governments also began to regulate airport operations in an effort to protect surrounding neighborhoods from noise. After all, the local government had to deal most directly with the messy political reality of angry constituents concerned about airport noise.

However, in a 1973 United States Supreme Court decision, *City of Burbank v. Lockheed Air Terminal, Inc.*, 411 U.S. 624, the Court limited the authority of local governments to enact noise control ordinances that applied to privately owned airports. In its *Burbank* decision, the Supreme Court invalidated the City of Burbank's noise control ordinance on the grounds that the Noise Control Act of 1972 preempted the local ordinance. Federal preemption of state and local law arises out of the clause of the US Constitution that declares that the Constitution, and the laws of the US, shall be the supreme law of the land that states and local governments must follow. The Court concluded that the federal government's effort to balance safety and efficiency in the air transportation system required a "uniform and exclusive" system of federal regulations, rather than a cluster of uncoordinated local ordinances regulating aircraft noise.

An important part of the facts in the case was that the airport at issue, the Hollywood-Burbank Airport, was privately owned. The Court noted that many airports are municipally owned and thereby created the "proprietor exception," drawing a distinction between local governments exercising control as proprietors of airports and local governments that did not own airports. As proprietors of airports, local governments could, for example, set limits on the hours of operation of noisier aircraft. While local governments could not place limitations on the hours of operation to reduce noise at an airport the local government did not own, they could legitimately regulate land use around airports to limit the encroachment of noise-sensitive residential development around airports. As airport operators, local governments were therefore not totally foreclosed from trying to address land use compatibility issues. This seemed to add some relief given the holding in the *Griggs* case whereby local governments that owned airports were the entities responsible for paying compensation for unreasonable interferences with the property rights of neighboring landowners. These cases helped encourage local governments to be proactive in trying to address land use compatibility issues around their airports.

Protecting airports from incompatible uses

In addition to protecting noise-sensitive residential development from airports, local governments also often adopt height limitations around airports in an attempt to protect airports from tall structures that might interfere with the operations of the airport. A 2006 case decided by the Supreme Court of Nevada entitled *McCarran International Airport v. Sisolak*, 137 P.3d 1110, however, presents a new twist in thinking about these types of restrictions. While the applicability of the case is limited to the State of Nevada, the case presents an interesting interpretation of the *Causby* and *Griggs* cases.

Following the decision in the *Causby* case, the Congress redefined "navigable airspace" based on the rule prescribed by the then existing Civil Aeronautics Authority that the minimum safe altitude of flight was 1,000 feet over congested areas and 500 feet over non-congested areas, except where necessary for take-off or landing. Since 500 feet was the downward reach of navigable airspace, the space below that was private while the space above was public.

The *McCarran* case involved a landowner's claim to have a property interest in the airspace up to 500 feet near McCarran International Airport serving Las Vegas. During the 1980s, Steve Sisolak bought three adjacent parcels of land for investment purposes, which were each zoned for the development of a hotel, a casino, or apartments. The parcels lie 5,191 feet from the west end of runway at the county-owned McCarran International Airport. When Sisolak purchased the property, a county ordinance was in effect that regulated the height of structures in the vicinity of all public use airports. According to the county, the purpose of the ordinance was to prevent the establishment of obstructions that would pose air navigation hazards that could reduce the size of the areas available for the landing, take-off, and maneuvering of aircraft, thus tending to destroy or impair the utility and capacity of public use airports and the public investment in those airports.

In 1990 McCarran Airport began expanding and upgrading the runway at issue for use by commercial jet aircraft. As a result, the county enacted two ordinances that further affected Sisolak's property. The first ordinance placed Sisolak's property in the precision instrument runway approach zone, which subjected the property to a 50 to 1 slope restriction (limiting an owner's use of airspace 1 foot above ground level for every 50 feet from the runway). On Sisolak's property, this ordinance resulted in an actual height restriction of between 41 and 51 feet. The ordinance also required that a property owner notify the FAA of proposed construction, and provided that the Clark County Planning Commissioners held final authority to grant variances from the height restrictions. The second ordinance designated the Sisolak's property a critical departure area and placed it under an 80 to 1 slope restriction resulting in height restrictions of 3 to 10 feet above ground level. The ordinance provided for a variance procedure similar to that in the first ordinance.

In 2000 a developer submitted proposed building plans to the FAA and the county for approval. The developer wanted to build the "Forbidden City" (a four-story, 600-room resort hotel and casino), and requested a variance from the height restrictions to build up to 70 feet. The FAA granted a variance allowing the building of Forbidden City up to a height of 66 feet. The FAA concluded that such a building would not constitute an airport obstruction. The Planning Commission also approved the proposal. However, the approval lapsed because the developer failed to commence construction within the required one-year period. Sisolak did not complete the sale, and no other variance applications were submitted.

Shortly after the attempted sale, Sisolak filed a lawsuit against the county claiming the height restrictions constituted a taking of his property under the US and Nevada Constitutions. Sisolak asserted that, under state and federal law, he had a property interest in the airspace up to the 500 feet definition of "navigable airspace" by the federal government. He argued that the county, by passing various ordinances, denied him the use of the airspace above the property by appropriating it for public use, which constituted a per se taking. Claiming that the occupancy of his airspace substantially decreased the value of his land, Sisolak demanded just compensation. The evidence indicated that some flights did fly over Sisolak's property at an altitude lower than 500 feet.

Relying on the US Supreme Court's decisions in *Causby* and *Griggs*, the Nevada Supreme Court agreed that Sisolak had a protected property interest up to 500 feet. Public ownership of navigable airspace begins at 500 feet. Below 500 feet the landowner retains exclusive rights. The county height limitation ordinances resulted in the permanent physical invasion of airspace and excluded the landowner from being able to use that airspace.

Again, while the direct impact of this case is limited to Nevada, it does have local governments thinking about the potential need to provide compensation to property owners whose uses of land are restricted by height limitations around airports owned by local governments. Does the case signal a new trend that other states will follow? Will it be necessary to compensate landowners who can only build low buildings because an airport is nearby?

Parting considerations

All of the above cases deal with basic issues of "fairness" related to the costs of operating the aviation system in the US. In the eyes of the courts, we do not want individuals living near airports to bear an unfair burden of the costs of living near airports which are needed for economically thriving communities. To make the world fairer for these landowners, we provide some level of compensation to assuage the burden. While the *Griggs* case places responsibility for providing compensation on the owner of airports, it is interesting to think about where the compensation really comes from. Even though the federal government is not responsible under *Griggs* for providing compensation, for years the federal government has provided grants to airport owners for land acquisition around airports. Even though the airlines are not responsible for paying compensation under *Griggs*, the costs of acquiring land and avigation easements around airports will be reflected in the lease rates that airlines pay to airport owners. The airlines pass these costs on to the passengers—the airport users. Since 1992 airports have also been allowed to collect passenger facility charges from passengers as part of the ticket price for commercial airlines. In the end, these funding mechanisms attempt to spread the costs of operating the nation's aviation system among a broader public rather than placing an unfair burden on a few landowners located near airports. It is debatable whether the *McCarran* case presents the landowner with an unexpected windfall or whether the landowner is indeed unreasonably burdened by a location close to an airport.

Over a century after the first aircraft flight, one might expect that society would have resolved all the issues related to compatibility issues related to airplanes, airports, and non-aviation related land uses. As we have seen, however, the issue continues to evolve.

References

City of Burbank v. Lockheed Air Terminal, Inc [1973] 411 U.S. 624.

Griggs v. Allegheny County [1962] 369 U.S. 84.

McCarran International Airport v. Sisolak [2006] 137 P.3d 1110 (Nev. 2006).

United States v. Causby [1946] 328 U.S. 256.

23

STRINGS ATTACHED?

The ethics of air service development incentives

Russell W. Mills[1]

The US aviation industry has undergone major changes since the 2000s. First the events of September 11, 2001 led to a drop-off in commercial flights, and the imposition of new security procedures have affected the costs of providing air service and lengthened passenger travel times. In addition, economic recessions in 2001 and 2007–2009 reduced demand for air travel, while the price of jet fuel increased substantially. These factors have caused billions of dollars in losses to US airlines, leading to bankruptcies, liquidations, mergers, and acquisitions. The airline industry responded by reducing capacity, retiring inefficient aircraft types, and raising revenue through the imposition of new and expanded ancillary fees. These responses allowed the industry to return to profitability in the past few years (Spitz et al. 2015).

The changes in the airline industry have not had a uniform effect on all airports. The airports serving smaller communities have been particularly affected by the changes, resulting in reduced service levels, less airline competition, and poorer service quality. For instance, from 2007 to 2012, available domestic flights at the largest 29 US airports fell by 8.2 percent, as compared to a 21.7 percent decline in flights at smaller airports (Wittman and Swelbar 2013). Small communities have also been harmed by the retiring of small regional jets that have historically served these markets. Higher fuel prices have hurt the economics of operating smaller, 37–50-seat regional jets, and carriers have responded by cutting flights operated by these aircraft at many small community airports (Wittman 2014). Some carriers are attempting to remove these aircraft from their fleets entirely, replacing them with larger, 51–76-seat regional jets, often at reduced frequencies. Reduced air service is a major concern for smaller communities—several applied and academic studies have found positive, significant relationships between increased access to commercial air service and factors related to economic development in local communities (Goetz 1992; Brueckner 2003; Green 2007).

The importance of air service to local economic development, coupled with airline industry changes, has led many communities to attempt to retain or attract new service by offering financial incentives directly to air carriers. These air service development (ASD) incentives are designed to reduce the air carrier's risk of serving a small community with an unproven market by insuring the financial success of new service provided by either a new entrant or an incumbent carrier. As Wittman notes (2014: 2), "Incentives can take many forms, and innovative airport managers and consultants have created a wide variety of incentive packages to attempt to lure airlines to their airport." Often ASD incentive packages require investments by local businesses and governments of several million dollars. The perceived link between air service and economic development is so vital that in 2001 the US Congress created the Small Community Air Service Development (SCASD) grant program that provides federal funding to local communities to supplement their existing incentive programs.

Despite the ubiquitous nature of ASD programs, it remains unclear whether they are effective at bringing sustained air service to a region. In some cases, carriers exited markets when incentives or subsidies ended, while in others incentive packages of over $3 million have not been enough to persuade an air carrier to start service. More importantly, many communities have not engaged in a robust discussion about the ethical questions and dilemmas involved in subsidizing air service. This chapter will explore many of the ethical issues surrounding the use of air service development incentives. First it will present an overview of common ASD incentives and programs. Next, case studies of incentive programs will illustrate the wide range of views on these programs. Finally, the chapter concludes by analyzing the ethics of incentives, not only in air service development but also from a wider perspective.

ASD incentives and air carrier decision-making

Local economic development officials have brought to bear various approaches and tools in trying to attract air carriers to offer service in their communities. One of the most common approaches to attracting air service has been the use of incentives. Although several studies have found that these incentives do not guarantee achieving greater levels of air service, many carriers have come to expect that communities will offer some package of incentives to entice the carrier to begin service. A study of ASD techniques found that air carriers now view small communities that are seeking service as partners, and therefore require them to participate in the financial risk of developing the new service (Martin et al. 2009). As a result, many local, and in some cases state, economic development officials design and implement their own incentive programs to try to attract air carriers to their communities.

When discussing ASD incentive programs, it is important to understand how airlines make decisions on starting or ending a flight in a community. At the most basic level the success and potential sustainability of a flight are products of the revenue it generates minus the cost to operate that flight. Revenue is determined by the number of passengers and the amount those passengers pay for their tickets and ancillary services. For most legacy carriers and some low-cost carriers (LCCs), a particular flight is likely to generate greater numbers of passengers if it offers connectivity to the carrier's route network, rather than serving only passengers traveling between the flight's Origin and Destination (O&D) locations. The mix of O&D and connecting passengers aboard a flight allows the airline to determine what percentage of a flight's revenue is attributable to the individual flight (referred to as segment profitability) and the revenue derived from supplying passengers across the airline's network through connections to other flights (known as beyond profitability). Segment profitability is calculated by determining the revenue in excess of costs generated by the portion of passengers on a flight who are O&D, while beyond profitability is the portion of revenue over costs generated by passengers who connect to other flights. When deciding whether to enter a new market, airline route planners have desired financial returns for segment and beyond profitability that they use to determine the overall profitability of a flight, and then compare that profitability to other existing or potential flights in their network (Spitz et al. 2015; Stanley 2012; Oimet 2010).

The overall profitability of a flight is also a function of the costs of operating the flight, which include fuel costs, labor costs, and airport fees (for example, terminal rent, fuel flowage fees, baggage handling fees, and landing fees). Other data considered by air carriers include an analysis of the current actions of competitors, community economic or tourism profiles, and strategic considerations and the likely responses of competitors (Mead & Hunt 2006). Many airport incentive programs are designed to reduce or eliminate many of the on-airport costs. However, the ability of a flight to generate sustainable revenue is often the primary driver of an airline's decision to start new service or end existing service. Many community incentive programs that focus on revenue guarantees are designed to offset early market entry costs experienced by carriers. Potential flights that are longer distances, less populated, and overfly competing hubs along the way are viewed by many legacy carriers as high risk and are less likely to be considered (Spitz et al. 2015).

Broadly speaking ASD incentive programs can be broken down into two categories: airport or community incentive programs (Table 23.1).

Airport incentive programs

Airport incentive programs are offered to an air carrier directly by an airport or its governing board. Importantly, airports are constrained in the types of incentives they can offer by Federal Aviation Administration (FAA) rules and Airport Improvement Program (AIP) grant assurances. Specifically, airports must abide by the following requirements when offering incentives to air carriers (FAA 2010):

- Cannot target certain types of carriers (for example, low cost carriers) or particular air carriers/aircraft types (this is to prevent cross-subsidizing of carriers)
- One-year time limit on incentives offered only to new entrants
- Two-year time limit on incentives offered to both incumbents and new entrants
- Airport revenue may be used to:
  - Promote competition
  - Increase air service
  - Raise public and industry awareness of airport facilities and services
  - Pay for a share of promotional expenses designed to increase travel using the airport
- Airport revenue may not be used for:
  - Destination or tourism marketing
  - General economic development/marketing not related to the airport
  - Direct subsidies to air carriers
  - Guarantees of passenger revenue, ticket sales or seats filled
  - Influence of ticket prices.

TABLE 23.1 Common air service development incentives

Airport Incentive Programs	
Reduced/waived fees and rents	Reduction or abolishment of landing fees, fuel flowage fees, departure charges, overnight aircraft parking fees, and terminal rent fees including baggage handling fees, ticket counter fees, and gate and ramp services.
Advertising or marketing assistance	Cash or in-kind resources for advertising new flights, carriers, or destinations; used to prime the market and inform them of new options available.
Start-up cost offset	Provision of equipment, training, personnel, and services designed to offset costs to air carrier including ground handling equipment and services, information technology support for ticketing systems.
Community Incentive Programs	
Minimum revenue guarantee	Agreements between air carriers and communities that establish a target amount of revenue the carrier will receive for operating service on a route over a specified period. Communities commit to a certain revenue level agreed upon with the carrier; if the carrier does not achieve the desired revenue level, community funds are used to bridge the gap.
Travel bank	Local businesses or individuals deposit funds in a bank account that can be used only for purchasing tickets on the specified air carrier during a given period.

Due to these constraints, airports typically offer incentives to air carriers to start or continue new service through reduced fees, waived or reduced rents, ground or baggage handling, and advertising and marketing assistance.

When an air carrier provides new service at an airport, there are a host of costs associated with the establishment of a new ground station at an airport, including training of new personnel. Additionally, airports charge for the use of their property, including landing fees, terminal rental fees (baggage claim, ticket counter, gate, and ramp services), and federal inspection charges. Start-up costs for an air carrier can be upward of $200,000 (Martin et al. 2009). The waiving of airport landing fees and rents tends to be one of the most common forms of incentives offered by local communities because they can be administered through the airport governing body and often require little involvement from community partners. Importantly, cost waivers by themselves will not differentiate an airport seeking service, as most air carriers regard some level of cost subsidy as a requirement for entering a market. As such many airports openly advertise their landing fee incentives on their official websites. Finally, because cost waivers do not typically require substantial community involvement, they are a weak signal to air carriers of community commitment to new service.

Community incentive programs

Another common incentive offered to air carriers is through the marketing and advertising of new service. In smaller communities it is often a challenge for air carriers to effectively market their new service to a new group of travelers who may have loyalties to another carrier or may travel to a nearby airport for their air travel needs. Therefore, local communities often offer to provide marketing or promotional assistance as part of incentive packages. While the FAA's grant assurances allow airports to directly provide marketing or promotional support for new service, airport staffs usually lack the expertise necessary to implement an effective marketing campaign. Therefore, local economic development agencies, chambers of commerce, or other private partners may provide financial assistance in the form of marketing and advertising as part of a community incentive program. Other community partners, particularly the media and billboard owners, may provide in-kind assistance through free or reduced advertising rates. In-kind contributions for marketing and advertising support provide substantial value to the local airport, demonstrating local support of the air carrier while also showing a high level of community buy-in to the new service. More recently, social media such as Facebook and Twitter have become ubiquitous and inexpensive tools that airports and local communities can use to market and promote new air service in their communities (Spitz et al. 2015).

Community ASD incentives differ from airport incentives in that they are not subject to FAA rules and AIP grant assurances. This means that a community can offer a direct subsidy to an air carrier for a specific route. Typically, a local economic development agency, a chamber of commerce, or a convention and visitors' bureau (CVB) operates community ASD incentive programs. Importantly airport sponsors must not be involved in the decision making of the community ASD incentive program, and airport revenue cannot be used in these incentive programs. Community incentives are less common than airport incentives and are often used to attract or retain specific service, such as service to a particular destination. The two most common community ASD incentives are minimum revenue guarantees (MRGs) and travel banks.

MRGs are agreements between air carriers and communities that establish a target amount of revenue the air carrier will receive for operating service on a route over a specified time period. Communities commit to a certain revenue level agreed upon with the carrier, and if the carrier does not achieve the desired revenue level, the community funds are used to bridge the gap. Importantly, if the new service does achieve the desired level of revenue, it is an indication of the route's success and indicates to the carrier that the new service may be self-sustaining without future subsidy. Many communities place performance requirements on the carrier as conditions for the payout of the revenue guarantees. Such requirements can include on-time performance benchmarks or limits on the number of cancelled flights. While revenue guarantees are an effective incentive, they also have some challenges. Many carriers may be hesitant to take a revenue guarantee if they know the route will be unprofitable because of the negative brand image it may leave in the community along with the cost of repositioning assets. Also, many airports and community members have noted that raising money from the community for revenue guarantees was difficult (Martin et al. 2009).

Guaranteed ticket purchases, or travel banks, ensure that the air carrier being targeted will have a certain level of passenger traffic worth a certain volume of revenue. Local businesses or individuals deposit funds in a bank account that can be used only for purchasing tickets on the specified air carrier during a given time period. Typically, local or state governments will match funds provided by businesses in the form of additional revenue guarantees. From the perspective of local governments, the benefit of a travel bank compared to revenue guarantees is that it provides an indicator to the airline of the community's commitment to use the proposed service. More importantly, it provides an indicator of the interest of the business community, which often offers the greatest source of funds to the travel bank and is generally the type of client the air carrier is most interested in securing for new service (typically high-yield customers). Additionally, travel banks also signal the level of commitment to existing carriers at an airport. A lack of support for a travel bank to attract a new carrier may signal to the air carrier that there is already a great deal of customer loyalty to an existing carrier at the airport or at a nearby airport.

While there are many positives to travel banks, there are also some challenges with implementing them as part of an incentive program. First, travel banks require a great deal of local initiative to organize and implement. They often are most successful when implemented through a grassroots organization led by a significant community champion. Second, airline acceptance of travel banks is not uniform. A 2005 GAO report found that most airline officials viewed travel banks unfavorably due to the difficulty and unreliability of their implementation.

The Small Community Air Service Development (SCASD) program

It is important to note that the federal government also plays a significant role in assisting local communities and airports in the attraction and retention of air service. The SCASD program is a competitive discretionary grant program administered by the US DOT's Office of Aviation Analysis (OAA) that allocates funds to provide temporary financial assistance to small communities in order to gain or improve access to the national air transportation system. Local communities, businesses, economic development corporations (EDCs), chambers of commerce, and CVBs use SCASD grants to supplement community-funded incentive packages that include minimum revenue guarantees, marketing support, reduced airport fees, or travel bank pledges (Wittman 2014).

The SCASD grant began as a pilot program in 2000, authorized under the Wendell H. Ford Aviation Investment and Reform Act for the 21st Century (AIR-21), P.L. 106-181. While no funds were appropriated in the first year of the pilot program's authorization, Congress provided $20 million for the SCASD program in 2002. The program was made permanent in the Vision 100-Century of Aviation Reauthorization Act of 2003 (Vision 100) and was reauthorized through FY 2015. If communities are awarded a SCASD grant, they enter into a reimbursable agreement with the DOT for the portion of total expenses that are federally funded. Importantly, while the DOT allocates funds to communities, there are no guarantees that they will be used. A new route introduced by a carrier may be so successful that it does not require subsidies to achieve the agreed-upon profitability margin. Conversely, an air carrier may view a route or community as so unprofitable that it will not enter into an agreement for new service even with millions of dollars in revenue guarantees available. In this case, the unused funds are returned to the DOT. Since its inception in 2002, the SCASD program has received over 1,100 applications for funding and allocated 364 grants totaling over $164 million (Table 23.2).

Case studies of air service development incentives

Fort Wayne International

In October 2014 US Airways (now American Airlines) began round-trip service from the Fort Wayne International Airport (FWA) in Indiana to two of its major hubs—Philadelphia International Airport (PHL) and Charlotte Douglas International Airport (CLT) in North Carolina. The addition of these two new destinations for Fort Wayne represented a victory for a community that had been trying to add east coast service to support the travel needs of local businesses and

TABLE 23.2 SCASD Awards (2002–2014)

Year	*Number of Applicants*	*Number of Grants Awarded*	*% of Applicants Receiving Award*	*Total Amount Awarded*	*Average Amount Awarded*
2002	175	40	22.86	$19,985,056	$499,626
2003	170	36	21.18	$19,849,807	$567,137
2004	108	40	37.04	$19,853,546	$496,338
2005	84	37	44.05	$18,952,685	$512,234
2006	75	28	37.33	$9,692,600	$387,704
2007	76	26	34.21	$8,975,678	$345,218
2008	66	16	24.24	$6,499,000	$433,266
2009	84	19	22.62	$6,445,450	$339,234
2010	74	19	25.68	$6,993,000	$368,052
2011	70	29	41.43	$14,984,000	$515,448
2012	61	33	54.10	$13,917,000	$421,727
2013	59	25	42.37	$11,484,375	$459,375
2014	52	16	30.77	$7,000,000	$437,500
Total	1,154	364	31.54	$164,632,197	$444,835

Note: The number of SCASD applications has decreased since the first three years of the program due to cuts and funding and the implementation of a ban on communities holding more than one grant at a time.

citizens for years (Rochford 2014). While Fort Wayne previously had eastbound hub service through Pittsburgh, Cleveland, and Cincinnati, a series of industry changes—including airline mergers, spiking fuel prices, and the elimination of smaller regional jets—left the airport without service to an eastbound hub. Prior attempts by the community and the Fort Wayne-Allen County Airport Authority to attract an air carrier had failed to result in new service despite several meetings with carriers at national conferences and at their headquarters. Multiple carriers, including US Airways, told the airport authority that while a Fort Wayne route to an east coast hub would be profitable in the long run, they were not willing to take the short-term risk of dedicating a valuable asset (namely, an aircraft) to the community without a more robust market for service.

The airport authority joined with the City of Fort Wayne and the Greater Fort Wayne Chamber of Commerce to assemble over $2.5 million in incentives to persuade US Airways to establish flights to Charlotte and Philadelphia (Miller 2013). The incentive package included a $600,000 SCASD grant to provide a minimum revenue guarantee to US Airways for non-stop daily service to its hub in Philadelphia. The $600,000 in funds from the DOT was matched by $1,400,000 in funding from the City of Fort Wayne and the Greater Fort Wayne Chamber of Commerce for a revenue guarantee and $60,000 in marketing support. In addition to the $2 million in minimum revenue guarantees the Fort Wayne-Allen

County Airport Authority waived over $500,000 in airport fees. The executive director of the Fort Wayne-Allen County Airport Authority, Scott Hinderman noted, "We believe there is risk coming into a new market and new community. That is the purpose of a revenue guarantee" (Sumers 2014). Despite the over $2.5 million in incentives offered by Fort Wayne to US Airways, Hinderman said, "The market has to be there before a carrier will come in. They are not coming into a market that would project to fail even if there is a minimum revenue guarantee" (Sumers 2014).

Charles M. Schulz Sonoma County Airport

Charles M. Schulz Sonoma County Airport (STS), located in Santa Rosa, California, is a non-hub commercial service airport that serves a six-catchment area of 1.1 million residents. STS faces significant competition from San Francisco International Airport (SFO), Oakland International Airport (OAK), and Sacramento International Airport (SMF). In 2012 STS captured only 8 percent of total bookings in the catchment area, with SFO retaining 78 percent and Oakland receiving 7 percent. While STS is only 65–75 miles from both SFO and OAK, drive times can range up to 3–5 hours on the congested Highway 101 corridor. This problem results in many residents from the Sonoma County area driving to SFO or OAK the evening before morning flights to avoid potential delays.

The recent history of air service development at STS begins in 2001, when United Airlines directed SkyWest Airlines to end its flights from STS to Los Angeles and San Francisco despite strong performance. The decision to abandon the successful SkyWest commuter routes was based on internal airline policy factors and was not related to the effects of the September 11 terrorist attacks. One local official noted that SkyWest's exit from STS was a "come to Jesus moment—we didn't even know it was there until it was gone." From October 2001 through March 2007 STS did not have commercial service. After a year of unsuccessful efforts to lure an air carrier to provide service at STS, the County Board of Supervisors created the Sonoma County Airport Airline Attraction Committee (AAC) in October 2002. The AAC was comprised of representatives from the County Board of Supervisors, local businesses, the Sonoma County Tourism Bureau, and the Santa Rosa Chamber of Commerce and had three subcommittees: one to secure advance ticket purchases from local businesses; one to secure in-kind advertising support from media outlets; and another providing feedback to ideas for new service.

The AAC and the airport's consultant identified Los Angeles and Seattle as their top destinations, and Alaska/Horizon Air as the preferred carrier. Alaska/Horizon was chosen as the preferred carrier primarily due to its use of the Bombardier Q-400 aircraft. The Q-400 is a high-performance turboprop aircraft that seats 76 passengers and can easily take off and land at STS's short runway. In 2004 the AAC and the airport submitted a SCASD grant application to the DOT for $635,000 in revenue guarantee and marketing funds. Importantly the AAC and

STS had worked to secure $500,000 in travel bank commitments from local businesses and a letter from Horizon Airlines in support of their application. The DOT awarded STS the full $635,000 requested in the application. The airport and the AAC partnered with American Ag Credit in Santa Rosa to offer Alaska Airlines branded debit cards to be issued to businesses that committed funds to the travel bank. During the three weeks following the announcement, the AAC secured over $500,000 in funds for the travel bank. One airport official noted that Alaska Airlines was supportive of the effort rather than a revenue guarantee because it demonstrated community commitment and investment in the sustainability of the service.

While the community had raised $500,000 in travel bank purchases and secured the SCASD grant, it still needed to gain a firm commitment from Alaska Airlines officials. Airport officials and the AAC flew Alaska Airlines executives into Oakland International Airport at rush hour. As the executives sat in the town car for three hours, airport officials made their pitch to the executives that residents of the North Bay sit in this traffic when they fly from SFO and that there is demand for direct service from STS to avoid this inconvenience. When the Alaska executives departed STS, airport officials and AAC members rented a helicopter and flew the executives over the nearby Sears Point Raceway, where a NASCAR race was taking place. Executives noted that they were impressed by the number of people that came to the area for the event. In 2006 Alaska Airlines announced that Horizon Air would begin service in March 2007 with two daily non-stop flights to Los Angeles and one to Seattle.

While the community was willing to support the Alaska Air service through a travel bank and a federally funded minimum revenue guarantee, several local residents expressed outrage at the idea of the community funding an MRG to incentivize an air carrier. One local official noted, "What makes me nervous about the subsidies—it creates this interdependency that's not healthy" (Spitz et al. 2015: 118). When asked about businesses contributing to an MRG for eastbound service, a local business owner proclaimed: "I don't want to come across as belly aching too much but, I mean, it (eastbound service) would be beneficial but I'm not willing to sell my soul for it . . . I mean if you just want cash in your pocket, forget it" (Spitz et al. 2015: 118). These quotations illustrate the wide range of perceptions on different ASD incentives and how one incentive may be acceptable to a community while another may seem unattractive.

Newark Liberty International Airport and Atlantic City International Airport

The Port Authority of New York and New Jersey (the Port Authority) is the government entity responsible for operating several airports in both New York and New Jersey—including John F. Kennedy International (JFK), LaGuardia Airport (LGA), and Newark Liberty International Airport (EWR). The Chairman of the Port Authority from 2011 to 2014 was David Samson, the former Attorney

General of New Jersey. Between 2011 and 2012 then United Airlines CEO Jeffrey Smisek was working with Samson to: reduce the airline's terminal rent and operational fees; increase access to EWR through a $1.5 billion extension in train service between the airport at New York City on the Port Authority's PATH rail line; and secure Port Authority funding for the construction of a new wide-body aircraft maintenance facility (Kocieniewski and Voreacos 2015; Mayerowitz and Caruso 2015). During a dinner meeting in 2011 Samson asked Smisek and other top United Airlines officials if the airline would consider reinstating a flight between EWR and Columbia, South Carolina—50 miles from Samson's weekend vacation home in Aiken, SC. A month after Samson's request, United Airlines route planners rejected the proposed route because it would not turn a profit (Kocieniewski and Voreacos 2015).

Not deterred by this initial rejection, Samson continued to push United Airlines to add the flight to Columbia. Specifically, during two monthly board meetings Samson removed agenda items of importance to United. At the December 2011 board meeting the Port Authority approved $10 million to support the construction of United's $35 million hangar for wide-body aircraft maintenance. The next month, during a meeting in Chicago with United Airlines route planners, officials from Columbia Metropolitan Airport (CAE) were pleased and surprised to learn that the airline was interested in restarting service to Newark, particularly given the airline's insistence that the route would be unprofitable. The twice-weekly flight (Thursday evenings from Newark to Columbia and Monday mornings from Columbia to Newark) began on September 6, 2012. Two weeks later the Port Authority began a study to extend the PATH rail line to EWR. The Columbia flight quickly became known as the "Chairman's flight" and ended in March 2014, four days after Samson resigned from the Port Authority amid public outcry over the George Washington Bridge scandal (Kocieniewski and Voreacos 2015; Bachman 2015). In February 2015 the US Attorney's office in New Jersey announced that it was formally investigating both United Airlines and the Port Authority over the Chairman's flight. Due to the mounting pressure from the federal investigation, United CEO Smisek resigned in September 2015 (see textbox for the legal repercussions).

During the same negotiations with United Airlines, New Jersey Governor Chris Christie and the South Jersey Transport Authority, which owns Atlantic City International Airport (ACY), agreed to pay the Port Authority $500,000 per year to market and attract more flights to ACY to help save the city's dying casino industry (Mayerowitz and Caruso 2015). Upon taking control of ASD efforts, the Port Authority worked with United Airlines to explore air service options at ACY. The Port Authority developed an incentive program that focused on providing marketing support ($60,000 per flight) and reduced landing, fuel, and handing fees to attract a new carrier to the region (Wittkowski 2014). At the same time the Atlantic City Casino Reinvestment Development Authority (CRDA) created a $5 million minimum revenue guarantee program to help offset losses by carriers during the first two years of service.

On November 14, 2013 Governor Christie and United CEO Smisek announced that the air carrier would begin service from its hubs in Chicago and Houston to ACY starting in April 2014. Interestingly, the announcement of the new flights did not take place in Atlantic City, but rather at United's Newark terminal—a reminder of the ongoing negotiations over fees and the PATH rail line between the airline and the Port Authority. As part of the agreement to start service at ACY, United took advantage of the Port Authority's newly created incentive program featuring reduced fees and marketing support. As a condition of receiving the incentive package, United agreed to repay any marketing incentives if it pulled the new service within the first year. Three months after the announcement of the ACY flights, Christie and New York Governor Andrew Cuomo announced they were moving forward with the PATH rail extension from downtown Newark to the airport (Mayerowitz and Caruso 2015).

During the first seven months of service between Chicago and Houston, the ACY flights averaged 50 percent load factors—an indication of a lack of demand for travelers to Atlantic City (Mayerowitz and Caruso 2015). While they were successful in persuading the Port Authority to move forward with the PATH rail extension, United Airlines was still engaged in intense negotiations with the port over fees at its Newark hub. When negotiations over fees broke down in November of 2014, United announced it was ending the ACY flights in December because of the unprofitability of the routes (Mann 2014). Per the incentive agreement, the Port Authority could have obligated United to repay over $104,000 in marketing incentives used to market the flights. However, they chose not to pursue repayment given potential legal costs and possible reputational damage that would discourage other carriers from starting service at ACY (Boburg 2015).

THE LEGAL REPERCUSSIONS OF THE CHAIRMAN'S FLIGHT

Jeff Smisek's resignation was not the only consequence of the Chairman's flight. The legal repercussions included United Airlines paying over $4.6 million to settle related criminal and civil investigations without admitting wrongdoing. As for David Samson, the former Chairman of the Port Authority, he pleaded guilty to using his position to coerce United into starting the Chairman's flight. He could have received up to two years in prison, but was instead ordered to one year of home confinement, a $100,000 fine, and four years of probation. His lawyers pointed out Samson's poor health as a reason to avoid prison time. However, the prosecution had wanted prison time because "a stiff sentence for a high-ranking official would send a message" (Ax 2017). Although this case is now resolved, the question remains whether the punishment fits the crime.

Ethical considerations and air service development incentives

The outline of airline incentives, along with the three case studies presented above, illustrates the complex and often non-transparent process by which airport managers, businesses, and elected officials attempt to attract and retain air service in communities. The provision of incentives to air carriers by communities raises several critical ethical dilemmas to address. In her innovative assessment of the ethical issues around providing incentives to change behavior, Ruth Grant argues that incentives are comprised of three interconnected components:

- An extrinsic benefit or a bonus that is neither the natural or automatic consequence of an action nor deserved reward or compensation
- A discrete prompt expected to elicit a particular response
- An offer intentionally designed to alter the status quo by motivating a person to choose differently than he or she would likely to choose in its absence. (Grant 2012: 43)

Additionally, Grant argues that incentives necessarily involve relations of power between two parties. The exercise of power using incentives can take the shape of bargaining, persuasion, or coercion in certain cases. In order for the use of incentives to be ethical, according to Grant, they must be for an appropriate purpose and must be voluntary—not done out of desperation brought on by a power imbalance. It is clear from the cases explored above that the air service development programs used by many airports and communities today meet Grant's definition of an incentive. From minimum revenue guarantees to waivers of airport fees, the incentives provided by airports to attract new air service are an extrinsic benefit, expected to elicit a response (namely, start new service) and to alter the decision making of the air carrier.

The more interesting questions regarding the ethical dilemmas around air service development incentives are whether they meet Grant's criteria of an appropriate purpose and if they are voluntary. The purpose or reason given by many communities using incentives to pursue air carriers is that air service is a vital economic connection to the US and the world. Additionally, advocates argue that it increases the potential for economic development in the region as well as increasing access to the rest of the world for citizens of the community. However, as the case of the Chairman's flight indicates, there may be motives or purposes other than enhancing air service and economic development behind offering ASD incentives that could be considered inappropriate. The federal investigation coupled with the resignation of United's CEO, Jeff Smisek, suggest that the trading of new flights as a *quid pro quo* arrangement for a new hangar or reduced fees at another airport is an unethical use of incentives.

The second criterion mentioned by Grant is that the offering of incentives is voluntary. By a strict interpretation, the decision by airports to offer incentives to air carriers is entirely voluntary—some airports make the choice to offer incentives while others do not. However, one of the major debates surrounding the use of ASD incentives is whether air carriers "game the system" by moving from community to community and taking incentives where they find them. Similarly, many communities justify the use of ASD incentives by claiming that if we do not offer them, the air carrier will simply go to an airport that does. A key component of the relationship between airports and air carriers is that airports typically lack key information on air carrier decision making, including the overall cost of operating a specific flight at their airport. Therefore, when airports design and offer incentive packages to air carriers, they are often uncertain as to what amount of incentive would be necessary to change the decision-making calculus of the air carrier. For example, was it necessary for Fort Wayne to offer $2.5 million in incentives to US Airways? Would US Airways have started service with an incentive package of $1.5 million instead? It is reasonable to conclude that the power dynamic favors air carriers in the use of ASD incentives, much as it favors companies looking to gain tax incentives when they look for locations to build new facilities.

Note

1 The research presented here was funded by a Transportation Research Board Airport Cooperative Research Program grant (ACRP 03-29).

References

Ax, J. 2017 (Mar 6). Ally of New Jersey Governor Christie avoids jail for airline scheme. *Reuters*. Available at: www.reuters.com/article/us-new-jersey-bridgegate/ally-of-new-jersey-governor-christie-avoids-jail-for-airline-scheme-idUSKBN16D16D [accessed March 30, 2018].

Bachman, J. 2015 (Feb 25). United and Port Authority subpoenaed over Chairman's flight. *Bloomberg*.

Boburg, S. 2015 (Sep 28). No payback of incentives required for United Airlines over Atlantic City service. *North Jersey Record*.

Brueckner, J. 2003 (Jul). Airline traffic and urban economic development. *Urban Studies*. 40(8): 1455–1469.

Federal Aviation Administration (FAA). 2010 (Sep). *Air Carrier Incentive Program Guidebook: A Reference for Airport Sponsors*. Washington, DC: Federal Aviation Administration.

Government Accountability Office (GAO). 2005 (Nov). *Commercial Aviation: Initial Small Community Air Service Development Projects Have Achieved Mixed Results*. Washington, DC: US Government Accountability Office.

Goetz, A. 1992. Air passenger transportation and growth in the US urban system, 1950–1987. *Growth and Change*. 23(2): 217–223.

Grant, R. 2011. *Strings Attached: Untangling the Ethics of Incentives*. Princeton, NJ: Princeton University Press.

Green, R. 2007. Airports and economic development. *Real Estate Economics*. 35(1): 91–112.

Kocieniewski, D. and D. Voreacos. 2015 (Apr 28). The dinner proposal that subjected United to a federal corruption probe. *Bloomberg*.

Mann, T. 2014 (Nov 7). United Airlines cuts its service to Atlantic City. *Wall Street Journal.*

Martin, S. C. et al. 2009. *ACRP Report 18: Passenger Air Service Development Techniques* Washington, DC: Transportation Research Board.

Mayerowitz, S. and D. Caruso. 2015 (Sep 28). Tangled web of United, Port Authority and Christie pals extends to Atlantic City routes. *Associated Press.*

Mead & Hunt. 2006. *Northwest Regional Air Service Initiative Handbook: Small Community Air Service Development.* Issaquah, WA: Mead & Hunt, Inc.

Miller, J. 2013. Fort Wayne International Airport application under the Small Community Air Service Development Program.

Oimet, P. 2010 (Oct 25). Airline routes: how you can influence their development. Paper presented to the 49th ICCA Congress and Exhibition, Hyderabad, India. Available at: www.iccaworld.com/cnt/progmdocs/ME302%20-%20Paul.pdf [accessed October 31, 2014].

Rochford, B. 2014 (May 5). FWA gains Philadelphia, Charlotte flights. *Fort Wayne Business Weekly.*

Spitz, W., M. O'Connor, R. W. Mills, and S. Murray. 2015. *ACRP Report 142: The Effects of Airline Industry Changes on Small and Non-Hub Airports.* Washington, DC: Transportation Research Board.

Stanley, J. 2012 (May). Airline network analysis in a changing US industry. *R&A Point-to-Point (Ricondo & Associates).* Available at: www.ricondo.com/articles/PTP-airline-network-analysis-may-2012.pdf [accessed October 31, 2014].

Sumers, B. 2014 (Dec 19). Fort Wayne International Airport says incentive plan has worked. *Aviation Daily.*

Wittkowski, D. 2014 (Apr 17). United to take advantage of airport incentives for Atlantic City flights. *Press of Atlantic City.*

Wittman, M. 2014 (Sep). Public funding of airport incentives: the efficacy of the Small Community Air Service Development Grant (SCASDG) program. *Transport Policy.* 35: 220–228.

Wittman, M. and W. Swelbar. 2013 (May). Trends and market forces shaping small community air service in the United States. Cambridge, MA: MIT International Center for Air Transportation.

24

WHAT CAN AVIATION MAPS TEACH US ABOUT ETHICS?

Peter Nekola

How we see the world and make sense of it does much to inform our actions upon it, and how we have come to inform and justify those actions. Practices of aviation mapping have been perhaps surprisingly relevant to both forming ethical perspectives and informing ethical practices since the mid-1960s. Envisioning the earth's surface in terms of air networks and surface conditions has made it easier, in turn, to envision a global setting for international laws, charters, and conventions. Also, maps drawn from aerial data to air-age global standards have also enabled possibilities for studying issues of environmental management and sustainability, as well as poverty and development, while being less hampered by national or state boundaries or interests. Such maps, and their own history, pose useful questions concerning ethics and especially the nature of ethical agency.

Agency is a core concern of ethical thinking. One who *acts*, or is somehow responsible for an action, is considered an ethical actor or *agent*. The consideration of ethical agency is the investigation of questions such as: Who has responsibility for what? Who is capable of making ethical decisions? And under what circumstances? Though many theories of ethical reasoning have cast decision-making as an activity carried out by individual people, for much of human history social and cultural norms and practices have denied many individuals the status of ethical agents—just as they have denied many individuals rights in favor of classes, groups, or nations, as de facto ethical agents. Those same social and cultural norms have long conditioned how people have understood their own identities, their roles in society, and the nature of the earth's surface.

The legacy of the Second World War

The legacy of the Second World War casts dark shadows on the subject of both ethics and aviation. This war, perhaps more than any other, brought questions concerning the nature of ethical agency into public debate. The Second World War was born of the ruthless assertion of *national* agency—the idea that nations, not individuals, were the ultimate sources of ethical agency. Near the end of that conflict modern jet technology was initially employed on an industrial scale. In the world of nations-as-ethical-agents, aviation technology could be understood largely in terms of the strategic value and destructive potential that had driven it during wartime. But that world was changing quickly.

After the war, new international organizations were developed to secure and maintain the infrastructure of peace: most notably the United Nations (UN), chartered in 1945; its component agencies such as the International Civil Aviation Organization (ICAO), formally ratified in 1947; and its ethical conventions such as the Universal Declaration of Human Rights (UDHR), adopted the following year. Membership in these organizations and the signing of these declarations was limited to states, which in turn were often national in character, as was the political norm of postwar territorial organization. Human thought and experience, however, was itself changing: evolving and adapting to worldview-challenging technologies and opportunities, as well as new ways of framing ethical questions and considerations.

The "underlying pattern of the world"

In December 1945, a few months after the end of the war, Robert Platt, a geography professor at the University of Chicago, delivered a presidential address to the Association of American Geographers titled "Problems of Our Time," which was published a few months later in the association's journal. Studying such problems, according to Platt (1946: 2), would "involve the interlocking of human life over the whole earth, and simultaneously the localization and even to some degree the isolation of life in every individual place on the earth." Platt suggested a study of a single square mile of the earth's surface could help make sense of the world in the wake of one of the most disastrous conflicts in human memory, a time when geographical knowledge and reasoning could, in his view, help preserve and inform the peace. "In one square mile," Platt said, "there is evidence of these antithetical aspects of life—world unity and disunity, coherence and separation, interdependence and independence—opposite but coincidental aspects of life to the met and dealt with . . . by everyone who lives on earth" (Platt 1946: 2).

The square mile Platt chose was the Chicago Municipal Airport (in 1949 renamed Chicago Midway Airport). That mile, according to Platt, could not be understood without consideration of its place in "world air traffic," only just beginning to recover from "wartime fracture" but with new technology that could reduce the distance between people and places considerably (Platt 1946: 2). When "the most distant point on earth is less than sixty hours from your local airport," Platt's colleague Walter Ristow remarked earlier, "all geography becomes home geography" (Ristow 1944: 333). Geographical knowledge and reasoning, as Platt had understood them, involved studying *patterns* on the earth's surface, instead of strict adhesion to the study of such entities as nations and states. While some patterns on the landscape fit within national or state boundaries, many did not; in 1945 air traffic patterns in particular were more global than they had ever been. Chicago's airport, Platt noted, "has a place within the world pattern. Its facilities are duplicated throughout the pattern" (Platt 1946: 8). He juxtaposed the airport with an analysis of the surrounding residential and industrial neighborhoods, as well as the school next door, and continued with a discussion of "another permanent characteristic of human life: the social and physical localness of human beings in distinct and different localities" which could "raise respective problems", such as "How shall we arrange our far-reaching activities, our world-wide interdependence, now calling for split-second cooperation and presupposing universal agreement on objectives and means of attaining them?" (Platt 1946: 12).

These questions, in the years after the war, could "be faced with either optimism or pessimism" (Platt 1946: 13). For Platt, thinking geographically, as opposed to thinking nationally or territorially, was itself optimistic. It meant thinking of the world not in terms of national actors—making sense of the world by drawing boundaries and emphasizing perceived differences—but by studying new patterns that developed on and across, as well as above, the surface of the planet. Maps of air networks were fast becoming among the most ubiquitous ways to visualize such patterns, while aerial mapping was fast becoming among the most comprehensive and detailed. Such thinking offered "no trick formula for comprehensive understanding," but a position from which to "see the range of problems, to trace world patterns in perspective" (Platt 1946: 15). Platt's address concluded not on an overwhelmingly hopeful or optimistic note, but with a call for such optimism that new interpretations of the "underlying pattern of the world" could be of use in somehow peacefully reconciling humans' "social and physical localness" with what he called the "unbroken unity of world habitat." Global standardization in aviation itself, and in its business practices and its culture, gave reason for optimism, voiced in Platt's observation that:

> The features of aviation are nearly identical over the world, the parts interchangeable: The fields, the planes, the pilots, the routine of operations, the ritual of service, the etiquette of crew and passenger behavior . . . all cut from one piece of cloth . . . of one cultural variety. (Platt 1946: 8)

The text of Platt's address was not long; it appeared in the *Annals* at only 18 pages but was followed by 25 pages of images, most of them maps. Beginning with maps emphasizing Chicago Municipal Airport's proximity to the city and surface geology of the area where the airport lay, Platt's presentation proceeded through a succession of maps: one showing prevalent winds; another establishing the extent of forecasting ranges for atmospheric conditions; another for establishing air traffic conditions and their local monitoring range; still another for the building of runways to accommodate these conditions within the local context of Chicago's street grid. Airline traffic patterns spilled over borders, demanding new projections, among them new maps of air routes drawn to scales ranging from the local to the regional to the global. Platt's presentation included a succession of quickly expanding North American route maps for airlines such as American, Northwest, Pennsylvania Central, Eastern, Chicago and Southern, Braniff, and Delta, while noting how others such as Pan American World Airways (more commonly known as Pan Am) and Transcontinental and Western Airlines (later Trans World Airlines, or TWA) were experimenting with global projections to illustrate the increasingly global character of their networks.

Further images in Platt's address showed aerial photographs of different patterns on the earth's surface, some of them clearly shaped or developed on the terms of human land use; others emphasizing geological, hydrological, alluvial, agricultural, botanic, and arboreal surface patterns as aviation technology had made possible new ways to study the earth's surface and the patterns on and across it. The final page of illustrations accompanying "Problems of Our Time" included a view of a social studies class in the school across the street from the airport. Students were shown studying a map labeled "Sources of Raw Materials," also bearing the name of a nearby pipe-manufacturing company. It showed the United States alone, appearing to float in a sea of undifferentiated oceans in all directions. While such a map might have helped explain some economic aspects of the local steel industry at the time, Platt's point was that it would not for long.

The following decades would indeed see the effective globalization of extractive industries, manufacturing, and investment. Patterns of human activity on the surface were, like patterns of air traffic, impermanent. "The airway pattern is a fragile feature of the current unstable Occidental organization of the world" (Platt 1946: 8). He continued:

> But, even though this particular pattern may disappear, the problem of interrelations on a world scale is with us to stay and probably to increase, under conditions in which one place has contacts with other places, more and closer contacts with any or all places for better or for worse.

Platt hoped for better, and saw a great role for aviation technology—not just in the transportation it enabled but also in the new view of the world it offered.

The Second World War was a moment in time that saw some of the most egregious of nationalistic acts committed—acts which many elements of modern aviation technology had been developed to support or oppose. The idea that "nation" was the proper form of human social and political organization was not new when nationalists took power in 1922 in Italian-speaking areas and in German-speaking areas beginning in 1933. But the idea that a political unit, even a national one, should be only inhabited by a select group who passed a sort of nationality requirement was, if not entirely new, fairly untested until the twentieth century. The National Socialism or "Nazism" of the "Third Reich," or "Third German Empire," had brought it to a new and highly militarized level; and local populations that were deemed insufficiently "German" were killed, often systematically, in the name of the "German Nation"—ostensibly a single source of unified, national agency and identity for all its sufficiently "German" inhabitants. It was in the combat service of the Reich that the first mass-produced jet aircraft, the Messerschmitt ME 262, took flight in 1944.

Nationalism, even in its militaristic variants, was not exclusive to fascist or Nazi regimes. Writing of one nation attacking or invading another was common journalistic shorthand in newspapers of the day. Even many scholars employed language that unthinkingly cast nations as the primary units of political and social organization. Narratives of invading, conquering, or securing independence from other nations dominated school history texts and classrooms, animated by interest in the war and its outcomes. And new maps were published and purchased en masse after the war to satisfy that interest. Such maps presented the world as a newly adjusted patchwork of nations and borders, a kind of lens through which the world could be understood—in terms of what appeared to be its basic constituent parts.

Platt was interested not in parts but in patterns. He was, in 1945, offering a template for rethinking the nature of ethical agency in a new, air-age global context, informed by the greater and more humane potential of aviation technology. Aviation maps allowed the visualization of this potential: They emphasized not borders but linkages, not limitations but possibilities, offering a visual language for an age in which earth's inhabitants could come to understand themselves as more connected than divided.

The "air age" and ethical agency

The idea of an "air age" captured the public imagination in many parts of the world in the years after the Second World War. While commercial aviation had become an established means of travel as early as the 1920s, its size, service areas, and cost-accessibility expanded astronomically in the years after the war, and it became as much a cultural phenomenon as a technological one, inspiring a variety of voices, from advertisers to politicians, designers to educators to cast the postwar period as an "air age"—a welcome next chapter of human history for a population weary of wartime. That aviation served as an age-defining symbol of progress at this time spoke not just of the cultural power of technology, but of air travel as progressive and even liberatory, the prospect of a nation tired of war embracing its collective achievements and its international linkages and future possibilities. The new "air age" was to be a progressive, cosmopolitan era for the human race.

Aeronautical iconography appeared in unexpected places after the war, from advertisements for non-aeronautical products to the shape and feel of products themselves. Many radios, fans, automobiles, and other consumer goods embraced streamlined designs, adopted tailfins, chrome, and "wings," and even "air" in their brand names. The same iconography and language found its way into schools; and among the most popular new textbooks published in the US after the war was a geography text titled *Our Air-Age World* (Packard et al. 1945). Its cover bore an image of a slightly tilted globe, with the North Pole visible, landmasses in red, and the sea in dark blue; rough lines of latitude and longitude contributing to a design effect of orderliness. Map imagery such as this was among the most ubiquitous forms of popular air-age iconography and became in many ways symbolic of the era. New maps of air routes increasingly accessible to travelers visualized a new global transit network. Some maps were designed to fuel public imagination, such as the perspectival innovations drawn during the war by illustrator Richard Edes Harrison for the US magazine *Fortune*. More localized and technical maps became increasingly possible with the aid of aerial photography and aerial observation of atmospheric conditions.

Political dimensions of "air-age internationalism" or "air-age globalism" took the form of bestselling works like US politician Wendell Willkie's *One World*, in which he invoked the prospects of aviation for civilized life. "When I say that peace must be planned on a world basis," Willkie wrote, "I mean quite literally that it must embrace the earth. Continents and oceans are only parts of a whole, seen, as I have seen them, from the air" (Willkie 1943: 84). While examples of popular "air age" cartography such as Harrison's maps for *Fortune* sought to communicate an *international* perspective on geopolitical issues, it was working maps—from routes and networks to aerial contributions to environmental studies—that may in retrospect have been more philosophically important. Those maps are best described not as inter*national*, but as global, for their networks and studies do their jobs without requiring the recognition of nation as category.

While this proclaimed "air age" in the years after the Second World War has been celebrated as a technological and cultural phenomenon, it also had legal and ethical dimensions, such as the creation of ICAO and the beginning of "open skies" negotiations and agreements after the war. As Dawna Rhoades has argued earlier in Chapter 4 of this volume, such agreements pitted philosophy against national sovereignty and economic self-interest. Key to Rhoades's analysis is the suggestion that ethical reasoning, and ethical responsibilities, not only transcended ideas of national sovereignty but also could actually be impeded by claims to national sovereignty. In this way, the philosophical groundwork can be understood to have been laid for such agreements at least as early as Immanuel Kant's essays "Idea of a Universal History from a Cosmopolitan Point of View" (1784) and "Perpetual Peace" (1795).

As Mark Waymack discussed in Chapter 1, for Kant, ethical reasoning is not to be understood solely in terms of individual choice, but within the context of broader human patterns and possibilities for expanding knowledge and reason. Kant had understood the *requirements of reason itself* to be at work in negotiating "peace among nations" in the form of a "federation of free states." According to him, "The idea of international law presupposes the separate existence of many independent but neighboring states," as legal entities of some sort had to serve as negotiating bodies. "Nations" were forms that had developed as humans differentiated one another by language or religion, but these differences were not absolute; they had developed historically, and would continue to, as history itself was ongoing. These differences involve "a tendency to mutual hatred and pretexts for war," Kant argued. "But the progress of civilization and men's gradual approach to greater harmony in their principles finally leads to peaceful agreement" (Kant 1957a: 354, 367). According to Kant, "The greatest problem for the human race . . . is the achievement of a universal civic society which administers law among men." Toward this end, "A philosophical attempt to work out a universal history . . . directed to achieving the civic union of the human race must be regarded as possible" (Kant 1957b: 22, 29).

In Kant's view, humans were capable of learning, of knowing, of employing reason, and the natural conclusion of this would be working to achieve a more rational society. Human reason was not bound up in nationality; nor was ethical agency, in any necessary or fundamental way. Indeed, agency had more to do with human reason and the responsibilities that came with it, according to Kant, than any particular sense of national, linguistic, or religious identity. Human history was a long process. At various points in time humans had seen fit to organize themselves politically in ways corresponding to patterns of language or religion; but none of those arrangements had lasted particularly long in the larger arc of world history. The capacity for reason gave humans the *responsibility* to use it in social and political affairs. Peaceful agreement was, in the end, more rational than hatred or war, and humans were ultimately capable of, and responsible for, ensuring it. International agreements were a start.

But understanding the world *as a world*, and not simply as the sum of national parts, was key. That involved understanding ethical agency not as an absolute characteristic of national, cultural, or religions identity but as something that develops and changes as humans learn to use reason in everyday affairs. In understanding the world by working to represent its patterns and conditions, maps and geographical work would, in the postwar years, play a key role in such reasoning. Cartography in the air age, both popular and highly technical variants, was increasingly moving beyond simple but philosophically problematic maps of states and nations to more substantive study of earth's surface patterns and conditions. In practical terms, these visualization practices were in the works already in 1891 with the proposal of the International Map of the World, which in turn would become the basis for the World Aeronautical Chart.

Representing the world as world

In 1893, decades before the "air age," Berlin geographer Albrecht Penck had written in *The Geographical Journal* of the inadequacy of the territorial maps that dominated the cartography of his time. "The interests of civilized life," such as science, navigation, administration, and commerce, he wrote, "make good maps a necessity" (Penck 1893: 254). Earlier, in 1891, Penck had proposed a standardized set of topographical sheets that would represent, on their completion, the entire world on a detailed scale. The project he proposed came to be known as the International Map of the World (IMW). In the early years of the twentieth century, after securing agreement from mapping agencies around the world, a uniform polyconic projection was adopted that accounted for the curvature of the earth more precisely than any projections prior to that had. A standard was thus adopted which divided the earth into 2,500 quadrangles which would cohere almost seamlessly and cover the surface of the earth on a meticulously detailed 1 to 1-millionth scale. The IMW's accomplishments included standardizing methods and techniques for representing and studying the earth's surface. In a series of IMW meetings in 1909, the French government agreed to accept the Greenwich Meridian; the British government reciprocated by accepting the meter (Willis 1910: 128).

Also among the IMW's accomplishments was its service as a base map for the World Aeronautical Chart (WAC), developed after the Second World War to accommodate global air traffic by representing local conditions on and above the surface on a globally consistent scale. The WAC, perhaps even more than the IMW, put into practical use the standardized visualization of the earth's surface conditions in great local detail, without the impediment of territories and boundaries. It constituted an early practical argument for international laws and standards. The WAC was ultimately a practical tool. In its work to enable safe travel it made no attempt to "mirror nature," but simply to represent surface conditions. Yet such attention to conditions, on a uniform scale and in uniform style, allowed the WAC to serve as base map for, among other uses, weather and climate study. Indeed, the WAC in the mid-twentieth century laid much of the visual groundwork for how climate and ecological problems and disasters, real and potential, have been studied in the decades since (Cabot 1947). Further examples involve actively working to improve the human condition through WAC-based land use surveys that assess local economic and ecological conditions in parts of the developing world, both to facilitate sustainable development and to assess and coordinate relief efforts when necessary (Van Valkenburg 1950; Fitzgerald 1974; Vink 1975).

Conclusion

Bound up in its practical uses, the WAC, as well as the increasingly global airline routes that stretched across the surface it represented, could be understood to have also served an ethical purpose. These maps and charts explored the idea that the proper sphere of human ethical action could be understood in the networks that play across that surface, no longer restricted by national borders. They made the argument, visually and practically, that spheres of human agency, and ethical action, are not national, not partisan, but involve more fundamental aspects of the human condition: Our best understanding of conditions of life and work on earth's surface, growing as we continue to study those conditions. It may be yet more significant that this dimension of ethical thinking came not from isolated philosophical speculation but out of the very fabric of modern life: The charts were practically necessary for safe aerial navigation, but they ultimately offered insight into who we are and where we live as humans. This was Platt's primary point, and Kant's as well. Ethical agency is bound up in how we see the world, how we know the world. If we understand it in terms of patterns, and if we study those patterns rationally and responsibly as they develop and change, we may ourselves develop a more rational, responsible understanding of who we are and how we should behave.

While the air age is a fairly new chapter in the long course of human history, with new kinds of ethical problems, the philosophical perspective it offers may also hold seeds of possible solutions. Issues such as international security agreements, emissions protocols, corporate responsibility and consumer and worker protection practices, laws, and covenants are not just international but global in scope—as are the mapped networks and the global market for air travel, and as is the global climate. Ideas of national or group agency have developed historically along with emerging patterns on the earth's surface; they continue to evolve as such patterns do.

References

Cabot, E.C. 1947. The northern Alaskan coastal plain interpreted from aerial photographs. *Geographical Review*, 37:4, 639–648.

Fitzgerald, N. 1973. Mapping by the Food and Agriculture Organization of the United Nations, *Geographical Journal*, 139:2, 375–376.

Kant, I. 1957a [1795]. Perpetual peace. In *On History*, trans. Lewis White Beck. Indianapolis: Bobbs-Merrill.

Kant, I. 1957b [1794]. Idea for a universal history from a cosmopolitan point of view. In *On History*, trans. Lewis White Beck. Indianapolis: Bobbs-Merrill.

Packard, L.O., B. Overton, and B.D. Wood. 1945. *Our Air-age World: A Textbook in Global Geography*. New York: Macmillan.

Penck, A. 1893. The construction of a map of the world on a scale of 1:1,000,000. *Geographical Journal*, 1:3, 253–261.

Platt, R.S. 1946. Problems of our time. *Annals of the Association of American Geographers*, 36:1, 1–43.

Ristow, W.W. 1944. Air age geography: a critical appraisal and bibliography. *Journal of Geography*, 43:9, 331–343.

Van Valkenburg, S. 1950. The World Land Use Survey. *Economic Geography*, 26:1, 1–5.

Vink, A.P.A. 1975. *Land Use in Advancing Agriculture*. Berlin-Heidelberg: Springer-Verlag.

Willis, Bailey. 1910. The International Millionth Map of the World. *National Geographic Magazine*, 21, 125–132.

Willkie, Wendell. 1943. *One World*. New York: Simon and Schuster.

25

ETHICAL DIMENSIONS OF THE 1981 PATCO STRIKE

Michael S. Nolan

On August 3, 1981 more than 12,000 air traffic controllers, all members of the Professional Air Traffic Controllers Organization (PATCO), illegally conducted a work stoppage (strike) that has had profound implications for both aviation and the labor movement in the US. The actions taken that year reverberate throughout the nation to this day. In response to the walkout of the air traffic controllers, then-president Ronald Reagan stated that the striking controllers were in violation of federal law, and that they had 48 hours to return to work or they would be fired. Most of the controllers did not return to work and were subsequently fired. I was one of them.

There are many misunderstandings, both in the aviation community and elsewhere, concerning the reasoning behind many actions taken by both the government and the union, but little investigation into the overall ethical situation faced by controllers, union representatives, and FAA management. Many of these situations are glossed over during discussions of the strike, but they deserve to at least be considered and discussed.

In retrospect the PATCO strike of 1981 was a watershed event that needs to be fully understood for its impact on American politics and labor relations. Most people reading about, inconvenienced by, or peripherally related to the strike thought it was just about money. From the controllers' perspective it concerned removing themselves, their job, and their employer from restrictions imposed by being civil service employees (a goal that replacement controllers eventually achieved). From the FAA and government perspective it was a chance to redistribute the balance of power between unions and their employees. Almost everyone agrees that the PATCO strike of 1981 changed the course of union/management relations that exist to this day.

In 1980 Ronald Reagan was elected US president. Much of his platform was based upon a return to "better days." Americans were felt to be weary of the ineffectual, liberal politics of the past. Reagan's 1980 campaign slogan, "Let's Make America Great Again," expressed the spirit of the times. Although he had been active in union politics early in his acting career, Reagan had changed his views since then and based his campaign on opposition to many union activities and politics. The PATCO strike would be the first opportunity for Reagan to publicly display his resolve, both to citizens of the US and to others watching this new president during one of his first confrontations.

During labor's heyday in the middle of the 20th century, American workers struck frequently and effectively. According to the Department of Labor's Bureau of Labor Statistics, between 1950 and 1980 the US averaged more than 300 major work stoppages per year (each involving at least 1,000 workers); but between 1982 and 2000 the annual average was 46. In the 21st century the average number of major strikes has been less than 20 per year, less than a tenth of what it was in the late 1970s (DOL 2017). Since the PATCO strike union membership has fallen by close to 40 percent. Experts agree that the organized labor movement in the US has not been the same since this strike.

For employers, however, the strike permitted them to regain "control" of the situation. When firing the air traffic controllers, Reagan legitimized a little-used union-busting tactic of replacing strikers with "permanent replacements." According to US labor law, employers are not permitted to fire workers for striking; but they do have the right to "permanently replace" them. Labor historian Joseph McCartin of Georgetown University states that in the 1950s and 1960s there was only 1 documented use of permanent replacements for about every 80 major work stoppages. Yet in the decade after the PATCO strike, there was 1 documented use of permanent replacements for every 7 work stoppages (McCartin 2006: 217).

McCartin also notes that, prior to the PATCO strike, it was generally not considered acceptable for employers to replace striking workers, even though the law gave employers the right to do so. Kathleen Schalch (2006) quotes McCartin as saying:

> any kind of worker, it seemed, was vulnerable to replacement if they went out on strike, and the psychological impact of that, I think, was huge. . . . The loss of the strike as a weapon for American workers has some rather profound, long-range consequences.

According to Schalch:

> When PATCO went on strike in 1981, Ken Moffett was the chief federal mediator. He says the union wanted a shorter work week and higher pay. Moffett says the strikers believed if they were gone, the safety of the flying public would be at risk. But that wasn't entirely the case. Moffett calls the

strike a "calamity," not just for the fired air traffic controllers, but for unions everywhere. Back in 1981, labor negotiations centered on the size of workers' raises. Subsequently, management began going after all unions for concessions and laying people off, he says.

Unions and federal employees

Prior to the 1960s most federal employees in the US were prohibited from joining labor unions with the right to collectively bargain wages and working conditions. This restriction was lifted by President Kennedy in 1962. In 1968 PATCO was created. Prior to its formation, controllers were eligible to join a number of representative organizations that included both controllers and supervisors/managers. PATCO membership was limited to working air traffic controllers, not supervisory or management personnel. Among other government unions it had participated in a number of quasi-job actions over the years in an effort to win increased benefits for members. During that time it was common practice to "slow down" activities or "work to rule," which was essentially the same thing. Employee "sick-outs," union-organized activities where employees called in sick, were somewhat common. Wildcat strikes, which are limited to local areas and assumed not to be sponsored by the national union, also occurred with increasing frequency.

In 1970 New York letter carriers (also federal government employees) staged an illegal strike that eventually spread across the country. Before settling about a month later, the Nixon administration had mobilized the US Army to perform some unionized mail carrier functions. Afterward the Postal Reorganization Act of 1970 was passed, in some ways due to the strike. This Act abolished the Post Office Department and created the US Postal Service, a corporation-like independent agency of the federal government giving postal employees rights and benefits similar to those desired by the air traffic controllers. All of this activity came to a head in the late 1970s when the two most active and militant unions in the federal government, those representing the postal workers and the air traffic controllers, began negotiations with their respective employers.

President Reagan and the PATCO strike

Just after the election of Ronald Reagan in 1980 the letter carriers were again threatening to strike, demanding more autonomy and workplace changes. This potential job action coincided with the air traffic controllers' increasing demands for workplace changes and organizational autonomy similar to those rights won by the postal workers a decade before. The union representing the postal workers was much larger, with a strike potentially more disruptive to the country. Each union was looking towards the other, wondering and hoping that the other would go first. Neither was certain of the ensuing response of the

Reagan administration. Whichever union went on strike would become the first domestic test of this new president.

In retrospect the outcome was not so surprising. Reagan had campaigned on toughness. He would "stand up" to challenges, both foreign and domestic. He did not believe in "politics as usual." The question in many people's minds, both in the US and possibly more importantly outside the US, was did he really mean it? What we now know to be historic negotiations with the leadership of the now defunct Soviet Union were to begin soon, and the Reagan administration, which had campaigned against weakness in negotiations, did not want to send the wrong message to the Soviets.

Within the FAA and the aviation community it had been known for years that in the 1980s' contract negotiations between PATCO and the FAA would be contentious. Newly hired trainees undergoing training at the FAA's academy in Oklahoma City were being told by instructors in the late 1970s that a strike was inevitable and that they should be prepared. Shortly thereafter, labor negotiations began between the FAA and PATCO. The union representatives, negotiating simply from the limited perspective of their issues and constituents, looked only at the issues related to air traffic control. Air traffic controllers felt that the newly elected Reagan administration would look favorably upon their demands, as PATCO had been the only major union to publicly lend its support to the Republican candidate in the recent presidential campaign.

The administration, however, was looking at the larger picture of overall labor negotiations and presidential image. Negotiators did not want to place the federal government at a perceived disadvantage when their counterparts were negotiating with the larger postal workers union. And administration officials were highly cognizant of the image that the Reagan administration was sending to the US and the rest of the world (particularly the Soviet leadership) concerning their resolve and negotiating tactics. Reagan has stated in his memoirs that he was aware that PATCO had been one of the few unions to support his presidential bid. "I supported unions and the rights of workers to organize and bargain collectively," he wrote, "but no president could tolerate an illegal strike by federal employees" (Reagan 1990: 282).

PATCO was negotiating for a reduced work week, a $10,000 pay increase, and a better benefits package for retirement. The FAA essentially stated that these items were non-negotiable. Civil service employment rules covered these matters, and they did not have the authority to negotiate in these areas. Contract negotiations with the FAA eventually stalled. PATCO threatened to go on strike early in the summer, but did not receive enough votes from the rank and file. A majority of controllers who worked at large, busy facilities voted yes but were outnumbered by controllers from smaller facilities who voted no. At the same time the postal employees threatened a strike but voted to continue negotiations, waiting to see what happened to the air traffic controllers.

During this time every controller was warned individually by their supervisors that a strike was illegal and that they might be fired as a result of participating in an illegal job action. In early August PATCO held another strike vote, but the procedure was weighted to give more representation to the busier, and felt by many to be the more "important," facilities. The vote was tabulated, and on August 3, 1981 about 13,000 PATCO members went on strike—in violation of federal law and the oath that they had taken upon accepting their jobs. Other government unions, particularly teachers and public safety workers, had occasionally declared strikes and very seldom were the strikers actually fired.

In the case of the air traffic controllers, President Reagan declared the strike a "peril to national safety" and ordered the controllers back to work or they would be summarily dismissed. In both public notices and private communication to each controller, the Reagan administration warned that a strike by federal employees was illegal, and that anyone who did not return to work within 48 hours would be terminated. The majority of the controller workforce refused to return to work and was dismissed en masse on August 5 with no public hearing or procedure. The striking controllers were replaced by controllers, supervisors, and other personnel not participating in the strike. The FAA had inflated the ranks of managers in the years leading to the strike in preparation for it. Military controllers were transferred to civilian facilities to supplement the workforce. Hiring of new controllers was accelerated.

The fired controllers were initially banned from any federal employment for life. The total ban on federal employment was lifted about one year later, but strikers were still restricted from ever being rehired by the FAA. Ironically, many gained federal employment with the US Postal Service. In 1993 President Clinton ended the prohibition on rehiring fired striking air traffic controllers. Over the next decade fewer than 1,000 were rehired. The controllers who remained at work, as well as their replacements, worked extended weeks for a number of years, and air traffic was restricted at many major US airports.

During the summer of 1987 increasingly disaffected controllers formed a new union, the National Air Traffic Controllers Association (NATCA), which was certified as the sole bargaining unit for air traffic controllers employed by the FAA. Part of NATCA's charter is a prohibition against conducting illegal strikes.

In retrospect the PATCO strike marked the turning point where many workers in this country lost the ability to wield their most potent weapon: the strike. Between 1947 and 1980 the Bureau of Labor Statistics annually reported at least 180 strikes involving more than 1,000 workers each. Since the PATCO strike, the number of such strikes has never reached even half of that level. Strikes by federal employees were illegal, but PATCO mistakenly figured that the nation could not survive for long without air traffic controllers. McCartin states:

> In the years after 1981, a number of prominent private sector employers followed Reagan's lead and permanently replaced their own strikers. The stiffened resistance to collective bargaining that became evident in the 1980s accelerated organized labor's decline. They would be wise to ponder an even more deeply rooted problem facing labor today—one highlighted by

> this week's painful anniversary. Since 1981 the strike has nearly disappeared from labor's arsenal. Unless unions can recover that weapon, they may not reverse their slumping membership figures. (McCartin 2006)

Ethical considerations

The short-term ramifications of the strike have been widely discussed. But what were the ethical dimensions of the strike? Many decisions needed to be made by both individuals and organizations. Controllers, managers, union representatives, elected officials—all had to make defendable ethical decisions. In general these decisions include the following.

Employees

Illegal strikes had been conducted in the past with little or no repercussions. There are many "illegal" activities on the books in the US that are selectively or rarely (if ever) enforced. Should this action have been one of those situations? Should employees hold the nation "hostage" in order for their demands to be met? Are there any other alternatives for employees in this type of situation?

Union leadership

The union leaders knew the potential consequences of a strike. They also allowed the employees to vote to strike. Then again the union was trying to achieve better salaries and benefits for its employees.

The federal government

Although FAA management officially opposed the strike, at the local level managers were less opposed, and in many cases supported it since benefits won by the controllers would be conferred upon them as well. At the FAA Academy, with full knowledge of the administrators, union representatives were advising trainees of the upcoming strike year before it occurred. In this author's personal experience, during the summer of 1981 I was "advised" by my supervisor that a strike was illegal. After signing a document to that effect, the supervisor shook my hand and said: "Good luck, I hope it works."

The Reagan administration made the members of PATCO an "example" to the rest of the nation, when apparently the demands being made were obviously not that excessive. PATCO was attempting to improve work conditions and pay. In addition, within five years of the strike, virtually every demand made by PATCO was granted by the FAA.

Finally, concerning the federal government, another question is whether or not the employees were basically led to strike by their own government. During the 1980 presidential campaign, it appears Reagan was going to support the ATC workers' cause. In a letter dated October 20, 1980 to Robert Poli, then president of PATCO, Reagan stated:

> I have been briefed by members of my staff as to the deplorable state of our nation's air traffic control system. They have told me that too few people working unreasonable hours with obsolete equipment has placed the nation's air travelers in unwarranted danger. In an area so clearly related to public safety the Carter administration has failed to act responsibly.
>
> You can rest assured that if I am elected President, I will take whatever steps are necessary to provide our air traffic controllers with the most modern equipment available and to adjust staff levels and work days so that they are commensurate with achieving a maximum degree of public safety.
>
> I pledge to you that my administration will work very closely with you to bring about a spirit of cooperation between the President and the air traffic controllers.
>
> (Hearings before the House 1981: 85)

The military

Out of all the players, perhaps the involvement of the military is the ethical consideration that is the most overlooked. The military appears to have been used to "make a political point" and/or to break the union. Without the use of military controllers, it is generally agreed that the striking controllers would have needed to be rehired.

Other organizations

Other aviation-related unions and professional organizations could have easily stepped in and forced an end to the strike. By tacitly agreeing to the actions of the federal government, other unions, particularly for pilots, made it possible for the strike to be broken. A simple statement by the pilots that they would not fly, or a statement to the effect that the strike had led to unsafe conditions, would have forced the FAA to renegotiate. The pilots' union made neither. Was this an ethical violation of what a union stands for? Unions are supposed to band together to protect the rights of all, not the individual.

References

Department of Labor. 2017. Table 1. Work stoppages involving 1,000 or more workers, 1947–2017. Bureau of Labor Statistics. Economic news release. Available at: www.bls.gov/news.release/wkstp.t01.htm [accessed February 13, 2018].

Hearings before the House of Representatives' Committee on Public Works and Transportation. 1981. *Aviation Safety: Air Traffic Control (PATCO Walkout)*. Washington, DC: United States Government Printing Office.

McCartin, J.A. 2006. A historian's perspective on the PATCO strike, its legacy and lessons. *Employee Responsibilities and Rights Journal*. 18(3), 215–222.

Reagan, R. 1990. *An American Life*. New York: Simon and Schuster.

Schalch, K. 2006. 1981 strike leaves legacy for American workers. *National Public Radio* [Online, August 3] Available at: www.npr.org/templates/story/story.php?storyId=5604656 [accessed: July 20, 2008].

26

AIR TRAFFIC CONTROL

A critical system in transition

Bill Parrot

While the previous chapter reflected on the PATCO strike of 1981, we will now address the future of ATC and the main challenges it faces. As with many industries in post-9/11 America, aviation transportation has undergone a metamorphosis of epic proportions. The US air traffic control system has seen and will continue to experience significant modification, evolution, and structural redesign as a most critical and integral part of the aviation industry. External influences coupled with internal pressures will combine to shape these discernible changes, resulting in a leaner, more efficient, and hopefully safer ATC operation. Pursuant to the industry's unquestioned commitment to safety, adequate oversight and focused supervision will remain absolutely essential.

Geopolitical influences dominated by Middle East instability, extremely volatile oil prices, and the War on Terror most certainly will continue to affect all aspects of commercial aviation. Financially strapped airlines have consolidated and downsized to cope with unsustainable overhead charges. Economic priorities shaping federal budgets—within the context of extensive military operations in Iraq, Afghanistan, and worldwide—have placed significant pressure upon all federal departments to streamline, reorganize, and cut costs dramatically. In addition to these explicit domestic constraints, the FAA must continue to comply with international agreements formulated through ICAO specific to international air traffic control procedures. With the FAA budget under intense scrutiny, the air traffic control system has experienced, and will continue to experience, wide-reaching budgetary reductions and adjustments.

Labor management relations between the FAA and air traffic controllers have descended to a new low as a result of these mandated changes. Pressure to reduce controllers' salaries mainly through the elimination of incentive pay, efforts to increase productivity primarily through increased controller time on position, and a newly introduced two-tier wage scale for all new-hire controllers have resulted in an extremely strained relationship between the FAA and the controllers' union,

NATCA. Without the authority to call for a work stoppage in the form of a strike, NATCA's only recourse during negotiation impasse concerns efforts to educate the traveling public and airline officials as to the critical state of the ATC system. Of late little sympathy has been forthcoming from airline employees, already reeling from numerous draconian pay cuts and work rule concessions. Additionally, the traveling public's interest rarely rises to a fever pitch in these types of disputes, absent cataclysmic nationwide delays or a viable perception of real compromises in air safety.

Issues in a deregulated environment

Deregulation of the US commercial airline industry in 1978 permanently eliminated the Civil Aeronautics Board (CAB), previously tasked with approving new routes, establishing ticket prices, and thereby formulating airline behavior nationwide. This significant development introduced a dramatically higher level of uncertainty related to airline business decisions and overall marketing strategies. Individual airlines selected city pairs between which to fly with relatively no regulatory authority dictating fares, frequency, or duration of service.

The "Hub and Spoke" system became the darling of the new era amongst airline planners maximizing the daily utilization of their aircraft. Busy hub airports were virtually saturated by peaks of traffic designed to squeeze that last bit of additional competitive advantage from the airline business model. "Complexes" of connecting passengers flooded terminals and created congested ramps with taxiing aircraft. ATC facilities, especially control towers at the busiest airports, found themselves inundated with traffic during select periods throughout the business day. Adverse weather, in the form of summer thunderstorms with damaging winds and hail, or winter snowstorms closing busy runways for snow plows and deicing equipment, proved to be devastating for these peak traffic periods, causing logistical nightmares for airline operations planners. Increased numbers of weather-related diversions to alternate airports during peak arrival periods would paralyze an airline's major hub operations. ATC's workload would thereby become substantially unmanageable.

More recently the FAA's best efforts to encourage more rational domestic flight schedules by the air carriers in an "unregulated" environment have managed to garner some level of cooperation amongst the major players, thereby alleviating some of the strain on the system. While the unprecedented levels of uncertainty in the recent past have been somewhat mitigated today by technological advances as well as more realistic flight scheduling and voluntary downsizing at the legacy air carriers, the ATC system continues to operate at peak levels amidst the unending search for improved efficiency and flexibility.

These peak traffic levels have primarily been the by-product of airline behavior and economic decisions. Air carrier scheduling and equipment purchase moves have exacerbated an already complex air transportation model. Most major air carriers have ceded much of their short-haul domestic market to the regional carriers through partnerships and alliances. Regional carriers typically operate smaller, 50–70-seat jet aircraft into cities previously served by aircraft accommodating 100–150 passengers. With consumer demand representing the major impetus for change, this trend towards an even larger aircraft fleet eventually will affect an additional burden upon the airspace grid.

The privatization debate

For more than a decade the debate over the privatization of the air traffic control system has remained a grave concern for the career controller population. Private contractors have already assumed management responsibilities at several visual flight rules (VFR) tower facilities nationwide, resulting in significant cost savings for FAA air traffic managers. Special interest groups have fervently argued both sides of the issue, seeking to garner public support for their specific agendas. Additionally, political leaders regularly rally their respective constituencies in pursuit of greater public awareness. Privatized European and Canadian air traffic models stand in stark contrast to the US system while serving as the backdrop upon which the debate continues. As long as budget constraints drive the cost-cutting bottom line for the FAA, privatization will be a significant part of the future of air traffic management in this country.

With the Trump administration, the topic of ATC system privatization in the US has returned to the front page, blogs, and visual news sources. Much has transpired since 2014 to shape and drive the debate towards its natural conclusion.

Opposition to privatizing the ATC system is widespread amongst organizations that represent General Aviation (GA) pilots, including: the Aircraft Owners and Pilots Association (AOPA), the Experimental Aircraft Association (EAA), and the National Business Aviation Association (NBAA). These organizations are uniformly and understandably concerned about user fees placing the burden of financial support upon their members. They argue that privatization would prove to be the death-knell of general aviation in the US (see Burns 2017). Discussions in the House and Senate involve sympathetic members on both sides focused intently upon the details of the proposal and its affect upon their constituents.

Unified support of the move, highlighted across all the major US carriers except for Delta Airlines, has galvanized behind carefully worded advocacy from NATCA, which represents a diverse, sometimes divided, membership. Nationwide air traffic facility workforce shortages—exacerbated by antiquated, inefficient, and costly hiring and training rubrics at the FAA—have resulted in increased scrutiny of that organization by elected officials and the public at large. The Department of Transportation Secretary, Elaine Chao, has conferred with both Canadian officials and the European Union (EU) to explore the transition, and to date appears supportive of privatization. As of 2017, with one political party in control of both the US House and Senate along with the White House, the momentum most certainly appears to be in the direction of this significant transformation of a highly complex and fluid air traffic management system.

THE PROS AND CONS OF ATC PRIVATIZATION

Florida Tech (2018) highlights some of the pros and cons of privatizing ATC. The *pros* include the following: 1) most major airlines support it; 2) the air traffic controllers' union, NATCA, is in favor as long as work protections, retirement, and pay levels are not negatively impacted; 3) it works in other countries, such as Canada; and 4) it might speed up the implementation of NextGen.

Regarding the *cons*: 1) fliers may have to pay more. One 2015 study shows that user fees may increase by 20 percent; 2) many private jet owners and operators are opposed; 3) the US system may be too big and complex for it. For instance, in 2014 Canada handled 1.3 million departures compared to 9.6 million in the US; 3) it might be disruptive because removing control from the FAA could stall the implementation of NextGen; and 4) the US Government Accountability Office shows that the current ATC system is proven to be safest in the world.

The role of new technology

Arguments that privatization will serve to accelerate the sluggish and meandering implementation of NextGen, the next-generation air traffic model, fuel the endless discussions towards a bill in the fall of 2017 aimed at a long-term budget for the FAA in the form of reauthorization (FAA 2017). Continuous, uninterrupted funding and facilitation of this next generation of air traffic control, featuring expanded satellite data in automatic dependent surveillance broadcast (ADS-B), will most definitely increase user situational awareness and arguably enhance safety for all participants. Timely and sometimes heated debates over which segments of the aviation community will ultimately bear much of the financial burden of this monumental change continue today.

Technological advances both in the ATC facilities and the cockpits of our air carrier aircraft have added a new dimension to the air traffic controller job description. Airborne traffic alert and collision avoidance system (TCAS) equipment mandates that the cockpit crew maneuver the aircraft away from an intruder that potentially violates their airspace without guidance from ATC. In this scenario, onboard technology overrides any ATC authority or assistance. New hardware recently introduced into the enroute centers (Air Route Traffic Control Centers) has essentially eliminated the need for a third controller at each radar position during peak traffic periods. This same equipment will eventually set the stage for "free flight," a new millennium air traffic management concept wherein every aircraft receives clearance directly to its destination, with the ATC system in a standby mode to alleviate down line conflicts. Increasingly, technological advancements seek to reduce the level of decision-making performed by human beings on both sides of the aviation equation.

Looking toward the future

Projections of ever-increasing passenger travel in the 2020s highlight the enormous dilemma facing the FAA's air traffic managers. With a finite amount of airspace and limited concrete at the busy terminals, new strategies will be necessary to meet the demand. Recently instituted programs including reduced vertical separation minima, permitting aircraft to utilize previously unusable cruise altitudes at the higher flight levels, coupled with technologically advanced ground and airborne radar displays will extract additional capacity from the system. Additional software and hardware innovations will be required to bring the concept of "free flight" to fruition and expand upon successes realized with Global Positioning System (GPS) satellite navigation. Increasingly, technology will determine the rate at which the system's capacity may expand.

Human factors related to the air traffic model will continue to occupy center stage with daunting challenges for the future. Heavy pressure related to system capacity, brought to bear upon FAA decision-makers by operators fighting for survival in Chapter 11 bankruptcy, must be addressed within the broad context of overall safety. As hardware/software breakthroughs seek to increase productivity, the human interface will be modified and adjusted. Controller pilot communications will necessarily become more automated for efficiency, and thereby require a reduced level of human interaction. Predictable morale problems within the controller workforce, directly attributable to the new two-tier wage scale and reduced incentive pay, will demand increased vigilance on the part of FAA management.

At some point perhaps budget constraints and management policies, seeking to do "more with less," may prove to be causal factors in an aircraft accident. While ultimately attributed to aircrew error exacerbated by unrelated conversations in the cockpit at a critical moment during takeoff preparations, the NTSB's search for probable cause in the August 2006 air carrier accident at Lexington's Blue Grass Airport served to highlight negatively the minimal air traffic controller staffing at the Lexington tower facility (NTSB 2007). Such tragic events serve to redirect public attention to real controller grievances specific to the workforce and training.

Additionally, significant pressures to reduce controller compensation and benefits may significantly alter the quality of new-hire applicants entering the ATC hiring process. Deep cuts in wages and job-related economic incentives will necessarily complicate the FAA's sincere efforts to attract self-motivated, qualified candidates.

At present the air traffic control system derives much of its success, characterized by an exemplary safety record, from multiple layers of redundancy. Controller pilot responsibilities and cross checks combined with improved technology will continue to improve situational awareness while ensuring a safe and relatively smooth operation. As elevated burdens upon the system arise, these redundant layers of error avoidance will be increasingly tested while accommodating the ever-expanding travel demands of the new millennium.

Adding significantly to the uncertainty facing the National Airspace System (NAS) future design, alternatives proposed in 2014 within Congressional initiatives debate in earnest the real possibility of privatization of ATC functions. In view of curiously inefficient and arguably ill-advised changes to ATC specialist hiring and training, the FAA has become the center of controversy across a broad swath of bipartisan discussion on Capitol Hill and throughout the news media. Legislation introduced in 2015, HR1964 (known as the Air Traffic Controllers Hiring Act), seeks to move the FAA towards much-improved transparency vis-à-vis the recruitment, selection, and training of future air traffic specialists (Congress.gov 2015). Coupled closely to the discussions, an ever-increasing call across the aisle in concert with the air traffic controllers' union, NATCA, appears to both focus on and support the dramatic move towards privatization similar to the European and Canadian models. Many elected officials prominently exercising leadership positions in both parties appear to be most committed to just such a move in response to the FAA's apparent intransigence regarding NextGen modernization of the system at large.

As has always been the case, the air traffic control system derives much of its success, characterized by an exemplary safety record, from multiple layers of redundancy. Controller–pilot responsibilities and cross checks combined with improved technology will continue to improve situational awareness while ensuring a safe and relatively smooth operation. As elevated burdens upon the system arise, these redundant layers of error avoidance will be increasingly tested while accommodating the ever-expanding travel demands of the new millennium. Ultimately the responsibility of change closely joined with rapidly advancing technology will fall to the FAA, an organization consistently pressured to remain current while openly committed to the rapid increase in safety and efficiency.

References

Burns, A. 2017 (June 6). What the aviation industry is saying about Trump's Air Traffic Control Reform Initiative. *Flying Magazine*. Available at: www.flyingmag.com/what-aviation-industry-is-saying-about-trumps-air-traffic-control-reform-initiative [accessed August 22, 2017].

Congress.gov. 2015. H.R. 1964: the Air Traffic Controllers Hiring Act. Available at: www.congress.gov/bill/114th-congress/house-bill/1964 [accessed August 17, 2017].

Federal Aviation Administration (FAA). 2017 (Apr). NextGen updated: 2017. Available at: www.faa.gov/nextgen/update/ [accessed August 22, 2017].

Florida Tech. 2018. The pros and cons of privatized air traffic control. Available at: www.floridatechonline.com/blog/aviation-management/pros-and-cons-of-privatized-air-traffic-control/ [accessed February 18, 2018].

National Transportation Safety Board (NTSB). 2007. *Attempted Takeoff from Wrong Runway, Comair Flight 5191, Bombardier CL-600-2B19, Lexington, Kentucky, August 27, 2006.* Aircraft Accident Report. NTSB/AAR-07/05. PB2007-910406. Washington, DC: NTSB.

INDEX

Page numbers in **bold** refer to figures, page numbers in *italic* refer to tables.

Made in the USA
Coppell, TX
03 October 2024

38018676R00201